Corrosion and Scale Handbook

Corrosion and Scale Handbook

J. R. Becker

PennWell

Copyright © 1998 by
PennWell Corporation
1421 South Sheridan Road
Tulsa, Oklahoma 74112-6600 USA

800-752-9764
+1-918-831-9421
sales@pennwell.com
www.pennwellbooks.com
www.pennwell.com

Marketing Manager: Julie Simmons
National Account Executive: Barbara McGee

Director: Mary McGee
Production/Operations Manager: Sheila Brock

Library of Congress Cataloging-in-Publication Data Available on Request

Becker, J.R.
Corrosion & Scale Handbook
ISBN 0-87814-749-7
ISBN13 978-0-87814-749-6
CIP DATA if available

Printed in the United States of America

2 3 4 5 6   11 10 09 08 07

Contents

Figure and Table Captions

Figures

Tables

Introduction

The handling of crude oil and the attendant problems of corrosion and scale are a major concern for oil companies worldwide. The variety and complexity of the fluids transported from the producing formation to the customer add to the difficulty of providing adequate methods of treatment to alleviate these problems. Each year oil companies spend millions of dollars to protect their investments in equipment and production, and some of the highest costs are incurred in the prevention of corrosion and scale.

Much of the literature available on the topics of oilfield corrosion and scale avoids the complexities associated with the multiple phases present in these systems. These discussions are generally confined to ones describing these phenomena in relation to the phase in which they are observed, or more particularly, the water phase. These phenomena are generally manifested in the water phase. However, it is also true that aggregate physical structures of oil and water play an extremely important role in determining how these phenomena affect the systems under study.

The following discussion will explain some of the phenomena manifested in oil and water systems by corrosion and scale, taking into account some of the macro-aggregate systems commonly present in crude oils. With this goal in mind, it is necessary to develop an understanding of some of the various interfacial structures present in these systems.

How these structures interact to affect the processes of corrosion and scale formation will also be addressed.

Although corrosion and scale are vastly different processes, the development of these topics in connection with macro-aggregate structures (e.g., emulsions, asphaltenes, and waxes) is treated concurrently throughout. This method of discussion is intended to avoid the obvious redundancy resulting from the necessity of illustrating the macro-aggregate interactions with both processes of scale formation and corrosion individually. Throughout the following discussions it will be helpful to keep in mind that three major surfaces will be considered: 1) solid, 2) liquid, and 3) gas. It will also be helpful to keep in mind that an idealized surface is determined by a minimum of two distinct physical regions in space (the demarcation line between one physical phase and another).

The distinction between two physical phases is by no means ideal, and as a consequence the chemical and physical processes taking place within and without these diffuse regions are greatly affected. In the following discussions much effort has been made to present these topics with intuitive clarity, while presenting a set of arguments with a minimum of mathematical treatment.

As mentioned above, corrosion taking place in crude oil systems is one of the most costly problems faced by oil companies; consequently, companies incur great expense to control it effectively. Corrodible surfaces are ubiquitous throughout production, transport, and refining systems. The equipment they represent constitutes a considerable initial capital expenditure. The protection of this equipment, therefore, is critical to the profitability and successful operation of these companies. Thus of all the problems faced by these companies, corrosion and its control rank as one of the most important.

While scale and its control rank considerably lower in importance, its presence can also prove costly in terms of equipment. Many production areas contain considerable amounts of scaling fluids, and the resulting scale formation can close off transporting equipment, requiring its replacement. Thus, scale control is an important part of any oilfield operation. The understanding of these processes as well as their impact on petroleum companies is, therefore, a primary aim of this book.

Some of the methods and mechanisms of corrosion protection will be presented in this book, while others will be omitted by choice. One such omitted method involves the protection of metal surfaces by polymeric coatings (paint, resins, and other thermosetting plastics). These materials and methods have had considerable success in certain applications, but tend to suffer problems in the harsh environments encountered in petroleum fluid systems.

Phenolic resins have been used for some time as coatings for tubing and transfer lines in production systems, but these materials are susceptible to mechanical damage when placed in such systems. These resins are brittle, and when they are subjected to mechanical handling, they are often damaged. The damage can be as minor as a scratch that penetrates the coating surface to the bare metal, or it may be a large surface coating loss.

These mechanical defects are difficult to detect once the metal products have been placed in service. Although surveys are conducted using calipers in tubing strings, a very small defect in tubing coating may go undetected. Corrosion processes can then lead to catastrophic tubing failures and significant production losses as a result. Thus, chemical corrosion inhibitors are frequently added to these coated systems as an additional precaution against such failures. Certainly plastic coating methods offer significant corrosion protection advantages over bare metal surfaces, and when their cost is justified, they should be used. However, coatings should not be relied upon as the sole protection against corrosion.

Early in the present discussion of corrosion phenomena it will be pointed out that mild steel components comprise most of the metal products used by petroleum companies. The reason given for the preponderance of mild steel usage by these companies is largely due to cost performance. The usage of costly alloys can only be justified in certain rare instances. However, the steel industry continues to make significant advances in the production of high-grade steel products and increased efficiency leading to lower costs. Thus the cost of certain alloys has dropped, but cost of mild steel also has dropped. Therefore, mild steel usage by petroleum companies will continue for the foreseeable future.

Corrosion and scale problems are always going to be a fact of life, no matter how well we understand these processes. They are part of the order that arises from chaos, or organization that results as a natural quest for function. They often trouble us, but as we delve further into their nature, we realize their essential qualities and appreciate their necessity. Without corrosion, our world would be littered with the remnants of eons of life and industry, not only from man, but also from all previous forms of life. We can look at corrosion as part of the cycle of existence, and realize that it is a means of achieving a higher state of entropy (the measure of disorder). This will fall again to scale, which forms from the chaos. Each process is necessary; in fact, each process depends on the other.

Biological life organizes from the chaos of its surroundings. In the process of living, it creates waste that degrades and collects in formations underground. It, too, eventually becomes part of this collection of matter. Fluids pass through the environment and carry with them the remnants of biologic and other forms of matter. They trickle through the accumulations of these other forms, taking this, leaving that, and all the time changing. It may be possible to slow some of these processes, but in the end, they will never be stopped.

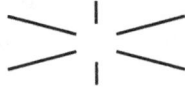

1
Petroleum-Related Corrosion and Scale

Petroleum producing, transporting, and refining companies spend hundreds of millions of dollars each year to locate and develop petroleum and petroleum derivatives. At the same time, these companies expend considerable millions on the maintenance of those facilities already in place. Much of the maintenance expense involves the protection of the vast array of transport conduit and machinery necessary for the movement of petroleum. Worldwide, the several thousands of producing wells each require tubing strings ranging from several hundreds to several thousands of feet. Pumps, transfer lines, tanks, and treating equipment are additional requirements to lift, transport, store, and treat the petroleum produced from these wells.

Generally, the high cost of special alloys, which offer a high resistance to the process of corrosion, precludes their use. Consequently, other means of protection are necessary. Several forms of protection are employed by petroleum companies, ranging from electrolytic (cathodic protection) to chemical surface passivation. Chemical means have been the most widely used methods of corrosion protection employed by petroleum companies. The reliance on these products by petroleum companies results in the purchase of millions of dollars of specialty chemical corrosion inhibitors per year. Nevertheless, corrosion-related failures occur quite frequently.

The consequences of not protecting these transport, storage, and treatment facilities can be catastrophic to the successful operation of a petroleum company. There have been numerous examples of petroleum spills and gas releases resulting from corrosion-induced failure of pipelines, tank bottoms, and pumps. These failures not only result in the loss of the product crude oil, but they also constitute a true hazard to the environment. Thus, the real cost of these failures is measured both in terms of economic and environmental impact.

Many countries in which crude oil is produced have relatively few environmental laws regulating the production, transport, and refining of petroleum. However, a growing awareness of environmental sensitivity is rapidly reforming government policies. The incidence of poisonous gases in petroleum (mainly hydrogen sulfide and low molecular weight mercaptans) represents a double hazard. The corrosive nature of hydrogen sulfide accelerates the processes leading to failure of the transporting conduit. It also endangers the safety of those living creatures that might be unlucky enough to be near when a breach occurs.

Regardless of however casually a government might regard the environmental impact of a petroleum spill, it will likely respond punitively to those incidents that result in the loss of life. Thus the liability inherent in petroleum operations, in addition to ethical and moral responsibilities, is considerable. Fluid flow restrictions resulting in increased pressure and, therefore, greater probability of conduit failure are also a major concern to petroleum companies. Organic constrictions (paraffin and asphaltene) and inorganic constrictions (scale and silicate fines) can be as much a source of problems as corrosion to the producer. Although the incidence of these problems is less pervasive than that of corrosion, they nevertheless represent a serious concern.

Macro-Aggregate Corrosion and Scale Formation

Many factors contribute to the problems of scale and corrosion. These include conditions of temperature and pressure, dissolved ion concentration in co-produced aqueous fractions, and conditions of drastically variable pH. The majority of crude oil produced from reservoirs is found in close association with water. These petroleum-related waters may have percolated through a vast array of mineral deposits on their way to the formation. This percolation through these deposits results in the concentration of soluble salts in the water phase. Subsequent (or concurrent) biodegradation and seepage of organic matter place the degradation product or "crude oil" in close proximity to the water.

Water is essential for biological functions, including those of organic decay. The microbial action responsible for organic decay can take place under conditions of high oxygen content, low oxygen content, or very little or no oxygen content (anaerobic conditions). However, microbial activity sometimes occurs. In this case, complex arrangements of multifunctional aggregates of complex molecules (polypeptides, amino acids, fatty acids, sugars, and carbohydrates) are left within the commingled oil/water phases.

These "artifacts" act as natural surfactants to intersperse the crude oil and water phases, resulting in gradations of solubility. Thus, the presence of microbial degradation products and the resulting macro-aggregates produce a complex system of fluids. These fluids affect the behavior of crude oils and further complicate the processes of corrosion and scale formation.

Most corrosion and scale models are developed from systems that are free from the complexities of macro-aggregate structures commonly present in crude oil systems. These useful, idealized models are only partially correct when applied to the real world of crude oils. The presence of discontinuous phases complicates the mechanisms responsible for the effects observed in both of these processes.

During the long history of the reservoir, a condition of equilibrium is approached by the various components of fluids, ions, and gases present. When the reservoir is tapped by a well, this equilibrium situation is dramatically altered. Shear forces caused by pressure drops produce conditions that favor a different distribution of the phases present (liquid, solid, and gas). The behavior of components of these phases is, therefore, changed. These changes are manifested by ionic redistributions within secondary structures (e.g., emulsions and organic complexes) that may now be capable of causing corrosion and scale.

From the instant the fluids begin flowing from the reservoir, a new system of equilibrium is approached by the components within them. Thus, across every pressure drop, through each temperature change, and within every region of turbulence different criteria for equilibrium are presented to the system. These changing criteria alter the process of corrosion and scale formation dramatically. Consequently, simple models related to their mechanisms and occurrence fall short of describing the behavior in dynamic systems.

Corrosion and Scale Forms

The forms that corrosion and scale take are highly dependent on these macro-aggregate structures and how these structures interact with their environment. If the systems were simple and homogeneous phases, the corrosion and scale would also be expected to be simple and uniform. However, real world corrosion and scale occur in many forms. Several of these processes include:

- Pitting
- Stress cracking
- Blistering
- Undercorrosion
- Embrittlement
- Passivation

Likewise, scale formation in real systems is manifested by several forms, some of which are:

- Homogeneous crystals
- Nonhomogeneous crystals
- Continuous (uniform surface coverage)
- Discontinuous (nonuniform surface coverage)
- Uniform crystals
- Nonuniform crystals

Subsurface (reservoir) and tubing corrosion and scale formation, therefore, take on characteristics that are determined by the fluid forms assumed by the various phases present.

Macrostructures produced by phase separations involving emulsions, organic crystals, and combinations of the two often interact with solid surfaces (e.g., metal, rock, or suspended solids) by surface adsorption. The pattern and type of corrosion or scale deposits observed depend upon the affinity of the emulsions, crystals, or their combinations with the metal or rock surface. Further, the range of emulsions and crystal size and type provides a variety of modality for these processes. This variety is manifested in a variety of corrosion and scale forms within the same system. Thus blistering, stress cracking, and embrittlement can all be present within a given environment, as can the various forms of scale.

Oilfield Scale Control Application

The possibility of several forms of corrosion or scale within the same system leads to problems of choice of the products used to control these processes. Scale buildup in the producing formation can reduce the quantity of production significantly, and various treatment programs are implemented to interfere with this process.

A particularly widespread practice of production companies involves a process called a "squeeze application" or "job." This practice involves the down-hole application of substantially high concentrations of scale inhibitor chemical. It assumes that the scale inhibitor chemical will adsorb to the formation and gradually be released to the aqueous production. This gradual feedback is calculated to be of sufficient concentration to inhibit the process of scale formation within the producing rock formation and up the well tubing. The effectiveness of these treatments varies from application to application and formation to formation.

In most instances the point at which scaling begins determines the need for squeeze applications, but often these applications are performed because the wells lack application capillaries. Scaling can and does occur in the formation, well tubing, transfer lines, treatment equipment, and storage vessels. The point at which scaling occurs in this downstream path determines the location and method employed for scale chemical application.

Where production fluid temperatures are high, and scaling occurs after the wellhead, the chemical addition may be placed at the transfer line influent from the wellhead. The effectiveness of these scale-control chemicals and their application in the presence of multiple scales will be discussed in subsequent chapters.

Oilfield Corrosion Control Application

Corrosion control in a producing well is approached in much the same way as scale control. However, the mode of operation of corrosion inhibitors is considerably different than that of scale control chemicals. One similarity is the requirement for contact with the corroding surface, which parallels the need for scale chemical adsorption to the growing scale crystal. This requirement is then met by several different corrosion chemical application methods, sometimes including squeezes.

The majority of time, the oilfield well application of corrosion control chemicals involves continuous or batch treatments. Continuous treatment requires a capillary string, while batch treatment involves the periodic pumping of chemical by treating trucks. Corrosion inhibitor chemicals are adsorbed at actively corroding sites on the metal surfaces and interfere with the processes of electronation and de-electronation.

The emulsions and organic crystals (and combinations of each) effectively compete for these corrosion control chemicals, just as they do with scale control chemicals. Thus, the presence of macro-aggregates often interferes with the activity of chemicals added to control corrosion and scale processes. In addition to the competition for chemicals, the accumulation of solids resulting from paraffin and asphaltene provides discontinuous sites for accumulation of corrosive environments adjacent to the corroding surface. Scale formation can also cause these discontinuities, providing similar sites exceptionally favorable for the corrosion process.

Surface Equipment

From the wellhead, well fluids are transported to treatment and storage facilities, where continuous changes in physical environment result in differing forms of corrosion and scale. The turbulence, temperature, and pressure experienced by these fluids in the transport from the wellhead to treatment facilities and storage vessels are widely varied. These variable conditions will often favor the processes of corrosion and scale formation within the transfer lines; consequently, chemical additives are usually injected early in the transport stream. Macro-aggregates are present throughout these dynamic systems in the form of natural emulsifiers, emulsions, biomass, paraffin crystals, and asphaltenes.

Generally, treatment chemicals that address the emulsions and paraffins have also been added either in the well or after these fluids have reached the surface. Thus, additional physical changes resulting from their introduction produce differences in the processes of corrosion and scale formation within the system. Settling tanks and storage vessels also provide environments with greatly reduced conditions of turbulence and significantly longer exposure periods than those of the well tubing and transfer lines. The reduced turbulence and extended exposure periods affect the morphology of the crystal forms of scale and paraffin. Storage tanks thus act to provide an environment favorable to the formation of sludge, throughout which corrosion processes may take place.

These relatively quiescent environments also provide an area into which the vapor phases of these systems can escape. Thus, corrosive vapors collect in the overheads of these vessels and produce corrosion damage quite different than that taking place in the liquid phase below. The provision of a vapor phase space is also responsible for concentration gradient changes that take place in the fluid phase. This shifts the equilibrium of the reactions involved in the formation of scale.

Aggregate Architecture in Scale and Corrosion

Throughout nature, the orderly assemblage of macrostructures from combined component atoms and molecules provides structural frameworks. These frameworks facilitate mechanisms involved in complex physical interactions ranging from photon emissions to nervous system networks and beyond. These multistructural systems provide form to the amorphous phases of matter and involve several forms or interfacial interactions wherein processes, including corrosion and scale formation, take place.

Biological systems represent some of the extreme limits of these assemblages. Within these systems, amino acids join to form polypeptides, polypeptides combine to form proteins, sugars and nucleotides join to form RNA and DNA, and fats and glycerides act to form emulsions. The combinations join to form cells that combine to form tissues, and so on. Simultaneously these structures provide pathways for selective processes of diffusion, electron transfer, ion transport, oxidation, and reduction.

While crude oil systems are less organized than biological systems, they possess much of the same character and tendencies found in these systems. The nonhomogeneous nature of phases present in crude oil systems illustrates the fact that they are the remnants of the breakdown of these biological systems. Many of the molecules involved in the architectural makeup of biological ancestors are present and unchanged in crude oils. The presence of these molecules provides avenues for the formation of higher order structures from combinations with dissimilar phases such as oil and water. Further, these aggregate structures also act as structural pathways for many of the same processes as those of the more highly ordered biological structures.

Considering the functions that these macro-aggregate structures perform, it should not be surprising that the mechanisms involved in the processes of corrosion and scale formation will be significantly affected by their presence. The determination of the effects these macro-aggregate systems have on these processes must be included in the descriptions of these phenomena. To this end, a discussion of these macro-aggregates, their formation, structural properties, and the effects they have on these processes is necessary.

Water in Oil Emulsions

Emulsions are among the most common macro-aggregates present in crude oil systems. The conditions that exist in crude oil formations, well tubing, transfer lines, and treating facilities approach conditions that are ideal for the formation of emulsions. Turbulence, pressure, and temperature combined with the presence of multiple phases and naturally occurring surfactants act in concert to form emulsions.

The natural surfactants present, as previously mentioned, are comprised of multifunctional molecules that remain from the process of biodegradation. These include fatty acids, glycerides, aliphatic substituted aromatics, and numerous other forms of bipolar molecules.

These bipolar species, or molecules, partition between polar and non-polar phases throughout the crude oil system, producing structures that separate and disperse the incompatible fluids.

The form that these dispersed phases take is determined both by the nature of the polar/nonpolar bulk fluid phases and the bipolar composition of the surfactant. The region separating these bulk fluid phases is designated as the interface, a diffuse region of space containing concentration gradients of the two bulk fluid phases along with the surfactant. This region of space and its composition determine, in large part, the nature of the corrosion and scaling processes that occur in crude oil systems.

Water and oil emulsions can be considered in conjunction with corrosion processes. The nature of the corrosion occurring at the surface of a metal will be considerably different than that observed when oil in water emulsions dominate. Likewise, if the phases are separated, or no emulsion is present, the nature of the observed corrosion will be different than either form of emulsion previously listed.

Scaling processes can be compared to those of corrosion by realizing that both processes involve ionic species present in the polar phase of the emulsion system. However, the mechanisms differ, since the types of ions transferred are different, and the compatibility of the diffuse regions (interfaces) with the ions is variable. Because of their smaller size, transport of hydrogen ions across these diffuse regions takes place much more readily than the transport of sulfate anions or calcium ions. Consequently, scale formation in these regions of diffuse composition takes place to a lesser extent than potential corrosion development.

However, due to the highly water-insoluble nature of the scales formed, these diffuse regions can favor the combination of scale forming anion/cation couples. In many cases the surfactant molecules act to facilitate the ionic interactions (e.g., fatty calcium salts can act as ionic carriers) within the diffuse interface. Figure 1-1 illustrates a possible mechanism for ionic transport in the diffuse interface region.

The sorbitan monooleate molecule (the surfactant molecule in Fig. 1-1) can interact via hydrogen bonding with the calcium cation. This interaction results in equilibrium between the aqueous phase and the diffuse interfacial region. Although these arrangements take

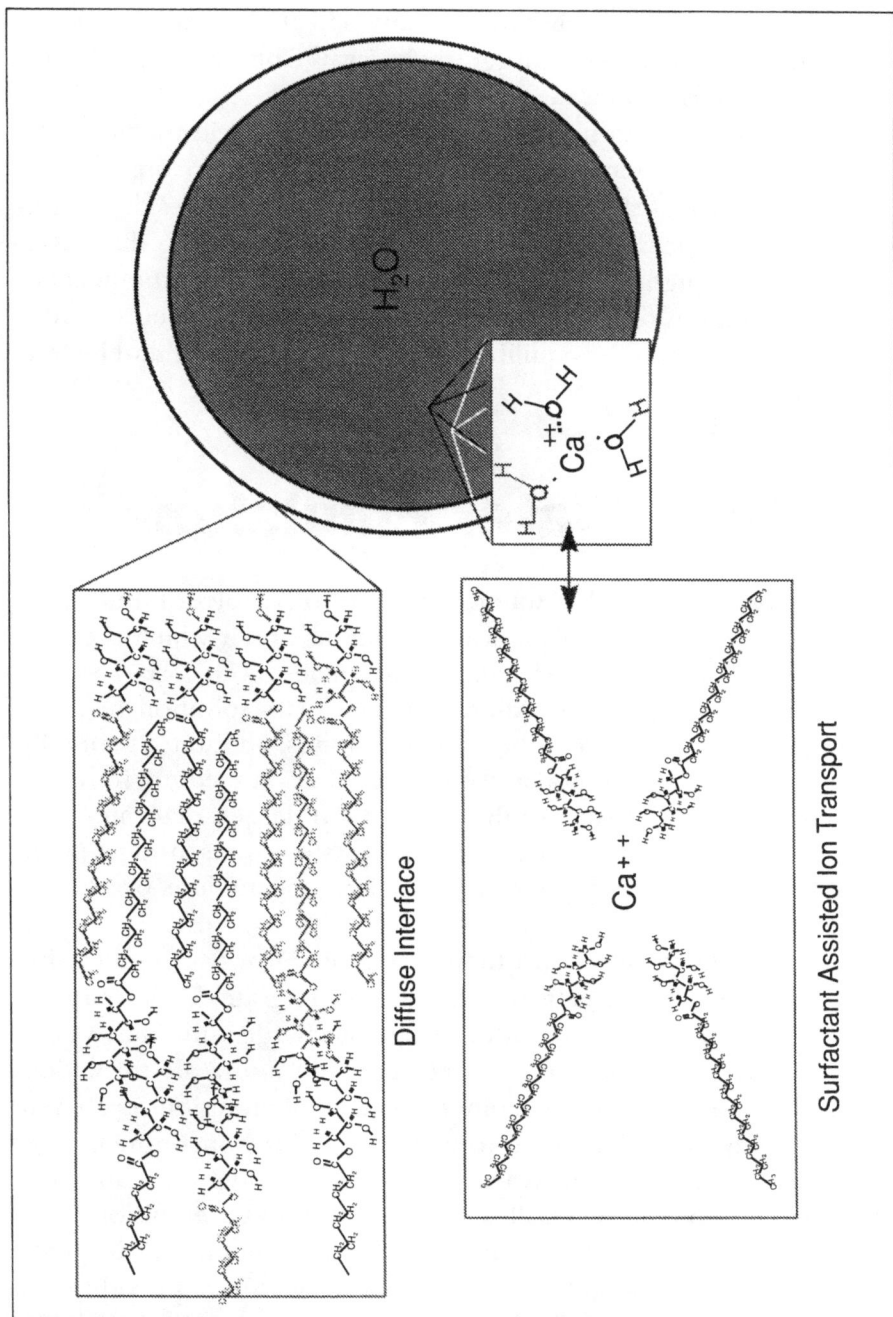

Fig. 1-1 Surfactant-assisted ionic transport, water in oil emulsion

place, the interaction established with fatty acid anions exhibits a higher stability. This results in larger shifts away from the equilibrium between the cation and water.

It is interesting to note that many of the fatty acid metallic salts are highly soluble (more appropriately, dispersible) in organic solvents. The bipolar surfactant acts to form micelle structures that envelop the cationic species and present a nonpolar exterior to the solvating phase. Although the charged cation is buried within the micelle, the partial content of water in the diffuse interface competes for the cation. This facilitates the equilibrium between the bipolar and hydrated complexes.

Oil in Water Emulsions

The presence of oil in water emulsions in petroleum fluids provides a very different set of physical conditions under which the processes of corrosion and scale formation take place. Both corrosion and scale processes involve the presence of ions. The solvation of these ions requires the presence of high polarity fluids, and water generally constitutes the most common and effective solvation medium for these ions. The nature and extent of the solvated ions largely determine the forms corrosion and scale take. Thus, fluid systems presenting a polar external phase to a surface contain a majority of the ions present in the system.

These polar solvents and their guest ions provide a conductive matrix through which electronic charges can migrate. The capability of a polar external phase to conduct electrical charge is intimately connected to the forms the processes of corrosion and scale formation take in the system. Oil-external-phase water in oil emulsions present a predominately nonpolar phase to a surface. This nonconductive or poorly conducting medium is less capable of providing a matrix for charge migration. Figure 1-2 illustrates an oil in water emulsion. The oil in water emulsion and the cation/surfactant complex, as illustrated in Figure 1-2 by the metallic salt of stearic acid, also provide sites that facilitate the attainment of equilibrium.

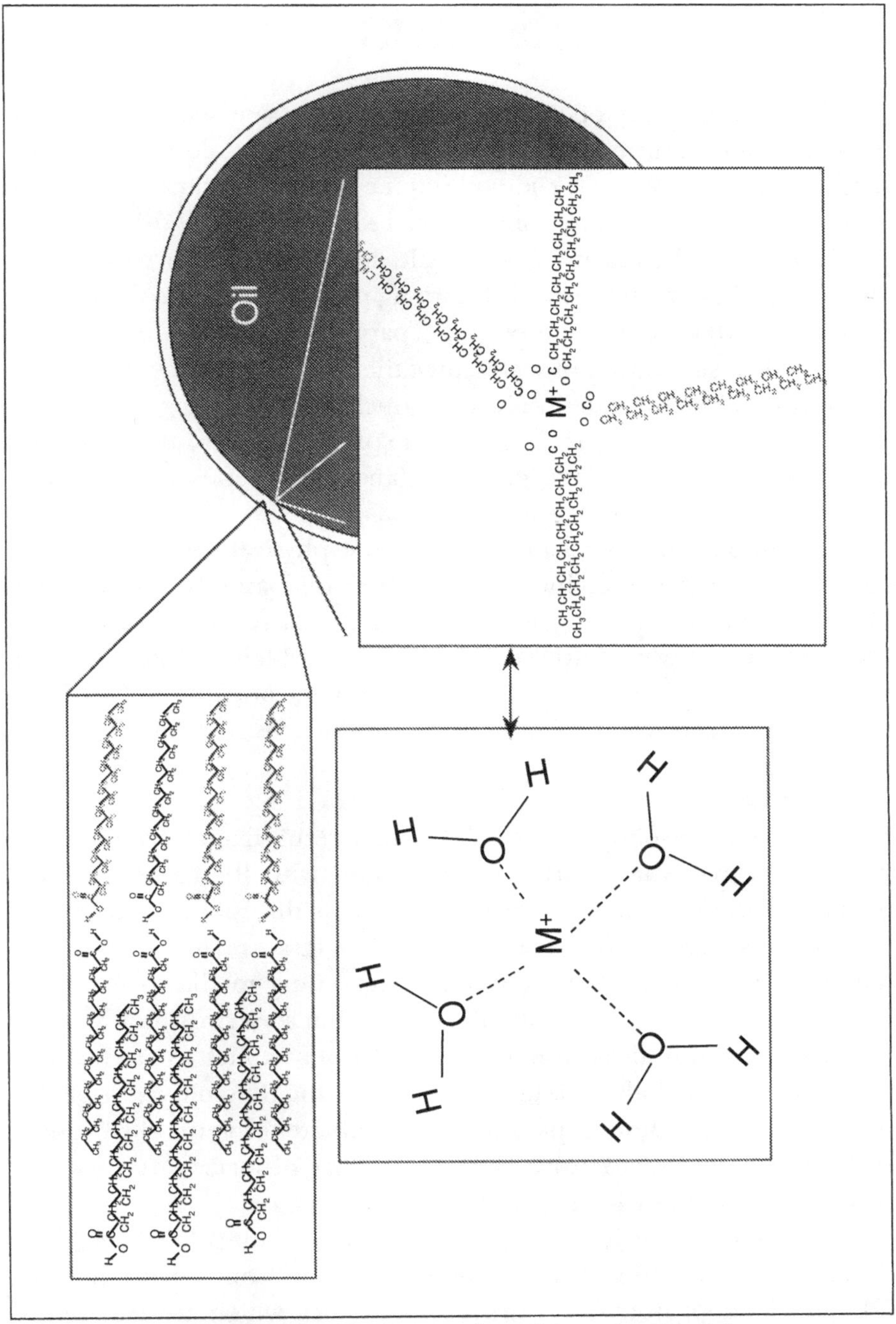

Fig. 1-2 Surfactant-assisted ionic transfer, oil in water emulsion

Solids

The interaction of scaling species and dispersed solids is an important mechanism in the processes leading to scale formation in petroleum fluid systems. Crude oil often contains substantial amounts of insoluble dispersed and nondispersed solids. These solids are either produced from the reservoir (sand, silt, and clays, etc.) or result from changes of shearing forces and streaming potential (e.g., asphaltenes), pressure, and/or temperature (e.g., paraffin crystals). These solids often seed crystallization of scale, since they provide surfaces that favor site growth of the saturated (as yet nonnucleated) scaling systems. This capacity for nucleating scale growth explains, in part, the frequent inclusion of silicates, clays, paraffins, and asphaltenes in the scales formed in crude oil fluids.

An important aspect determining the physical forms that scale assumes involves the co-incidence of the solid and adhered water. Migrating solids possessing solvation sheaths consisting of water have the necessary intermediary medium through which hydrated scaling species may achieve contact. It is this contact and the subsequent bridging between scaling sites that lead to the gross formation of scaling networks.

Corrosion processes are also facilitated by the presence of hydrate-sheathed solids. These solids either result from formation silicates and clays, which carry connate water with them, or they are formed in the production stream by aqueous diffusion. In the diffusion process, gaseous (unprotected) water migrates from water internal emulsions through the nonpolar intervening media to sites providing concentration gradients (those sites that seek hydration water). Physical discontinuities within the nonpolar phase, such as sand, paraffin crystals, or asphaltenes, provide surfaces for the collection of the diffusing water. Although paraffins and asphaltenes generally present surfaces of a nonpolar nature, the co-incidence of surfactants provides surface compatibility for the diffusing polar gases.

Although water is the predominant polar molecule undergoing diffusion within crude oil systems, it is not the only one. Hydrogen sulfide and its deprotonated hydrosulfide radical, which are frequently present in crude oil systems, are also capable of undergoing diffusion. Hydrogen gas and hydrogen radicals that result from electrolytic processes taking place within the crude oil system are also present. Carbon dioxide, carbon monoxide, and radical carbon monoxide are present in variable concentrations as diffusing gases in crude oil systems. These diffusing gases combine with the absorbed water of the solids present in the crude oil system and initiate complex ionic interactions leading to corrosion and scale formation.

Combined Emulsion Forms

Crude oils containing water generally possess both water in oil emulsions and oil in water emulsions. This co-incidence of both emulsion forms plays an important role in the nature of the corrosion and scale processes. Examination of the diffuse regions (interfaces) of each emulsion form provides a glimpse of possible mechanisms leading to scale products and corrosive media. Thus, the equilibria facilitated by the formation of bipolar metallic complexes also provide pathways by which complexed ions can be exchanged.

Processes known as "phase transfer" by chemists facilitate these exchanges. Phase transfer reactions are often performed in the laboratory by altering the solubility characteristics of an ionic species. This alteration allows it to be compatible with, and reactive toward, the solvent and reactant (respectively) in which it is dispersed. This phase transfer phenomenon can be used to rationalize many of the forms of scale that result from the admixture of oil external and water external emulsions. Given the structural arrangements of these two forms of emulsion, it is not difficult to visualize combined structures that represent a further progression to higher organization or lowered entropy. Figure 1-3 combines the two emulsion forms to show possible routes to scale products and corrosive species.

These increasingly complex structures (emulsions within emulsions) can, in theory, repeat their pattern of dissimilar phase inclusion until the limit of molecular dispersion is reached, if the concentration of surfactants is sufficient. Surfactant (bipolar) molecules in multiple

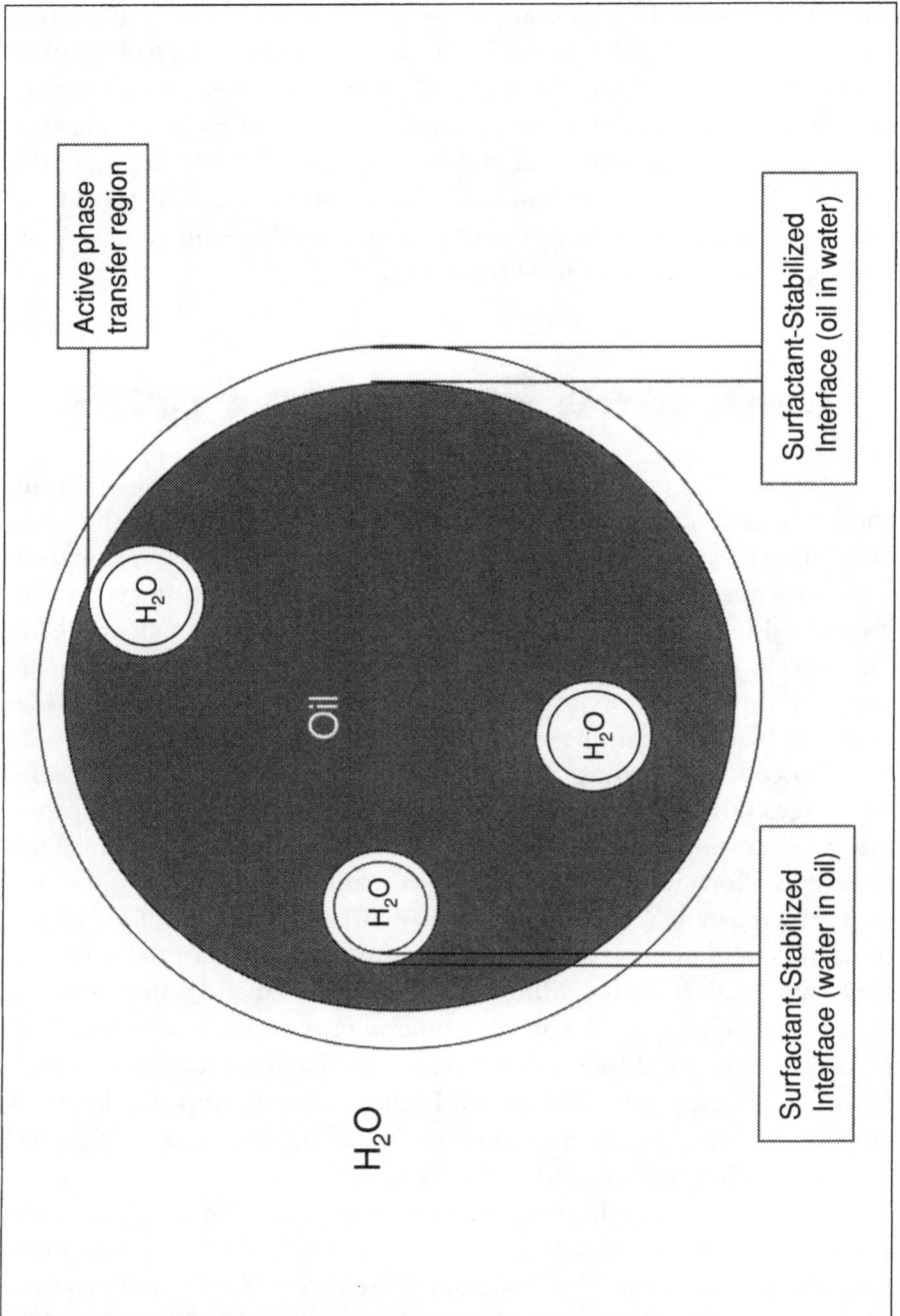

Fig. 1-3 Combined oil in water and water in oil emulsions

phase systems facilitate the spontaneous (energetically favored) formation of structures capable of supplying pathways for additional chemical and physical processes.

The degree of complexity of these functional structures ranges from quite simple to the very complex structures involved in biological functions. In biological systems, the bipolar surfactants often undergo molecular reactions to form polymeric products that impart strength to the containment vessel (emulsion). These structural systems then provide pathways for processes such as salt uptake, aqueous diffusion, and pH control. They afford the appropriate reaction environments for further molecular synthesis.

Summary

The gross manifestations of corrosion and scale forms in crude oil systems often differ from those expected from highly differentiated or resolved component phase systems. Pure ionic brine waters will produce scale and corrosion effects that exhibit considerably different gross character than those systems that contain emulsions, paraffin crystals, asphaltene particles, sand, and silt.

This discussion provides a basis for understanding the complications introduced by the presence of different surfaces in crude oil systems. However, it is a general discussion that intentionally avoids elaboration on the types of scale and corrosion involved. In the following chapters, the types of corrosion and scale processes will be brought into focus.

References

Barrow, Gordon M. 1966. *Physical Chemistry*. 2d ed. New York, St. Louis, San Francisco, Toronto, London, and Sydney: McGraw Hill Book Company.

Dickerson, R. F., H. B. Gray, and G. P. Haight, Jr. 1970. *Chemical Principles*. 1st ed. New York: W. A. Benjamin Inc.

Huheey, James E. 1978. *Inorganic Chemistry Principles of Structure and Reactivity.* *2d ed.* New York, Hagerstown, San Francisco, and London: Harper & Row.

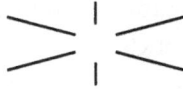

2
An Overview of Petroleum Fluids and Scale

Typical Forms

Oil field scale reflects the geologic composition of the formation from which the crude oil fluids are taken. A large percentage of petroleum-producing reservoirs occurs in regions that are rich in carbonates, sulfates, and silicates that provide considerable concentrations of anions. The abundance of cationic species present in surrounding geologic formation influences the forms the mineral layers will assume. High percentages of silica, calcium, and iron provide cationic species for combination with the anionic ions available.

Over the vast periods of geologic time, the processes of biodegradation and aqueous percolation through varying types of mineral layers have produced ionic brine waters. These brine waters are responsible for much of the crystal formation in the voids, which were created by processes that occurred in an earlier geologic timeframe. Water, ice, and wind erosion continue to reshape the surface of the earth, while periods of equilibrium are established and abolished.

In the geologically brief time during which the fluids contained in a producing formation are confined to regions of low fluid mobility, the system will have reached an apparent equilibrium. Thus, in the context of nongeologic time, the compositional changes have approached a state that appears to be constant. However, when the formation fluids are provided an avenue of escape, such as an oil well, the dynamics of the system change. As gases and liquids escape through the exit provided by the well, pressures and temperatures drop, and shearing forces change. With these physical changes a shift of established equilibrium occurs, and processes such as scaling and corrosion begin.

Solubility products of anion/cation interactions are extremely important in the processes leading to scale formation. The calcium carbonates, calcium sulfates, iron sulfides, silicon oxides, barium sulfates, and barium carbonates all have very low solubility in aqueous systems. Further, although barium cations are not as abundant as those of calcium and iron are, their extremely low solubility sulfate and carbonate forms often insure deposition. Scale forms found in producing wells tend to be those derived from the combinations of soluble salts of cations and anions that produce low solubility product salts and are geologically abundant.

Carbonate Scales

Of all the scales present in oil field systems, it is a safe bet that the most abundant are those consisting of carbonate. The successful precipitation of calcium carbonate depends upon the equilibrium:

$$Ca^{2+} + 2HCO_3^- \rightleftharpoons CaCO_3 + H_2O \ (K_{sp} = 10^{-8} \text{ mole/liter at } 20° \text{ C})$$

The widespread distribution of dissolved calcium bicarbonate contained in formation brine solutions favors the formation of calcium scales when appreciable amounts of calcium ions become available. The widespread occurrence of the soluble bicarbonate is due, in part, to the broad biological use of this soluble carbonate to produce rigid shells, exoskeletons, and bones. The form of deposit taken by the low solubility product (K_{sp}) calcium carbonate scale is highly variable, and depends on several factors.

Temperature and pressure are two physical factors that are extremely important in determining the scale form assumed by the low solubility salt. High temperatures and decreasing pressure favor scale formation, since these conditions favor the concentration of the ions within the solvent water by the gaseous diffusion of water from the solutions.

The period of time required for the gaseous water to exit the ionic solution is dependent on both conditions of temperature and pressure. However, this time factor also plays a role in the type of crystal structures formed. If microscopic examinations are performed on scale products formed under different conditions of temperature and pressure, gross morphological difference are visible. These visible differences are the result of the extent to which these salts interact to form multicomponent aggregates or crystal networks.

Generally, the more time required for the gaseous diffusion of the solvent water, the more extensive the crystal network becomes. Herein lies a subtle reason for the morphological differences observed in crystal forms as a result of the presence of surfaces in crude oil systems. Since increasing surface tension hinders gaseous diffusion, and emulsions and solids provide surfaces through or from which water must pass, the effluent and influent rates of gaseous water are hindered. Thus, the free diffusion of water from a system consisting of only water and salt will form crystals with different morphology than a system containing an emulsion or other surface.

Coordination numbers possessed by cationic species also determine the morphology of the crystals formed. Calcium and barium cations (2+ species) can possess up to a dozen available coordination sites. The combinations of these coordinate sites with ligand groups within the crystal provide the structural form for crystal growth.

Although coordination numbers greater than eight are seldom found, there is a potential for their inclusion into the growing crystals. The predominant forms of both barium and calcium carbonate scale reflect most common coordination complexes available to these cations, and generally range from six to eight ligand sites. The most common coordination numbers available to the barium and calcium cations are illustrated in Figure 2-1.

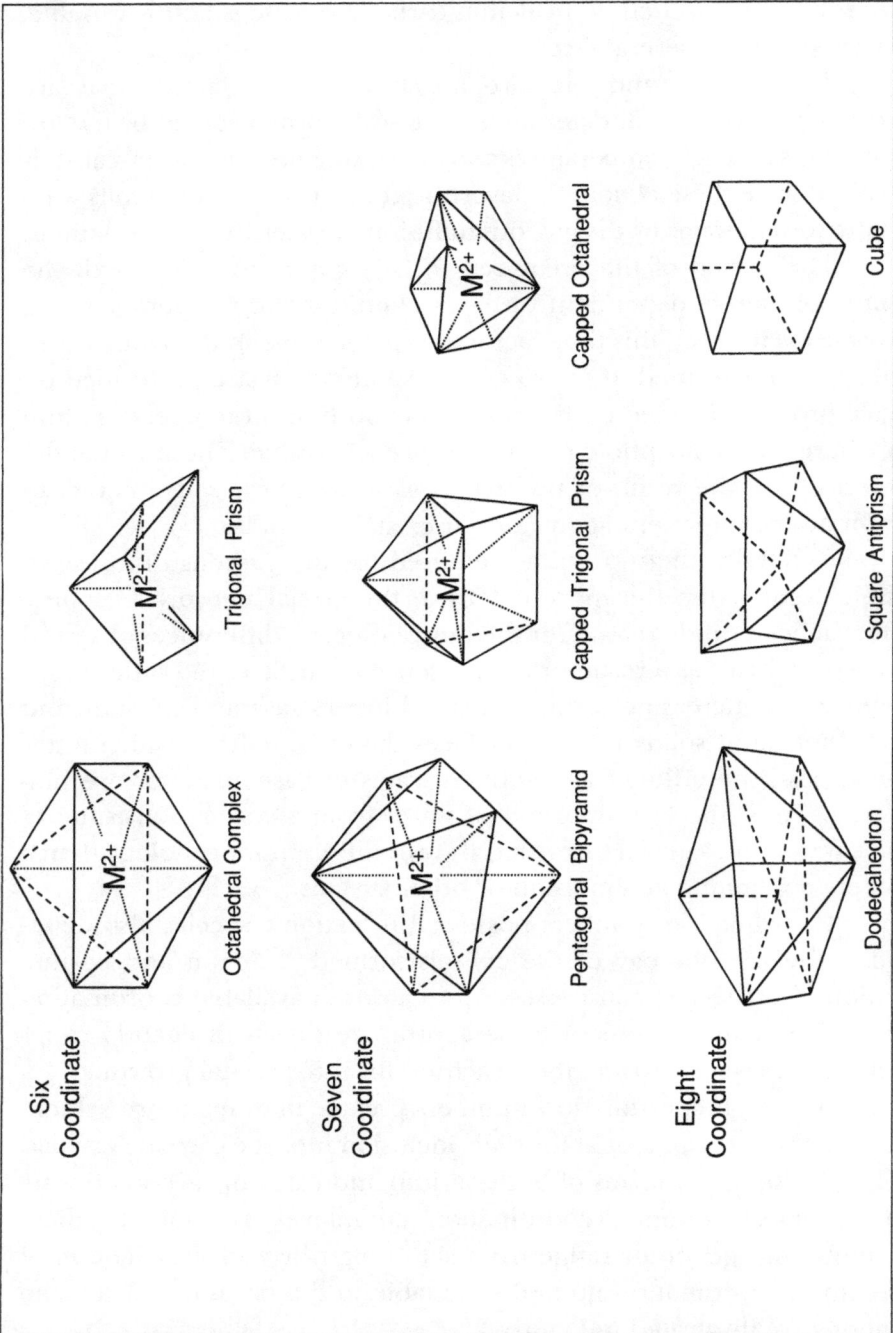

Fig. 2-1 Common crystal forms of six, seven, and eight coordinate cations

The crystals derived from the coordination numbers available to cations from Group 2A of the periodic table (including barium, calcium, and magnesium) are capable of a variety of morphologies. The carbonate anion is capable of occupying two vertices of the crystal forms listed above, leaving four, five, and six ligand sites available for combination with other ligand groups. Thus, Figure 2-2 shows some likely arrangements of the carbonate anion with the various coordination crystal arrangements in Figure 2-1.

In aqueous systems, the most common ligand group available is the water molecule. Thus, the extent to which coordination water is incorporated is a function of the coordination sites remaining available to the neutral metal salt. Therefore, the crystal forms of the carbonate salts of calcium and barium would be expected to include four, five, and six sites for the coordination of water. The hydrated cations are more stable than the isolated cations, and as a consequence, the hydrated forms are favored over the nonhydrated forms.

Sulfate Scales

The coordination numbers of the central metal are not altered by the change of the anion from carbonate to sulfate. However, the metallic salt resulting from the neutralization of charge by the larger and stronger sulfate anion causes a greater distortion of the coordinated complex. Thus, the crystal forms of the multiple combinations of complexed salts present different morphologies when the individual unit cells are deformed. The deformation of the complex also reduces the ability to accept water at the open ligand sites. This inability to accept coordination water changes the solubility characteristics of the salt complex.

A general trend of increasing water solubility is observed with increasing coordination by water at the available ligand sites of the metallic salts. Therefore, decreases in the number of available ligand sites result in lower water solubility. Coordination numbers (ligand sites) assumed by the central metal depend on the availability of energy levels possessed by the metal. They are determined by the physical

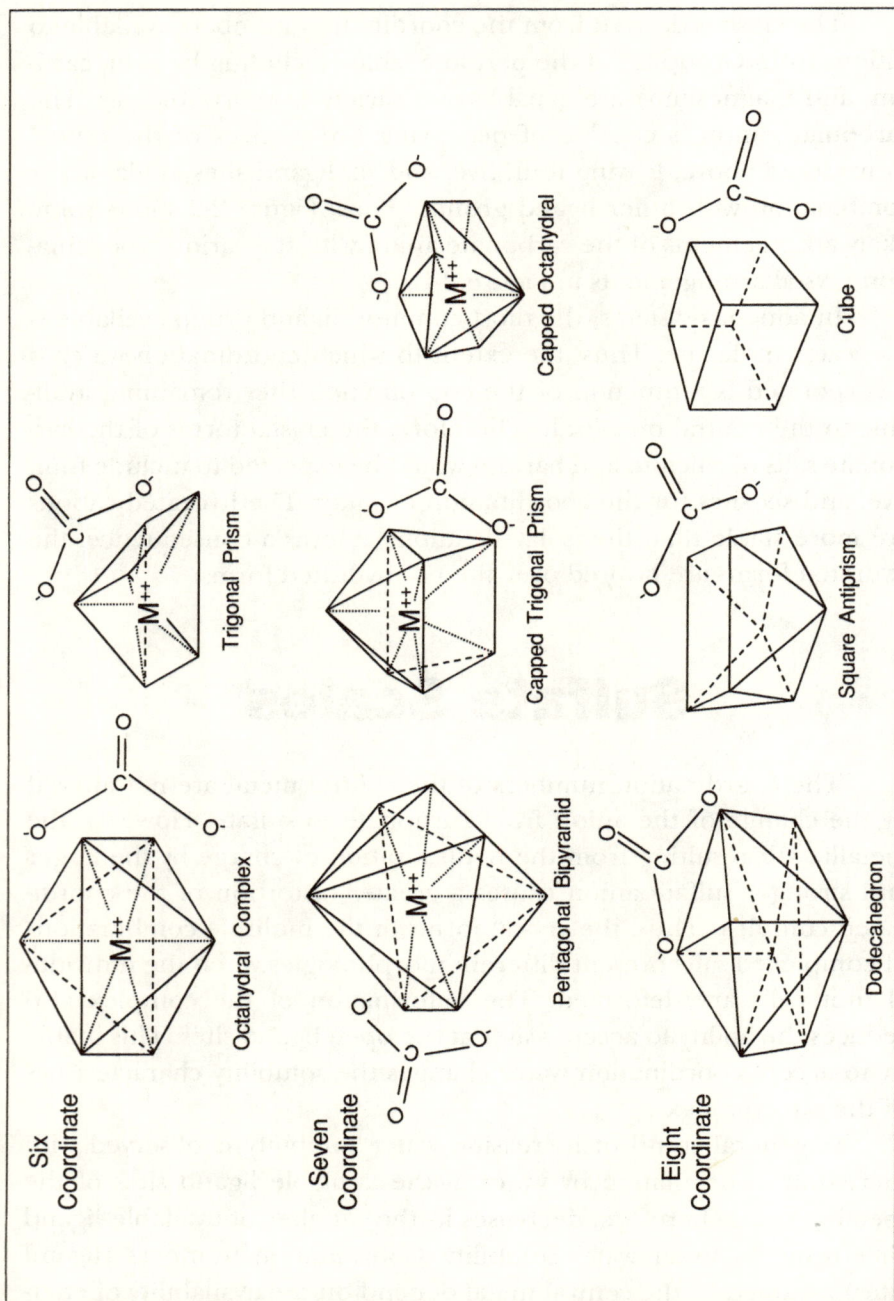

Fig. 2-2 Carbonate salts of six, seven, and eight coordinate cationic complexes

conditions to which the metal cation is exposed. Conditions of high thermal energy generally favor higher numbers of available energy levels for combination with potential ligand groups.

Since coordination sites consist of energy levels, they are subject to quantum restrictions that demand that these levels be discrete. Thus, the highest coordination number is determined by the highest quantum state reached by the available metallic orbitals. Electronic effects exerted by ionic ligand groups also affect changes in the nature of the complex and the coordination number assumed by the metallic cation. Thus, the environmental effects exerted by anionic species act to increase the energy of the central cation by providing an electrostatic potential difference. This potential difference can be high, as in the case of the strong sulfate anion, or somewhat lower, as is the case with the carbonate anion. Figure 2-3 indicates some of the most common coordination conditions taken on by the sulfate complexes with the divalent metals.

The distorted forms of the metal complexes are not shown in Figure 2-3. However, it should be realized that the stronger ionic forces possessed by the sulfate anion can result in a greater distortion of the geometry of the complex. Distortions of the complex can range from a complete conversion of its geometry to simple elongation or contraction along a given axis of symmetry. Figure 2-4 shows some examples of these geometrical distortions. Since distortions of the complex geometry affect the gross crystal morphology, the crystal types of scale formed from sulfate anion salts should be different than those formed from carbonates.

Comparison of the alkaline earth metal salt complexes found in some crude oil systems indicates that salts of large ions follow trends expected if changes in hydration energy dominate. Thus the solubility trends show the following order:

$$BaSO_4 < SrSO_4 < CaSO_4 < MgSO_4$$

Some data about the size and solubility behavior of some of the major types of scale encountered (at $25\,^\circ C$ and 1 atmosphere pressure) in the oil field are listed in Table 2-1.

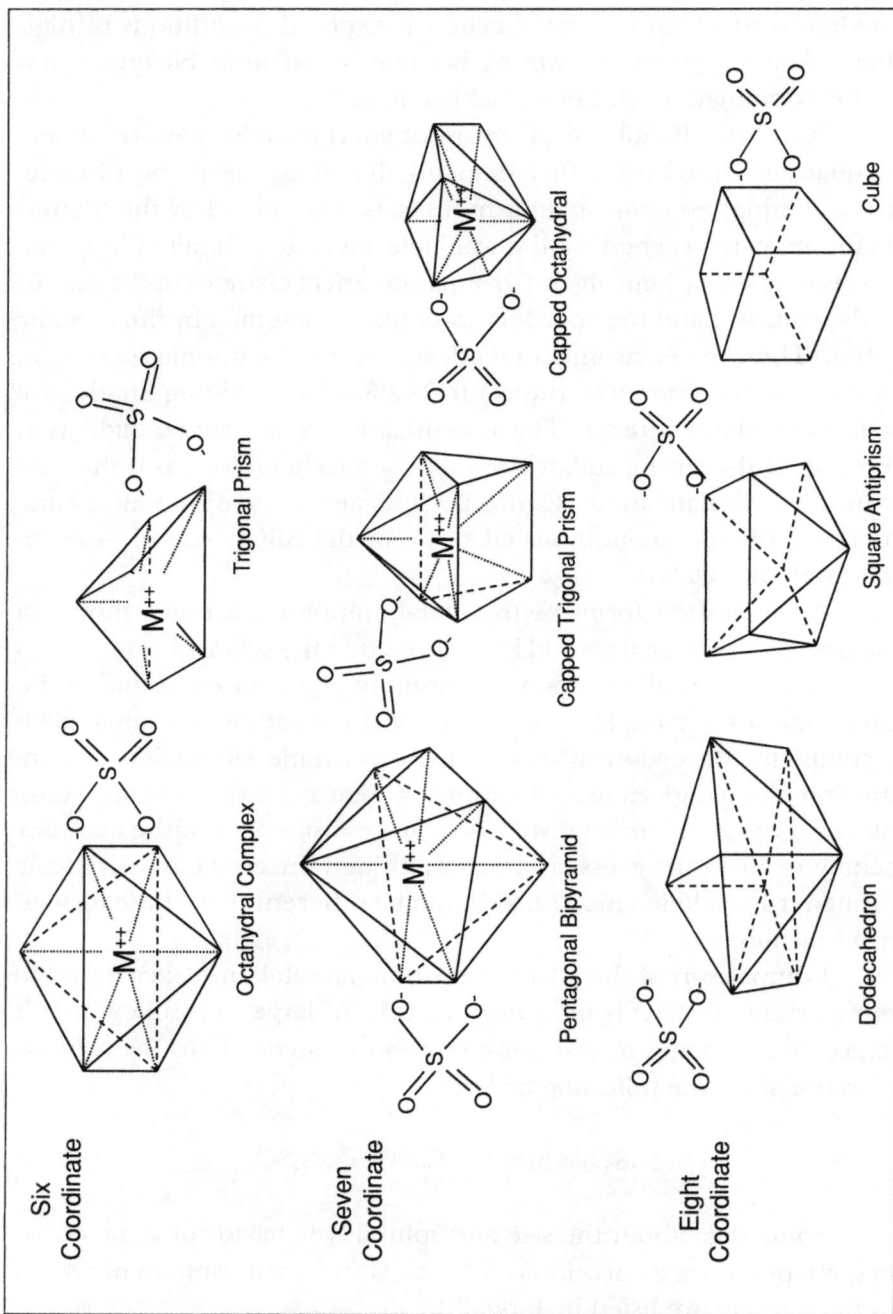

Fig. 2-3 Divalent metal sulfate complexes

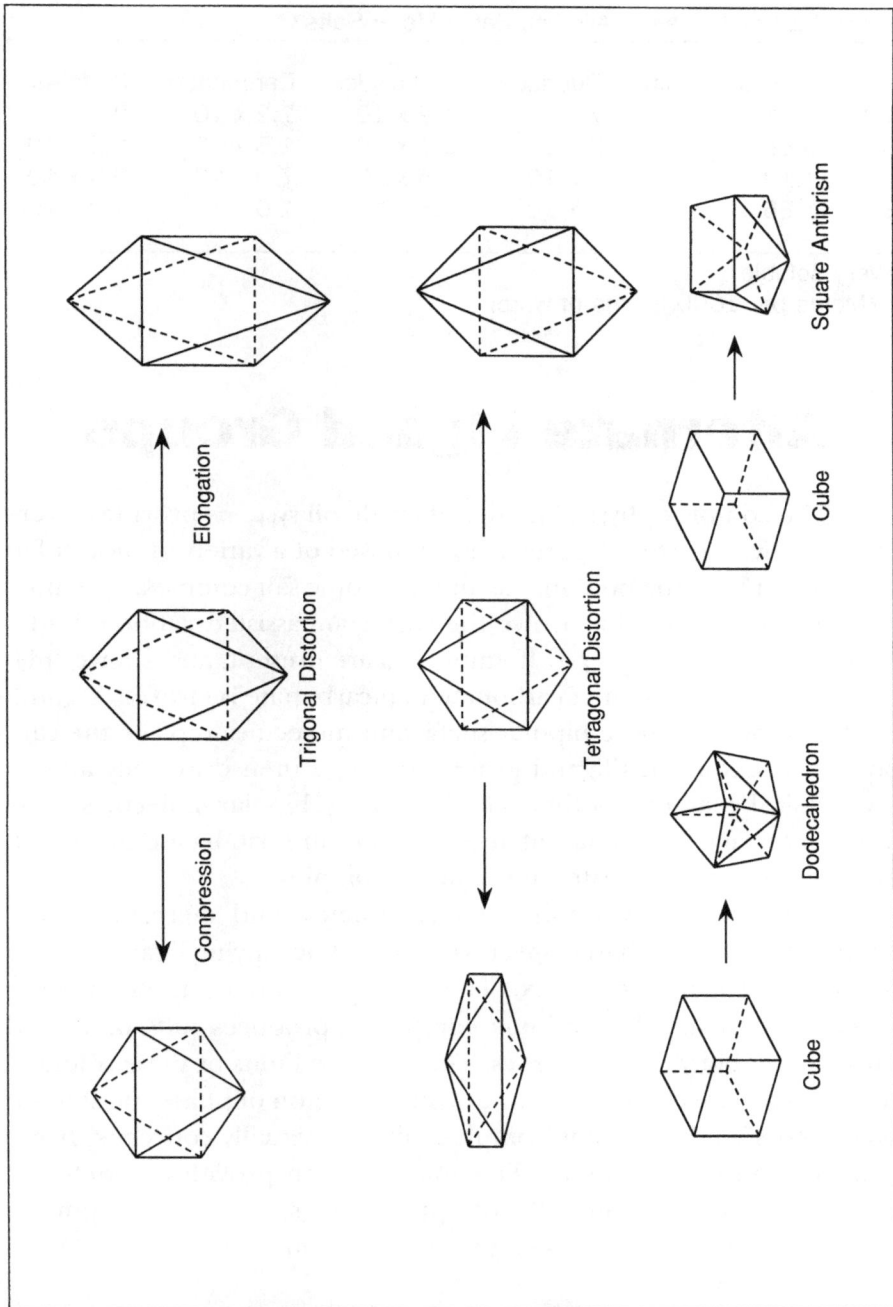

Fig. 2-4 Distortion of six and eight coordinate octahedral and cubic complexes

Table 2-1 Solubilities of Alkaline Earth Metal Salts**

M^{2+}	Radius (10^{-8}cm)	Fluorides	Hydroxides	Carbonates	Sulfates
Mg^{2+}	0.31	v.s.*	1.2×10^{-11}	1.2×10^{-3}	2.4
Ca^{2+}	0.65	2×10^{-4}	2.1×10^{-2}	1.5×10^{-4}	1.5×10^{-2}
Sr^{2+}	1.13	1×10^{-3}	6.5×10^{-2}	7.0×10^{-4}	5.0×10^{-3}
Ba^{2+}	1.35	1×10^{-2}	2.8×10^{-1}	1.0×10^{-4}	1.0×10^{-5}

*Very soluble
**Moles per 1000 Grams of water

Alternate Ligand Groups

The complex physical nature of crude oil systems provides several alternative structural aggregates composed of a variety of molecular species that have the potential to alter the ionic salt complex. One possible mechanism might involve the surfactant-assisted cationic transport discussed in chapter 1. If emulsions are present, and a lower pH prevails, the conversion of carbonate to bicarbonate is favored. Ligand exchange occurs where bipolar surfactant molecules replace the carbonate di-anion. This ligand group exchange then drastically affects the physical character of the scale. In this way, bipolar molecules within the interfacial boundary of an emulsion can form ligand groups at available coordination sites in the metal complex.

Fatty acylates, alcohols, amines, sulfates, and mercaptans are among some of the bipolar species capable of occupying ligand sites of the metallic cation complex. The incorporation of these bipolar species within metallic cationic complexes produces salts of widely variable solubility characteristics. The extreme limits of these solubility alterations are represented by the phenomenon of phase transfer. In phase transfer, the metallic complex salts are actually converted from polar to nonpolar solubility. This phenomenon provides models for scale system behavior and affords opportunities for the development of methods aimed at controlling scale formation.

Chelation Complexes

Sulfate anion displacement from the metallic salt complex is more difficult than that of carbonate. The stronger sulfate di-anion (higher pK_b or lower pK_a value) requires a drastically lower pH (higher concentration of protons) to protonate the $BaSO_4$. This is illustrated in the following reactions:

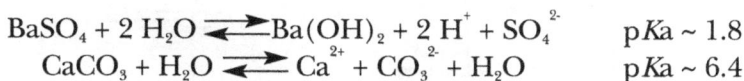

$$BaSO_4 + 2\ H_2O \rightleftharpoons Ba(OH)_2 + 2\ H^+ + SO_4^{2-} \qquad pKa \sim 1.8$$
$$CaCO_3 + H_2O \rightleftharpoons Ca^{2+} + CO_3^{2-} + H_2O \qquad pKa \sim 6.4$$

Thus the conditions for ligand transfer in the barium sulfate metal complex are less likely to exist in emulsion systems. Barium carbonate complexes are susceptible to similar ligand group transfer, but the larger barium cation tends to be slightly more stable due to its greater hydration energy. The possibility for ligand group transfer provides some clues as to how a successful treatment program might be devised to interfere with the formation of scale. The divalent cations of barium, calcium, magnesium, and strontium allow ionic salts to be formed using variable valence anions.

These anionic structures can range from monovalent to polyvalent forms, and depending on their molecular configuration, can produce complexed salts of variable solubility. Ethylenediaminetetraacetate (EDTA) is a good example of a multivalent anion that possesses two amine and four acetate groups capable of interaction with ligand sites of metal complexes. Carbonate scales are particularly vulnerable to attack by this anion, and its addition to systems containing metallic carbonates often results in reduction of the carbonate scale deposits. However, the success of this molecule in the control of scale is highly dependent upon the prevailing pH of the system.

Figure 2-5 shows two arrangements of the EDTA molecule with two forms of six-coordinate metallic complexes. There are several other polyfunctional anionic and nonionic molecules capable of occupying multiple ligand sites in metal complexes. A few of these anionic compounds are listed in Table 2-2.

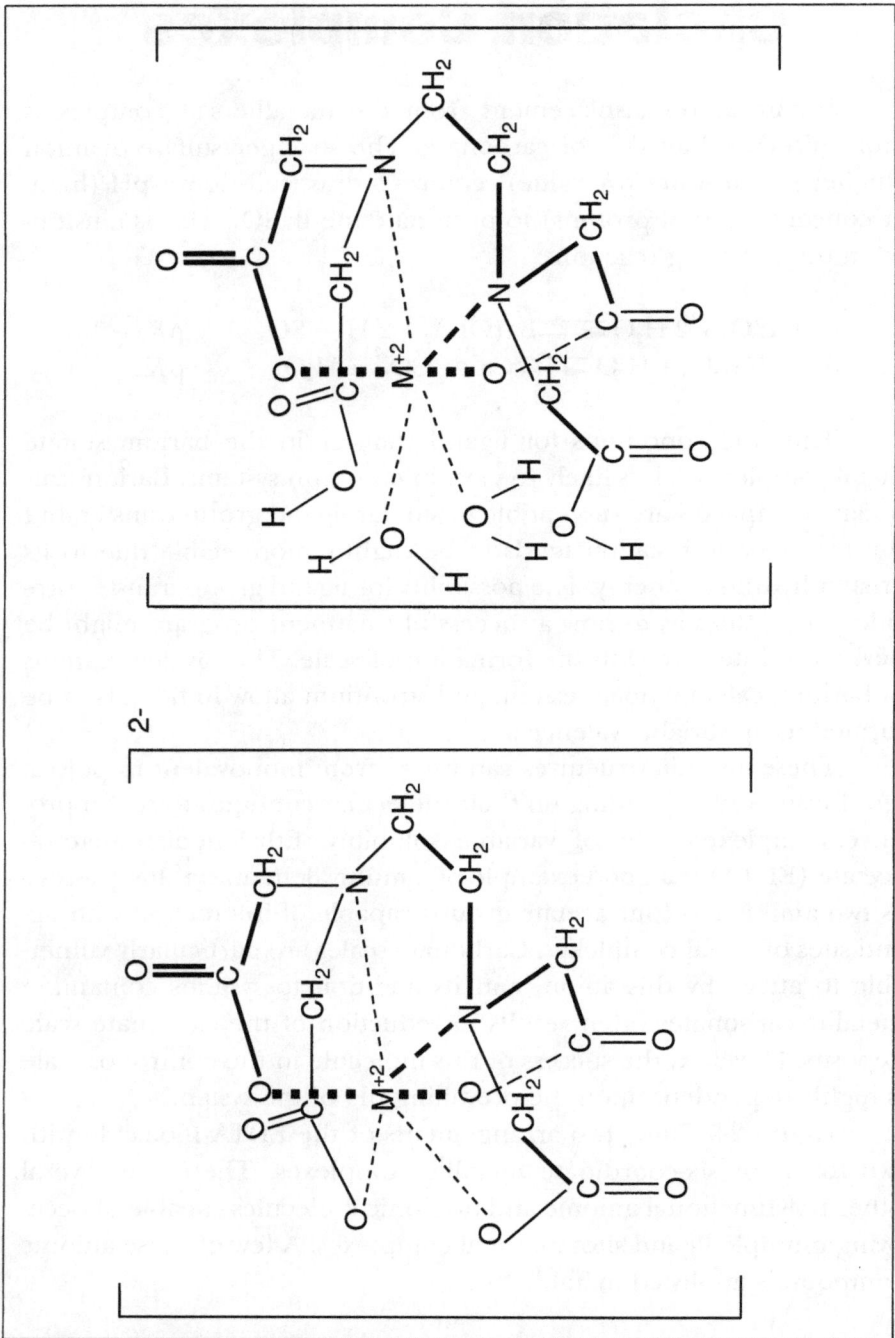

Fig. 2-5 Two EDTA complexes of a six coordinate metal cation

Table 2-2 shows some examples of multifunctional ligands, but there is a considerably larger list of available compounds that are capable of occupying ligand sites of metal complexes. Thus, polyacrylates, polymaleates, polyethers, and polyfumerates, as well as a host of others can be added to this listing. Upon closer examination of the changes in solubility of the metal complexes resulting from interaction with these multifunctional ligands, some important features emerge.

Table 2-2 Some Common Chelating Groups or Multifunctional Ligands

Symbol	Ligand Name	Formula	Bonds
en	Ethylenediamine	$NH_2CH_2\ CH_2\ NH_2$	2
pn	Propylenediamine	$NH_2CH_2CH(CH_3)\ NH_2$	2
dien	Diethylenetriamine	$NH_2CH_2\ CH_2\ NH\ CH_2\ CH_2\ NH_2$	3
trien	Triethylenetetraamine	$NH_2CH_2\ CH_2\ NH\ CH_2\ CH_2\ NH\ CH_2\ CH_2\ NH_2$	4
ox	Oxalato	$-OOC\text{-}COO-$	2

Size and composition of the nonpolar functions present in the ligand group determine the degree of dispersion in the solvent matrix. Thus, large aliphatic groups tend to drive the resulting complex into the organic phase, while smaller and higher polarity substituent groups favor the higher polarity phase (normally water). Here again, it becomes clear that the architecture or macro-aggregate structures, as exhibited by metallic complexes, lead to functional and behavioral variations in the resulting products.

Solvent Effects

Previous sections of this chapter dealt with the forms of carbonate and sulfate complexes formed by the divalent Group 2A elements in aqueous solvent systems. The forms taken by the metallic salt complexes are highly dependent on the solvent systems in which they are present. In the crude oil systems under discussion, the environmental conditions leading to the formation of these complex salts has been described as the slow, aqueous-assisted seepage of biodegrading substances to geologic reservoirs.

The intimate association and abundance of water attending this process favor the aqueous coordination products of the metallic complexes. Thus, the various coordination compounds present in the predominately aqueous solvent system will possess several water molecules as ligand groups. Some important considerations affecting the behavior of these complexes include factors such as the solvent phase concentration of different salts, conditions of temperature and pressure, hydrogen ion concentration, and hydroxyl ion concentration. Each of these factors can alter the behavior of scale formation and change the appearance of the scale products formed.

Salt Concentration

It should be clear that metal salt complexes of barium, magnesium, calcium, and strontium are capable of forming from a variety of anionic species. Salts of chlorides, fluorides, hydroxides, sulfates, carboxylates, carbonates, and sulfides are representative of the variety of anionic species capable of salt formation with these metals. It should also be clear that molecules containing electronegative groups (those having unshared electrons) are capable of taking position at coordination sites of the complex.

Water is a good example of a molecule that has unshared electrons capable of occupying coordination sites possessed by a central metal or metal salt. The sheer abundance of water present in most situations statistically favors its inclusion as a ligand group in the metallic salt complex. However, the increased interaction strength of electrostatic forces between anions and the metal cation compete with water for inclusion, even though their concentrations may be low. There are several factors that will determine the equilibrium condition of the solvent/solute system. These are:

- The extent to which ionic salts are soluble in aqueous phase
- The types of cations and anions present as ionic pairs
- The electrostatic attractive forces between cations and anions

Temperature and Pressure

Temperature and pressure also affect the solubility of the various salt forms present in the solvent system. The extent to which the salt is soluble is determined by the available energy level vacancies provided by these salts. The levels provided are functions of the temperature and pressure. Most salts exhibit higher solubility at elevated temperatures, but some salts exhibit lower solubility when temperatures exceed a particular point. Pressure effects at constant temperatures often result in vapor phase solvent escape, which acts to concentrate the salts present in the solvent system. The reason salts exhibit reduced solubility at higher temperatures is thus determined by concentration increases due to vapor release, or the failure to attain a higher energy level vacancy condition.

A good example of this second reason is provided by the behavior of barium sulfate scale. The aqueous complex of barium sulfate decreases as the temperature approaches approximately $80\,^{\circ}C$, where the coordination number decreases from eight to six, and water ligands depart.

pH

The pH of the solvent is an important factor affecting the nature and concentration of metallic salt complexes. In many instances the anion contained in a complex can form a weak acid with a hydrogen ion, as is true in the case of carbonic acid from calcium carbonate. This weak acid is considerably more soluble in water than the calcium carbonate parent complex and after a short time decomposes to carbon dioxide and water. Similar solvent effects can be attributed to the presence of hydroxyl anions in a solution of complex metal salts. Thus at high pH, the exchange rate of hydroxyl for carbonate anions as complex ligands is high. This replacement of carbonate with hydroxyl anions leads to the increased solubility of barium salts.

Approaches to Scale Treatment

The preceding sections of this chapter discussed some of the characteristics of the Group 2A alkaline salt complexes, so that some of the alteration and control methods used might be better understood. A large variety of chemical compounds have been developed to take advantage of behavioral changes resulting from ligand and ionic group replacement. These chemicals are frequently designed to include multiple functional groups that consist of polar and anionic species capable of interacting with the metal salt complex.

Often, advantage is taken of the "chelate effect," which adds to the stability of the resulting complex. This effect can be characterized by example. Ammonia and polyamines often displace ligand groups from sites of a metal salt complex in an equilibrium reaction. When the ammonia departs a ligand site, the probability that it will return to occupy the same site is extremely low. However, when one amine group of a polyamine is displaced from a ligand site, there is a good chance that it will return to the same site later. This phenomenon can be understood when one considers that the distance the displaced amine group must move to reoccupy the same site is very small. The configuration of the molecular bonds of the compound favor such a reorientation.

The ability to bridge is often engineered into a potential scale treatment chemical and involves the use of multivalent anionic compounds containing several anionic sites. Since many scales typically found in the oil field are composed of the bivalent 2A series elements, petroleum companies use chemicals possessing several anionic sites. These anionic compounds (with greater than an equivalent number of opposite charges possessed by the metal) have the potential to interact with more than one metal salt complex.

If the stability of the metal salt complex was the only factor determining the deposition of scale, it would seem that these two approaches to scale control would be reasonably adequate. However, complications arise because of variable solubility effects due to partitioning of these complexes between differing solvent environments. The bipolar nature of many anionic ligand group replacements frequently affords migration paths from one solvent to another within a crude oil/water system.

The dynamics introduced by these migration pathways can shift the equilibrium in unexpected directions, acting either as feed sources for the inflow of cationic metals or as sinks for their exit. Thus, issues of solubility and phase transfer must be addressed in order to effectively control scale-forming tendencies. Some functional design criteria for effective scale control can now be listed as follows:

- Complexes that possess highly stable ligand groups
- Complexes formed from molecules containing multiple ligand groups
- Molecular complexes that bridge multiple salt complexes
- Molecular complexes that contain variable phase solubility

Thus, synthetic chemical designs intended for use as scale control chemicals should, at a minimum, incorporate these criteria. Additionally, pH, temperature, and pressure sensitivity of the product complex can be considered subcategories.

Supplementary Ligand Groups

Phosphate esters, phosphonates, dithiophosphate esters, sulfamates, sulfonates, nitrates, nitrites, carbamates, thiocarbamates, amino acids, glycolates, thioglycolates, and many other polar anionic groups are capable of occupying ligand sites in metallic complexes. Figure 2-6 illustrates some of these structures. Chemists frequently incorporate one or more of these active ligand replacement groups into the synthetic design of scale treatment chemicals. Figure 2-6 also indicates that the R group can consist of alkyl, aryl, or alkyl aryl groups. This flexibility allows the synthesis chemist control over the solubility of the synthetic compound.

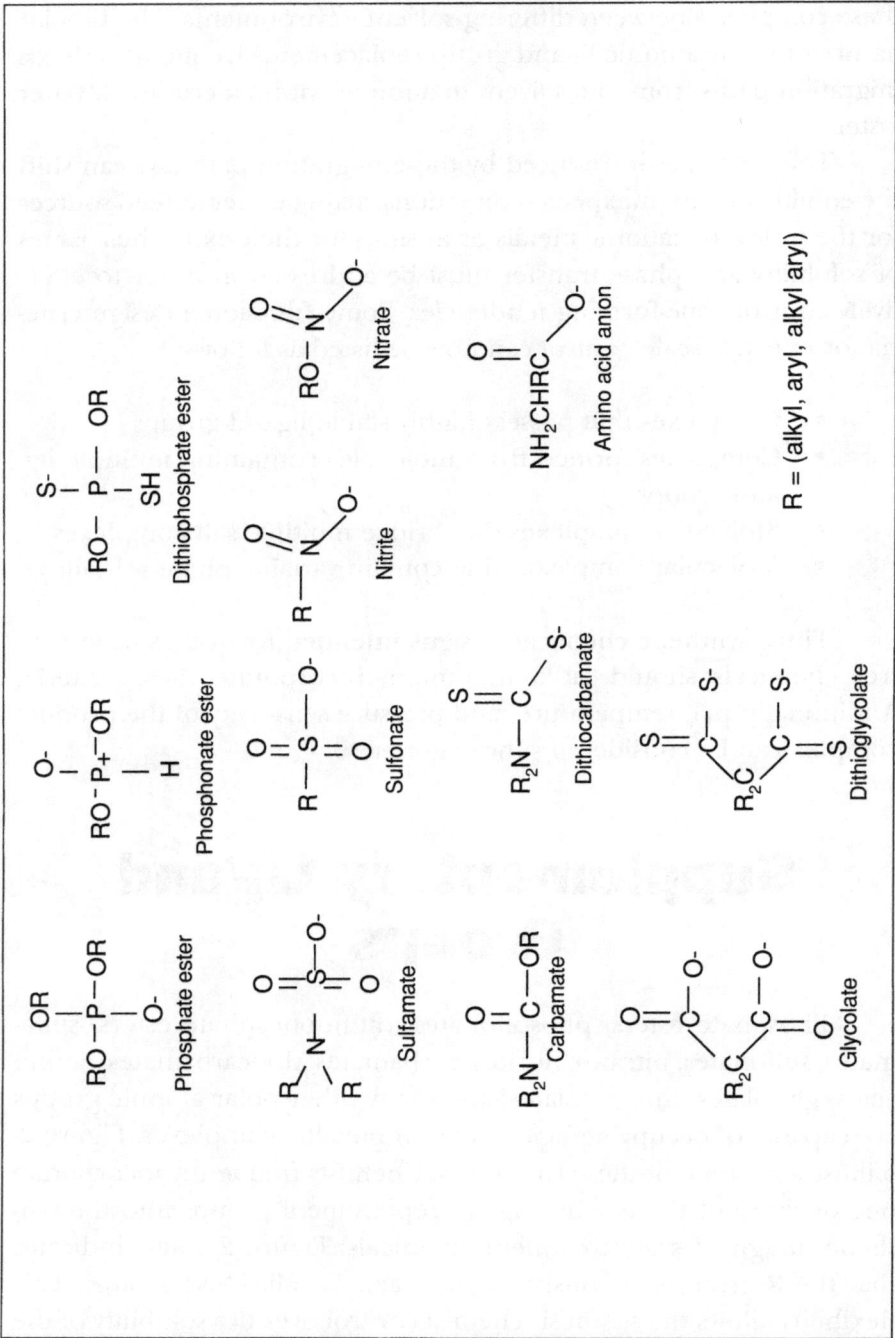

Fig. 2-6 Some additional anionic ligand group structures

In addition to the presence of anionic groups, the inclusion of electronegative substituents at positions within the synthetic molecule can enhance the chelate effect. This practice often involves the use of polyamines, polyethers, or combinations of each in one or more of the R groups of the compound. Additional discussions of the macro-molecular aspects of structure versus function will be presented in later chapters, but this brief sojourn into the area should pique the reader's interest for what lies ahead.

Summary

The preceding chapter provides a brief introduction to the common types of Group 2A metal salt complexes that occur frequently in oil field fluids. Geometric structures assumed by the six, seven, and eight coordinate forms were illustrated to show the spatial orientation of the metal cations. Carbonate and sulfate salts were used to illustrate some trends in solubility of the metal salts in connection with the size of the metal cation and the strength of the anions. Continued emphasis on the phase transfer phenomenon, presented in the preceding chapter, was pursued in connection with these metal complexes.

Representative chelate structures were introduced to illustrate some of their functional group properties and demonstrate their effect on metal complex stability when they are incorporated by the metal salt. Brief discussions of the effects of temperature, pressure, ionic concentration, pH, and solvent were related to the stability of the various complexes. These discussions led into the section dealing with the factors that determine the stability of complexes, listing a set of four criteria for achieving complex stability. Finally, a short discussion was conducted pointing out some approaches to building a successful chemical treatment for the control of scale.

References

Barrow, Gordon M. 1966. *Physical Chemistry.* 2d ed. New York, St. Louis, San Francisco, Toronto, London, and Sydney: McGraw Hill Book Company.

Dickerson, R. F., H. B. Gray, and G. P. Haight, Jr. 1970. *Chemical Principles*. 1st ed. New York: W. A. Benjamin, Inc.

Hamill, William H., and Russell R. Williams, Jr. 1966. *Principles of Physical Chemistry*. 2d ed. Englewood Cliffs, New Jersey: Prentice-Hall.

Huheey, James E. 1978. *Inorganic Chemistry Principles of Structure and Reactivity*. 2d ed. New York, Hagerstown, San Francisco, and London: Harper & Row.

Noller, Carl R. 1966. *Textbook of Organic Chemistry*. 3d ed. Philadelphia and London: W. B. Saunders Company.

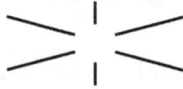

3
An Overview of Petroleum Fluids and Corrosion

Mild Steel

Oilfield corrosion processes that most adversely affect the economics of production, transport, and refining are primarily those that involve the destruction of iron. The low cost, ease of equipment fabrication, structural strength, and availability of mild steel make it one of the metals of choice of petroleum companies. Most mild steels are interstitial alloys of iron containing small amounts of carbon and other atoms. High-carbon steel consists of 0.75% to 1.5% carbon, situated in the octahedral holes of the iron lattice. Some of the other impurities that produce electrolytic cells, thereby enhancing the potential for corrosion, are displaced by the incorporation of carbon into the iron.

The smelting of iron ore almost always results in the inclusion of several other metals that are found in close association with iron. Thus, the impurities in the metallic iron remain throughout the fabrication process, unless special purification steps are taken or a displacement by agents such as carbon is performed. Obviously, purification methods involve costs that are, except in rare instances, in excess of those the petroleum companies are willing or able to pay.

Iron

Since mild steel is so widely used by petroleum companies, and iron is the major component of mild steel, it is prudent to examine the nature of iron salts in preparation for the study of how this steel is affected by corrosion processes. Iron predominately forms two cationic species, the ferrous ion (Fe^{2+}) and the ferric ion (Fe^{3+}). Ferrous cations can take on coordination numbers of four, six, and eight, while ferric ions add an additional coordination number of five. Figure 3–1 illustrates the geometric configurations that these coordination complexes represent.

From the geometry of the complex metal salts of iron, it is reasonable to suspect that the microscopic appearance of corrosion processes will resemble the geometry of the complexes. Thus the corrosion of metallic iron would be expected to progress from zero to the divalent or trivalent states of oxidation by successive anionic group occupation of available ligand sites. This expectation might be true if the corroding iron were pure (contained no inclusion elements), but carbon or mild steel alloy contains a considerable number of inclusion elements. Thus, discontinuities are observed when a microscopic examination of the corroded surface is performed. However, the bulk of corrosion sites observed do resemble the geometry of the salt complexes.

Corrosion of the mild steel equipment employed by most petroleum companies occurs under widely varying environmental conditions. Liquid, solid, and gas phases are numerous and diverse in oilfield transfer, storage, and processing areas. Consequently, the types of corrosion processes that take place are diverse. Much of the production from crude reservoirs contains associated water that is present in free and emulsified form. Gaseous components such as hydrogen sulfide, steam, hydrogen gas, carbon dioxide, carbon monoxide, and light hydrocarbons are frequently coproduced with the well liquids. Additionally, suspended solids including silt, paraffin, asphaltene, scale, and corrosion products are distributed throughout the system,

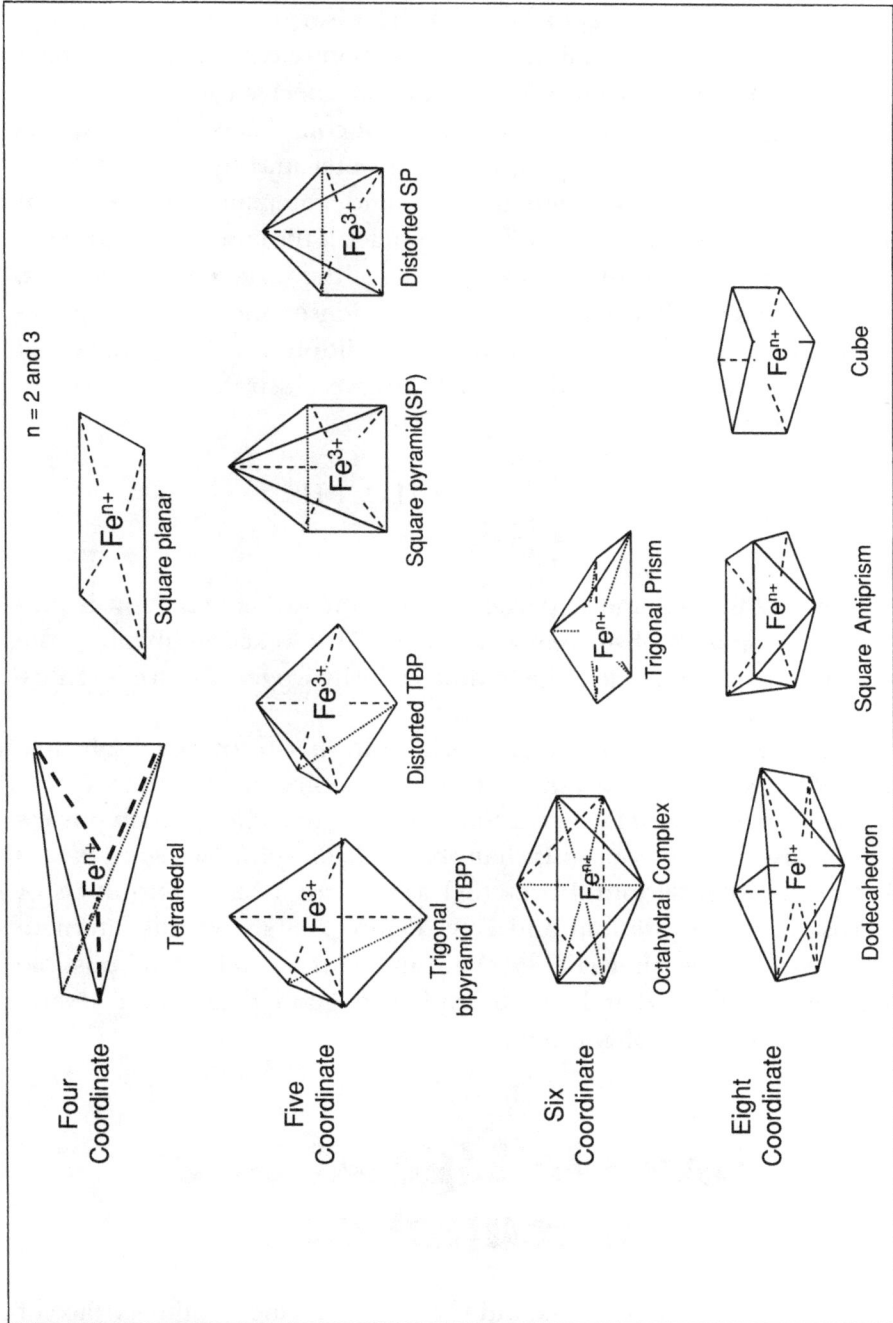

Fig. 3-1 Ferric and ferrous coordination complexes

providing heterogeneous surfaces for the adsorption of moisture.

Several gaseous and liquid disassociation reactions occur at these heterogeneous surfaces to produce anionic species capable of acting as ligand group occupants for metallic salt complexes. Thus, gaseous migration and surface condensations of water and hydrogen sulfide, water and carbon monoxide/dioxide, and water and hydrogen gas place these components in intimate contact. This contact occurs within emulsions or as absorbed components of the various solids present. The close association of these surface condensed molecules frequently provides reaction sites and favorable conditions for the formation of reaction products. Two of these reactions are illustrated in the following equations:

$$H_2O + H_2S \longrightarrow H_3O^+ + HS^-$$
$$H_2O + CO_2 \rightleftharpoons H_2CO_3 \rightleftharpoons 2H^+ + CO_3^{2-}$$

An additional and extremely important surface reaction is provided at corrosion by-product surfaces. This reaction involves the absorption and bond cleavage of molecular hydrogen. Figure 3–2 illustrates this process.

There is another cleavage mechanism that results in the formation of a proton and a radical from the molecular hydrogen. This mechanism is favored by metals that produce metal salts, which possess higher dissociation constants than that of hydrogen. Two such metals that are frequently found occluded as oxidation-clad by-products of corrosion are magnesium and zinc. Figure 3–3 shows this alternate mechanism. This alternate mechanism provides a host of possible reaction pathways that help to explain a good deal about several observed corrosion phenomena.

Anionic Ligand Site Occupation

The hydrosulfide anion and the proton formed at the surface of the oxide-occluded interstitial metal can provide pathways for the for-

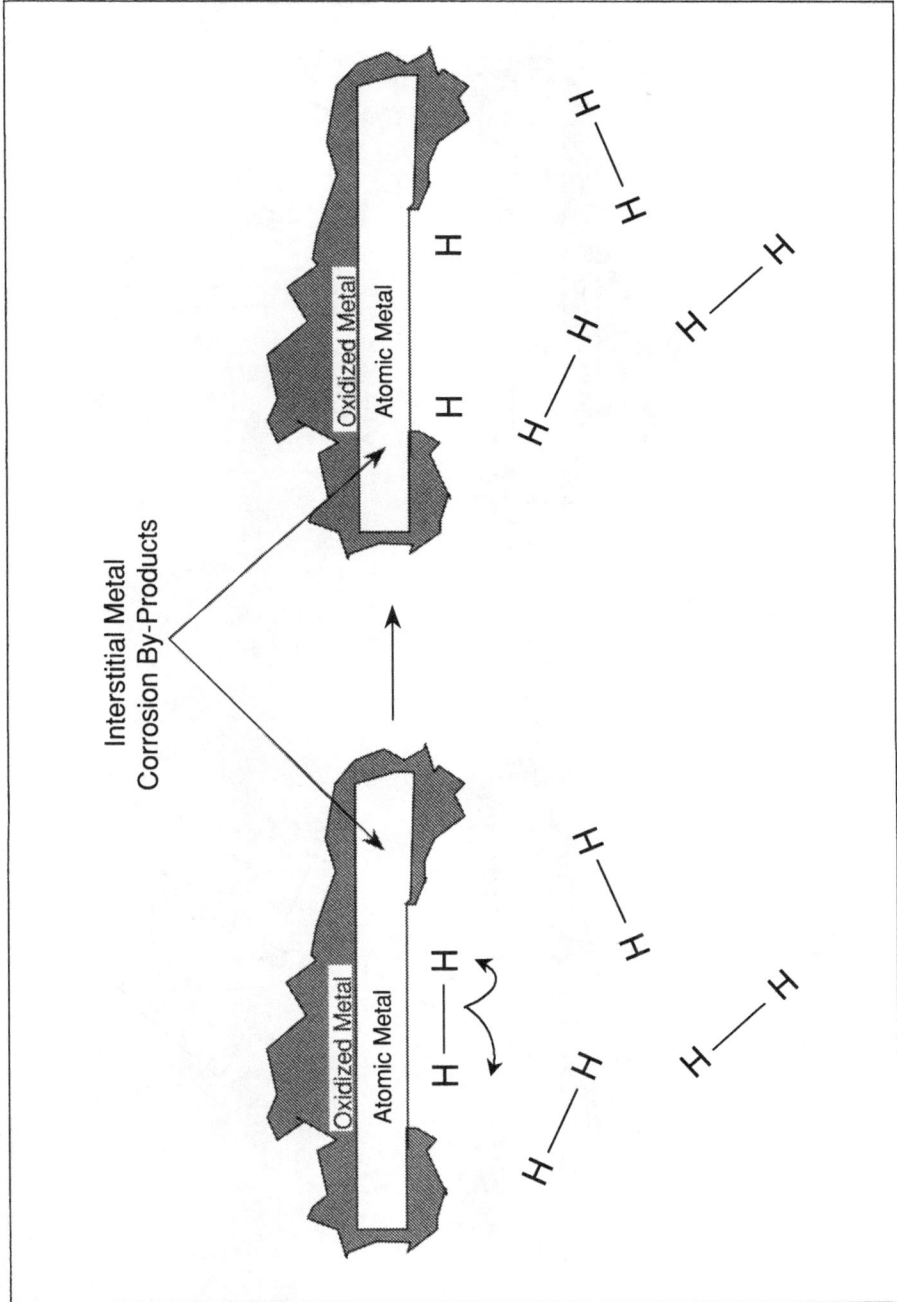

Fig. 3-2 Molecular hydrogen absorption and cleavage to form atomic hydrogen

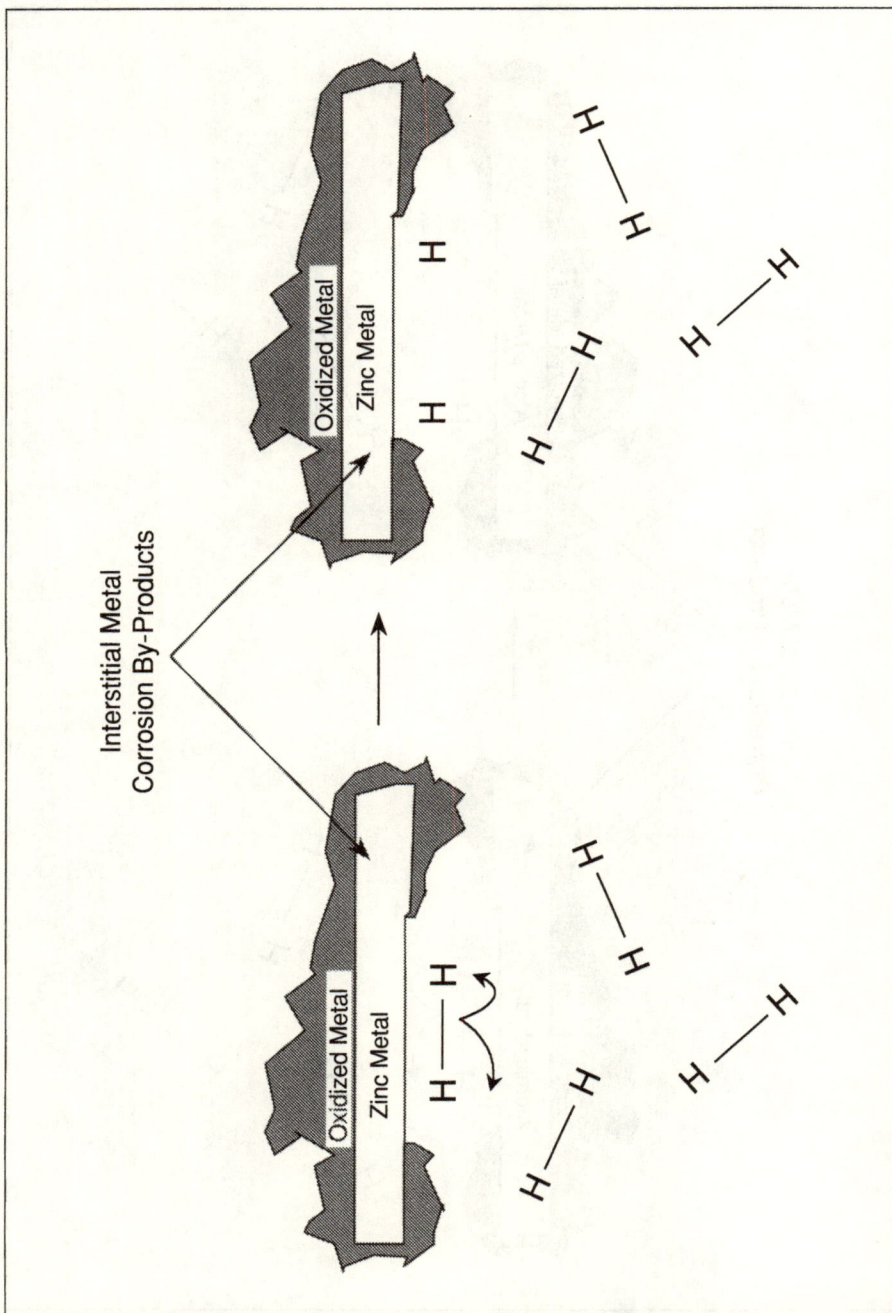

Fig. 3-3 Molecular hydrogen absorption and cleavage to form proton and radical

mation of additional anionic species as illustrated in the following equations:

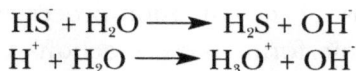

$$HS^- + H_2O \longrightarrow H_2S + OH^-$$
$$H^+ + H_2O \longrightarrow H_3O^+ + OH^-$$

Thus, an additional anion (hydroxyl ion) becomes available as a ligand capable of occupying coordination sites of the corroding metal. The hydroxyl and hydrosulfide anions possess sufficient base character to readily attack the iron, while the carbonate anion requires longer and hotter exposure conditions to ionize the metal. The process of metallic oxidation must take place prior to continued occupation of the coordination sites of the metallic salt. This restriction then requires that the atomic metal be oxidized to a minimum valence prior to ligand group occupation of coordinate sites.

Therefore, iron must be converted to ferrous ion (2+ oxidation) prior to ligand occupation of the coordination sites. Further oxidation to the ferric ion (3+ oxidation) then only succeeds in providing an additional coordination condition (five coordinate) to the metal salt. Thus, oxidation to a minimum valance condition can be seen as a rate-determining step (RDS) for the process of corrosion.

Once the oxidation process has successfully achieved the minimum valance condition, the coordination sites can be filled by a great variety of electron-rich ligand groups. The type of ligand group that assumes position in the metal salt complex is determined by two important factors:

- Group electronegativity
- Group concentration

These two factors can act in a concerted fashion (high values of both electronegativity and concentration are possessed), or they may act in competition. Generally the overwhelming presence of water found in crude oil systems favors the aqueous complexes of the metallic salt. Thus most of the oilfield corrosion salts involve aqueous complexes of ferrous and ferric hydrosulfide, hydroxide, and carbonate forms.

Radical Corrosion Processes

The focus of most of the corrosion processes literature is on the importance of the ionic forms these reactions take. Little has been mentioned about the role of free radicals in these processes. Free radical species are those chemical substances that possess a single unpaired electron in an expanded energy level condition. These species are, on average, short-lived, highly energetic compounds. Their presence in a system is often hard to detect because of their transitory nature. There is reason to believe that their occurrence is much more common than is indicated by the incidence of their detection. A good example of indirect evidence for the involvement of radicals is illustrated by the reaction of the low-temperature, low-pressure atmospheric formation of monoclinic sulfur (S_8) from hydrogen sulfide:

$$H_2S + Air \longrightarrow S_8$$

The results of this reaction are often seen in the oil field, where elemental (monoclinic) sulfur collects at gas release points. (These release points might include wellhead chemical injection points, sample valves, and leaking sucker rods.) The indicated direction of this reaction is overwhelmingly unfavorable from the point of view of thermodynamics. Unless other species present in the atmosphere are responsible for this reaction, there would appear to be no reason for its occurrence. Nitrogen is present in air to an even greater extent than oxygen, and oxides of nitrogen are also found. Consequently, it would seem reasonable that one of these species could be invoked to explain the reaction.

A simplified scheme for the inorganic photochemistry of smog might give some clues to the reason for the direction of the air reaction with hydrogen sulfide.

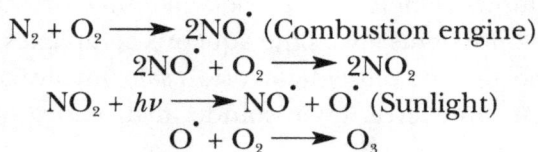

$$N_2 + O_2 \longrightarrow 2NO^{\bullet} \text{ (Combustion engine)}$$
$$2NO^{\bullet} + O_2 \longrightarrow 2NO_2$$
$$NO_2 + h\nu \longrightarrow NO^{\bullet} + O^{\bullet} \text{ (Sunlight)}$$
$$O^{\bullet} + O_2 \longrightarrow O_3$$

These reactions thus indicate the formation of free radicals in the engine combustion process and by irradiation from sunlight. These free radicals are so energetic that they can oxidize oxygen (O_2) to ozone (O_2). The combustion process requires considerable temperature and pressure to proceed. However, the radiation process takes place readily at normal conditions of pressure and temperature. By employing this scheme for the air reaction with hydrogen sulfide, it can be seen that a free radical reaction nicely explains the reaction's direction.

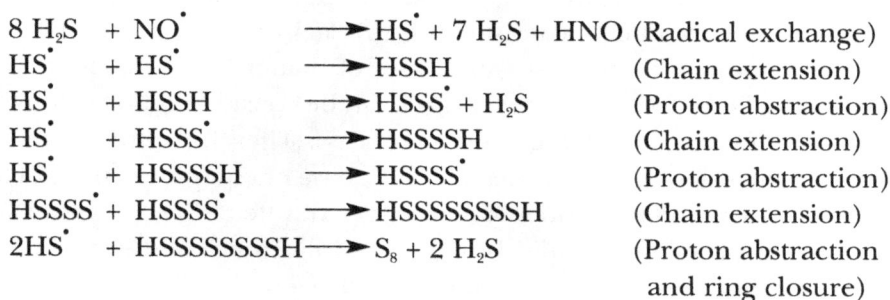

$$8\ H_2S\ +\ NO^{\bullet}\ \longrightarrow\ HS^{\bullet} + 7\ H_2S + HNO \quad \text{(Radical exchange)}$$
$$HS^{\bullet}\ +\ HS^{\bullet}\ \longrightarrow\ HSSH \quad \text{(Chain extension)}$$
$$HS^{\bullet}\ +\ HSSH\ \longrightarrow\ HSSS^{\bullet} + H_2S \quad \text{(Proton abstraction)}$$
$$HS^{\bullet}\ +\ HSSS^{\bullet}\ \longrightarrow\ HSSSSH \quad \text{(Chain extension)}$$
$$HS^{\bullet}\ +\ HSSSSH\ \longrightarrow\ HSSSS^{\bullet} \quad \text{(Proton abstraction)}$$
$$HSSSS^{\bullet} +\ HSSSS^{\bullet}\ \longrightarrow\ HSSSSSSSSH \quad \text{(Chain extension)}$$
$$2HS^{\bullet}\ +\ HSSSSSSSSH\ \longrightarrow\ S_8 + 2\ H_2S \quad \begin{array}{l}\text{(Proton abstraction}\\\text{and ring closure)}\end{array}$$

Thus, invoking a radical reaction mechanism, a reasonable scheme can be proposed for the mild conditions of air reaction leading to the formation of monoclinic sulfur. A free radical hydrogen can form with the simultaneous appearance of a proton and radical on the surface of an interstitially occluded metal. This mechanism is involved in the provision of important pathways for the formation of anionic ligand groups. Protonation and proton abstraction are illustrated by the following equations:

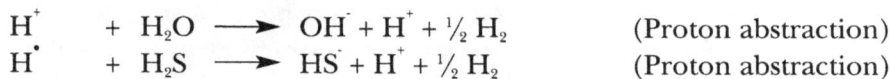

$$H^{+}\ +\ H_2O\ \longrightarrow\ OH^{\bullet} + H^{+} + \tfrac{1}{2}\ H_2 \quad \text{(Proton abstraction)}$$
$$H^{\bullet}\ +\ H_2S\ \longrightarrow\ HS^{\bullet} + H^{+} + \tfrac{1}{2}\ H_2 \quad \text{(Proton abstraction)}$$

In addition to the process of anion generation, the hydrogen radical constitutes a very small labile species capable of migrating through the metal lattices. When the hydrogen radicals encounter voids (metal crystal imperfections), they collect in them and undergo radical recombination reactions to form diatomic hydrogen molecules. The molecular hydrogen, thus formed, becomes trapped in these metal cavities (voids), since its molecular dimensions exclude its exit.

The continued migration of the hydrogen radicals into the cavity can result in extremely high concentrations of molecular hydrogen within the cavity. This process frequently continues until the pressure of the molecular hydrogen within the cavity is high enough to cause the cavity's rupture. The early stages of this process result in a corrosion phenomenon identified as *hydrogen blistering*, and the later stage (cavity rupture) is called *pitting*.

The fact that the hydrogen radicals possess no charge is an extremely important characteristic. It is this lack of charge that allows hydrogen radicals to populate all phases of matter (e.g., gas, liquid, and solid). Thus, hydrogen free radicals can be considered ubiquitous in systems undergoing the corrosion process. Therefore, the occurrence of corrosion in the overhead or gas phase of equipment used by oil companies can be considered, in large part, due to the ubiquitous nature of hydrogen free radicals. Figure 3–4 illustrates blistering and pitting phenomena exhibited by radical hydrogen.

The presence of hydrogen radicals in the gas phase combined with the vapor-phase water lead to other forms of surface corrosion, which will be examined in the next section.

Surface Moisture and Radicals

The presence of condensed moisture on metallic surfaces provides an aqueous medium for the proton abstraction processes that result in hydroxyl and hydrosulfide anion formation. The anions thus produced will oxidize the metal atoms from the metal lattice and expose the remaining ligand sites of the metallic salt. Water and/or hydrogen sulfide molecules then occupy the exposed ligand sites of the metal salt. Although different anions can be involved in this process (e.g., cyanide, carbonate, and sulfites), they occur far less frequently in petroleum-related gaseous corrosion processes.

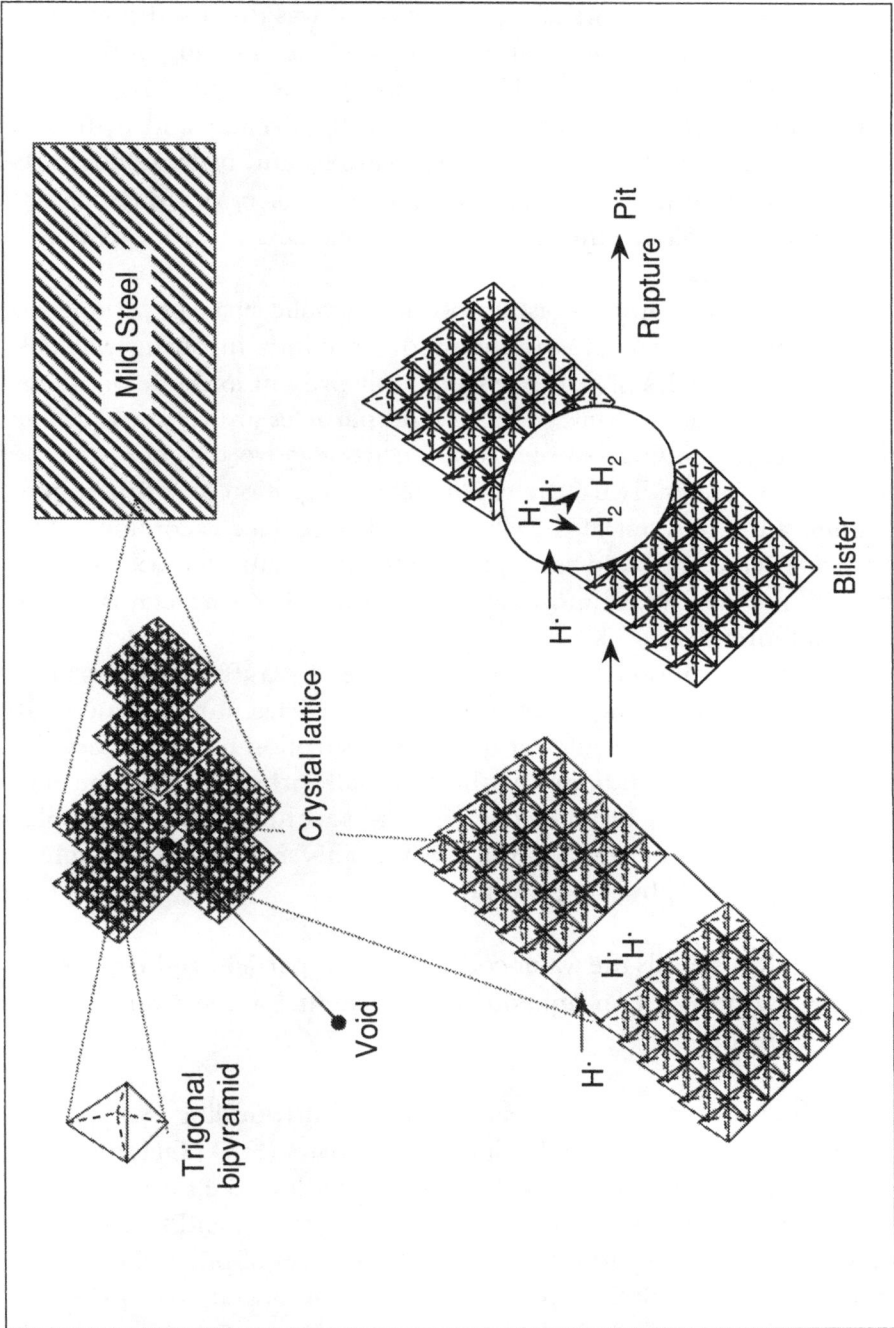

Fig. 3-4 Corrosion blistering and pitting by hydrogen radicals

Atmospheric factors also play a role, as was discussed previously in the mechanism for the creation of photochemical smog by the presence of oxides of nitrogen. Under atmospheric conditions, nitrous oxide radicals can also abstract hydrogen from water and hydrogen sulfide molecules, thereby producing hydroxyl and hydrosulfide radicals. Thus, atmospheric conditions (inclusion of environmental air) can be responsible for the creation of an oxidizing environment leading to corrosion.

The adsorption of moisture on nonmetallic surfaces produces a significantly different environment for reactions involving radicals. Suspended particles of sand are frequently present in crude oil fluids. The SiO_2 (silicate) crystals of these sand particles are very susceptible to hydrogen bonding between the electronegative oxygens and the hydrogen of water. Thus, the vapor phase migration and subsequent condensation of water at the silicate particle surface is common. The co-incident water-coated sand particle is often stabilized as a suspension by the migration and orientation of bipolar surfactants at the water-oil interface.

The adsorption of water at the surface of paraffin particles takes place by a different mechanism. Paraffin particles must be cocrystallized with bipolar molecules prior to the adsorption process. Once the first layer of water condenses on the externally projected polar head of the cocrystallized bipolar/paraffin, a second layer of bipolar molecules can add to the adsorbed water. This scheme then establishes two important physical conditions:

- It suspends the water-coated paraffin particle in the oil phase
- It provides an aqueous environment for ionic and radical reactions

The products of ionic reaction in the surrounding water phase are mainly comprised of hydronium/anion pairs (H_3O^+, OH^-, and HS^-) in equilibrium, while the radical reactions result in the formation and evolution of hydrogen radicals, molecular hydrogen, and hydrosulfide radicals. Asphaltene particles present a unique set of physical chemical conditions that result from their heterogeneous makeup (see Becker, J. R. 1997). Figure 3–5 illustrates a suspected structure for asphaltene.

Aliphatic hydrocarbon tails surround the structure illustrated in Figure 3–5. These must either interact in one of two ways. They can interact through London forces, with the nonpolar tail of a surfactant molecule providing a polar surface for water adsorption. The other way they can interact is to accept water molecules within the pi-bonding region. Both arrangements are probable, but the nature of aqueous reactions that occur will differ in each arrangement. Thus, anionic groups will deprotonate acids of the bipolar surfactant, if the fatty acid is the bipolar species interacting with the nonpolar aliphatic groups of the asphaltene.

However, if water molecules are included in the pi-bonding region of the asphaltene particle, the presence of the ferrous or ferric hemin complex produces a reducing environment. In this case the hydroxyl or hydrosulfide anions are reduced to form water and hydrogen sulfide. Free radical reactions that take place in the polar region of the fatty acid surface behave as those suggested above for paraffin. Those taking place in the pi-bonding region react with double bonds and other free radical traps (e.g., Fe^{+n} + radical = $Fe^{+(n-1)}$) in the complexing porphyrin.

Emulsions

Emulsions provide aqueous environments that are represented by both water internal and oil internal forms. The nature of the internal phase is determined by several factors. These include the hydrophilic/lipophylic balance (HLB) of the bipolar surfactant, the composition of the nonpolar phase, and the relative densities and viscosities of the internal and external phases. The HLB value of the surfactant is a function of the strength of its polar group interaction with the polar phase and its nonpolar group interaction with the nonpolar phase.

Crude oils contain variable amounts of bipolar surfactants that frequently exhibit ranges of HLB values. Additionally, crude oils are composed of variable density and viscosity nonpolar and polar phases. Thus, crude oil water production streams are very frequently com-

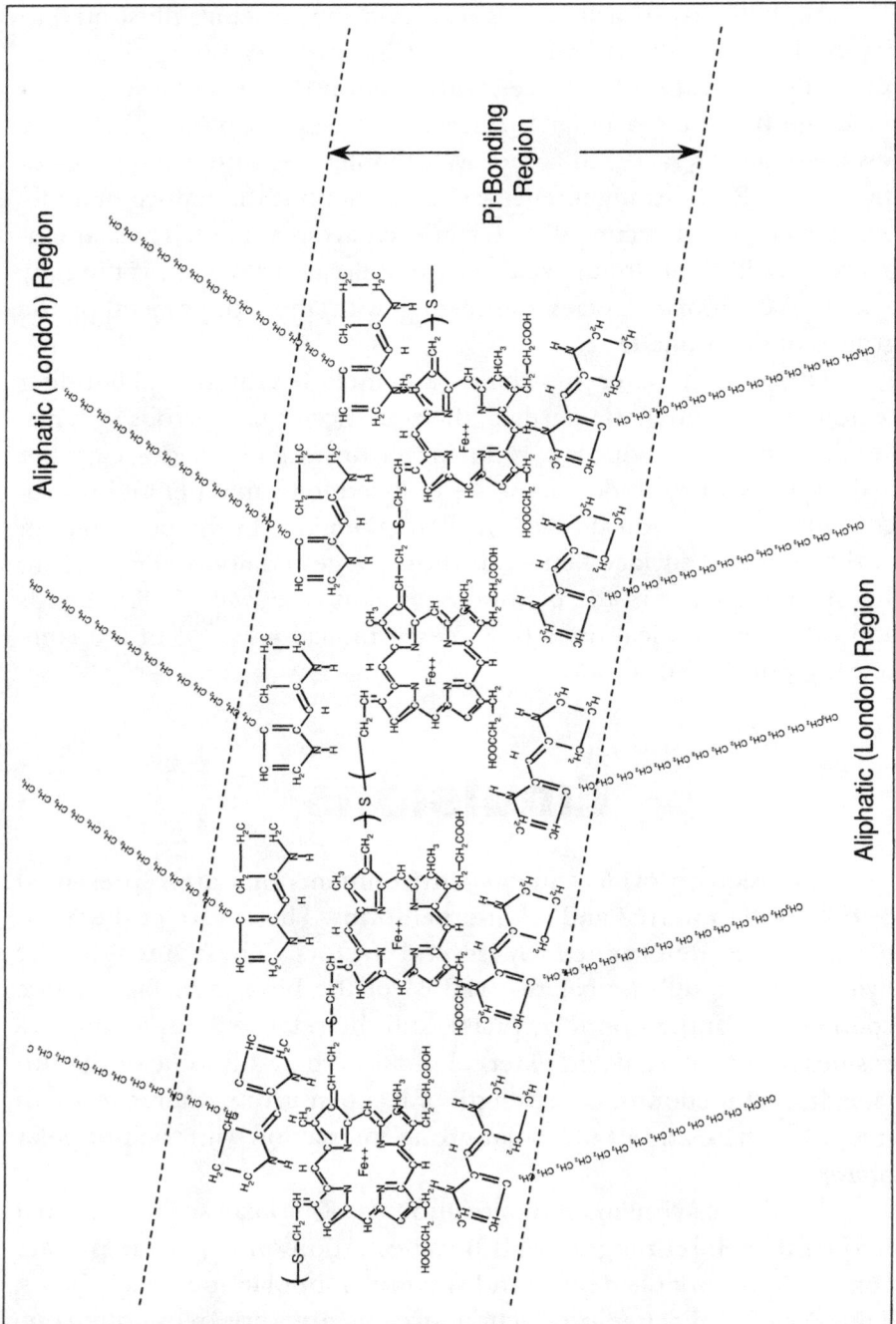

Fig. 3-5 Proposed asphaltene structure

posed of emulsions. Figure 3–6 illustrates a schematic view of both the oil external and water external emulsion forms.

The types of ions present in the water often affect the ionic and radical reactions taking place in the aqueous phase of an emulsion. The aqueous phase of an emulsion often contains high concentrations of ions, while condensed water at the surface of dispersed solids is nearly free of ions. The ionic reactions described above are unhindered, while the free radical reactions are redirected by metallic oxidation/reduction reactions similar to the reactions mentioned for the pi-bonding region of asphaltenes. Thus, in a general sense, the asphaltene pi-bonding region (high polarity, high ionic concentration) and the highly ionic water phase of emulsions can be considered free radical traps. The effectiveness of these radical traps depends upon the type of metal salts present and the oxidation state of the central metal.

Summary

The preceding chapter developed the concept that corrosion processes are largely determined by the salt forms derived from the corrosive destruction of the metal. It also discussed how corrosion occurs because of the oxidation of the metal by anionic ligand groups. It also discussed the consequent freeing up of the remaining coordination sites for occupation by nonionic ligands. A discussion of hydrogen radical and proton formation, at dissimilar interstitial metal impurities, illustrated molecular adsorption and heterolytic cleavage of hydrogen molecules.

A brief sojourn into the radical chemistry involved in monoclinic sulfur formation from hydrogen sulfide described how radical molecules could be used to explain observed phenomena. Gas phase corrosion is described as occurring because of the presence of hydrogen radicals in the condensed water phase at the metal-water interface. Finally, different types of reactions of the radical forms indicated that the condensed water phase surrounding a solid provides a source of hydrogen molecules due to radical/radical combinations. High ionic aqueous phases are generally efficient free radical traps.

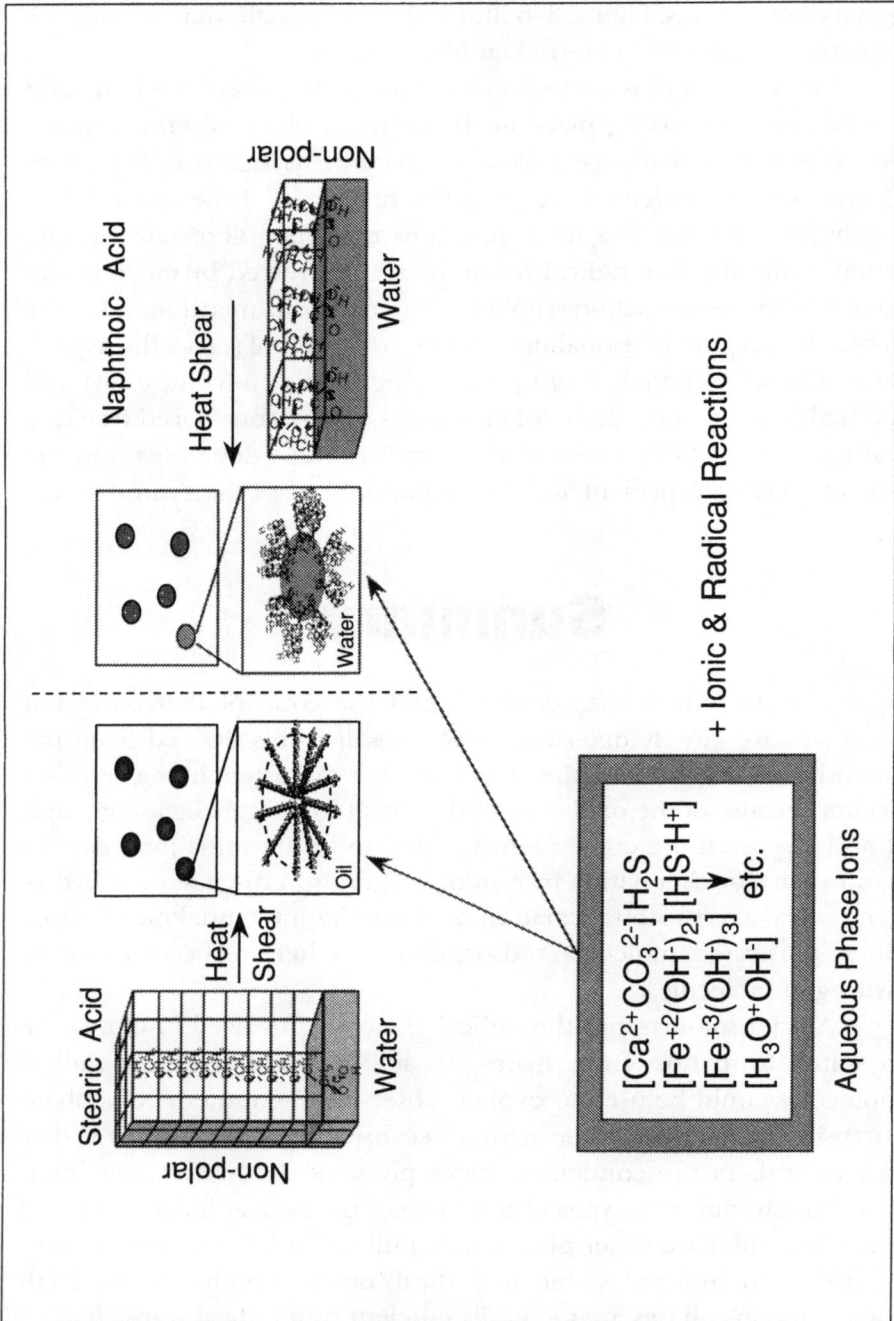

Fig. 3-6 Water external and oil external emulsions

References

Barrow, Gordon M. 1966. *Physical Chemistry.* 2d ed. New York, St. Louis, San Francisco, Toronto, London, and Sydney: McGraw Hill Book Company.

Becker, J. R. 1997. *Crude Oil Waxes, Emulsions, and Asphaltenes.* Tulsa: PennWell Publishing Company.

Bockris, J. O'M. and A. K. N. Reddy. 1973. *Modern Electrochemistry.* New York: Plenum Publishing Corporation.

Dickerson, R. F., H. B. Gray, and G. P. Haight, Jr. 1970. *Chemical Principles.* 1st ed. New York: W. A. Benjamin, Inc.

Hamill, William H. and Russell R. Williams, Jr. 1966. *Principles of Physical Chemistry.* 2d ed. Englewood Cliffs, New Jersey: Prentice Hall.

Huheey, James E. 1978. *Inorganic Chemistry Principles of Structure and Reactivity.* 2d ed. New York, Hagerstown, San Francisco, and London: Harper & Row.

Noller, Carl R. 1966. *Textbook of Organic Chemistry.* 3d ed. Philadelphia and London: W. B. Saunders Company.

References

_____

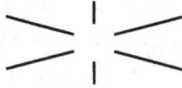

4
The Petroleum Industry and Mild Steel

The extensive use of mild steel in equipment design and construction by petroleum companies requires a closer look at the nature of this metal. The source and composition of its precursors and the methods of its production are important factors that need to be examined. The successful implementation of the mild steel products in petroleum industry applications depends upon several factors, but resistance to corrosion is among the most important. However, a balance between equipment cost and the acceptance of less-than-perfect performance must be struck. Thus the cost performance of mild steel has been judged by petroleum companies to be most practical.

Pig Iron

The separation of iron from its ores stands as one of man's most important technical advances. Iron is a moderately active metal that can be reduced by carbon monoxide from its most common form, iron oxide (Fe_2O_3), through a series of steps:

$$3\ Fe_2O_3\ +\ CO\ \longrightarrow\ 2\ Fe_3O_4\ +\ CO_2$$
$$Fe_3O_4\ +\ CO\ \longrightarrow\ 3\ FeO\ +\ CO_2$$
$$FeO\ +\ CO\ \longrightarrow\ Fe\ +\ CO_2$$

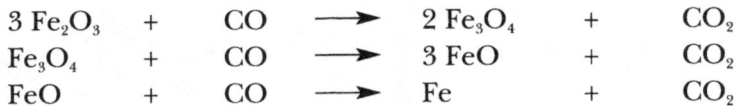

The primary reduction of iron is carried out in a blast furnace, where high temperatures are maintained. The primary impurity, SiO_2, is removed by adding lime (CaO) to form slag. The molten slag is considerably less dense than the molten iron and remains separated as the top layer as the molten products are collected. The iron is drawn off from the bottom of the mix. The pig iron produced in this process is so brittle and low in tensile strength that it is of little use.

Further processing is required to produce metals that have characteristics required for fabricated products. The poor quality of pig iron is mainly due to the impurities it contains: 3–5% carbon, and lower percentages of silicon, phosphorus, and sulfur. The further processing of pig iron into steel requires the removal of the carbon and nearly all of the phosphorus, sulfur, and silicon.

Mild Steel

The making of mild steel (carbon steel) from pig iron requires the use of higher temperature processes than those used in the production of pig iron. The incorporation of air feed into the process increases the temperature of the molten pig iron. This is accomplished by feeding high pressure air through the molten pig iron in a process known as blasting. More recently the air used in this process has been replaced with oxygen, which results in even higher temperatures.

After the steel has been produced, it is processed by the pouring of ingots, rolling, extrusion, and mold casting. Lower quality mild steel is cured in the ingots at atmospheric pressure. The higher quality steel is subjected to vacuum curing in the slab form to yield steel with fewer void spaces.

Each processing step adds to the final cost of the end-product steel. Consequently, petroleum companies most frequently use the lower cost, higher void containing steel products for tubing, transfer line, and tank construction. This grade of steel possesses useful properties of high tensile strength, high ductility, ease of machining, and moderate resistance to corrosion. The desirable physical properties combined with its nominal cost make mild steel an acceptable compromise for crude oil equipment manufacture.

Even with its moderate resistance to corrosion, petroleum companies still consider mild steel a better alternative than the exotic and costly alloys of stainless steel. These companies have determined a cost advantage for the employment of mild steel, even though it requires various means of controlling its corrosion. Thus, treatment programs of both chemical and physical types (cathodic protection) are economically justified.

One of the factors that determines the corrosion resistance of various grades of mild steel is its physical processing (cooling time, external process pressures, and uniformity of temperature). The physical processing, to a major degree, determines the coordination states of the metal undergoing crystallization. Thus, molten steel can be considered as a saturated solution of iron species ranging from coordination number four through eight, excluding seven.

The variation of coordination numbers will result from the states of thermal energy maintained by the system for a sufficient time for crystallization to occur. Thus, grain boundaries or different crystal forms of steel are frequently encountered in the processed metal. These grain boundaries can be considered structural discontinuities, which lead to different behaviors when subjected to oxidizing environments. Grain boundaries then lead to differential corrosion rates within metals of high purity.

Interestingly, voids act as cavity radiators, and cavity wall temperatures are always in excess of the bulk metal temperature. This phenomenon is described as *black body radiation*, and frequently leads to the formation of a grain boundary at the internal surface of the cavity. Because of the high viscosity and minimal agitation of the cooling bulk metal, temperature gradients exist throughout the sample, and grain boundaries are formed. Figure 4–1 illustrates some of the factors determining the corrosion resistance of metals, voids, and grain boundaries.

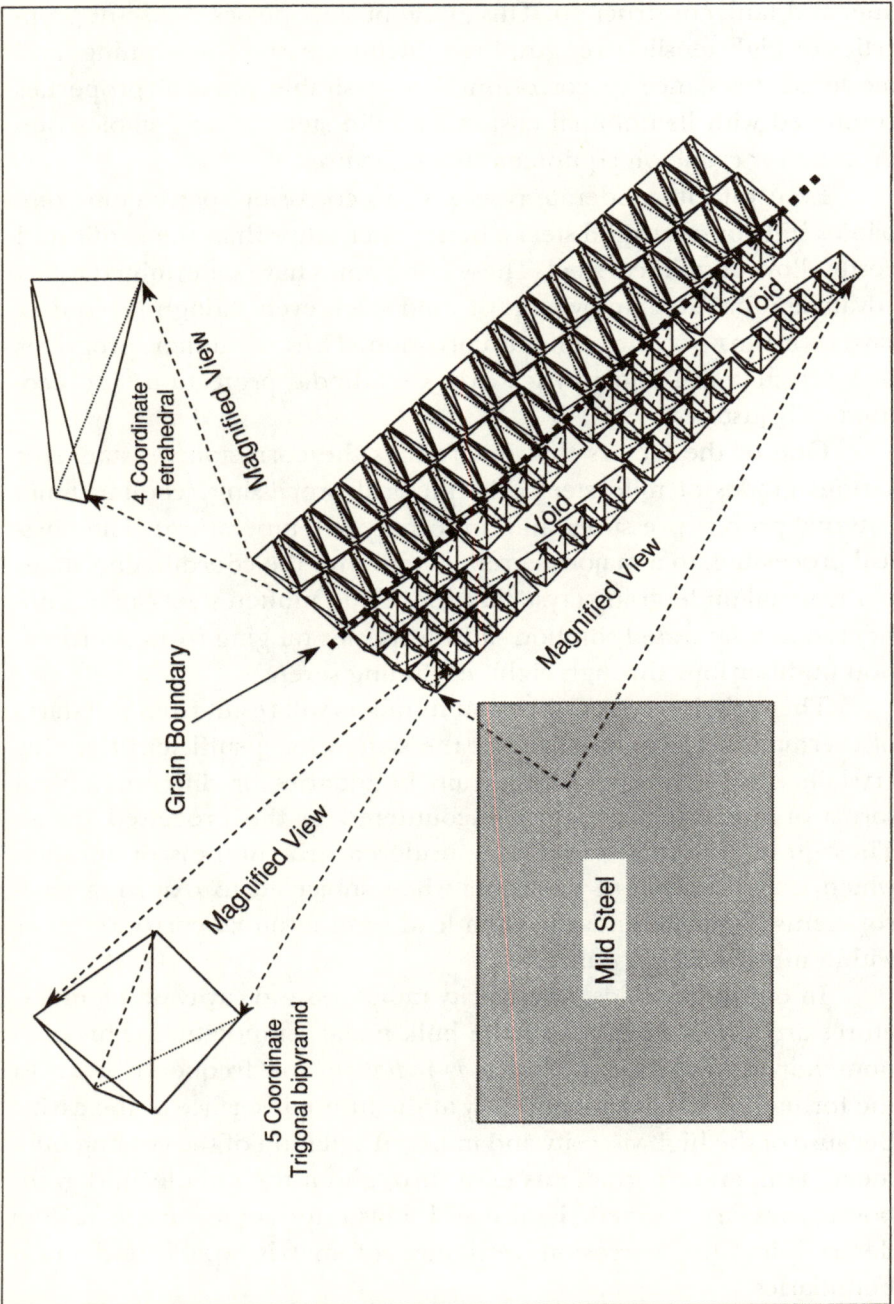

Fig. 4-1 Illustration of grain boundaries and voids resulting from four and five coordinate iron in mild steel

The grain boundaries contained within pure metals are formed because the geometry of the coordination states produces irregularities in the normal metallic crystal lattice. These irregularities frequently provide corridors for the migration of anions, radicals, and uncharged ligand groups. Grain boundaries also undergo differential crystal distortion when subjected to compression, elongation, and torsion stress forces. Combined corrosion and stress forces lead to stress cracking of the metal. Figure 4–1 illustrates a combination of two coordinate crystal forms. However, depending on the environment and cooling duration the metal experiences, examples of several coordination conditions can be present in a given sample of metal.

Steel Well Tubing

Well tubing provides some of the first metal/petroleum fluid contact, and as such is exposed to some of the most drastic conditions of pressure, temperature, pH, and fluid velocity. Well depths of 15,000 feet and deeper are not that uncommon in certain areas of the world. Temperatures in excess of 400°F and pressures greater than 3,000 pounds per square inch have been found at these well depths. The fluid velocities depend on such factors as the production rate, tubing diameter, fluid viscosity, water content, oil content, gas content, emulsion presence, and solids content.

Frequently, the co-produced water possesses high concentrations of ionic materials, which have been leached from the overburden minerals and deposited in the reservoir. Before the well is completed, these ionic materials or salts have achieved a condition of equilibrium and present little impedance to flow. However, when the well provides a point of exit, the resultant pressure drop results in a disturbance to the conditions of equilibrium. The temperature drops, and salts deposit as scale.

The pH of the reservoir water changes in response to ionic changes in equilibrium and the presence of the metallic tubing. The passage of reservoir fluids through the well tubing produces dynamic forces that cause rapid changes in the environments of the metal/fluid interface. Thus, the processes of corrosion and scale formation are changing continuously as the metal/fluid interface is replaced by fresh fluid.

Well tubing is usually fabricated by extrusion or welded seam method; each has advantages and disadvantages for the petroleum companies. The extruded tubing is usually considerably more costly than the welded seam product, but its resistance to corrosion is greater. The technology required for the extrusion method has been improved over the years, and these improvements have made it economically feasible for employment in well applications. However, welded seam tubing is still used in many instances, and its use will probably continue into the next century.

Welded seam tubing presents a continuous metal-metal boundary at the weld. Depending on the conditions of the ingot pour and cure, these metal-metal boundaries may consist of variable crystal coordination states or grain boundaries. The physical makeup of the tubing, combined with the prevailing conditions of the well, produce scale and corrosion effects that reflect the nature of both.

Steel Tubing Scale and Corrosion

The process of scale formation in well tubing is disturbed by the process of tubing corrosion. This results from the changing ionic character of the fluids occupying the fluid/metal interface. The aqueous electrolytes (ions) present in well fluids invariably include H_3O^+ and dissolved oxygen as electron acceptors. Thus the corresponding reactions are:

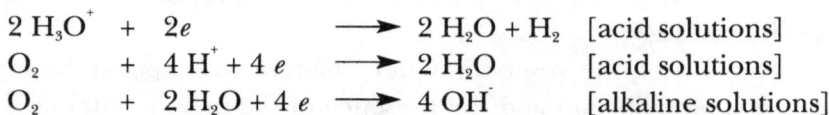

$$2\,H_3O^+ + 2e \longrightarrow 2\,H_2O + H_2 \quad \text{[acid solutions]}$$
$$O_2 + 4\,H^+ + 4\,e \longrightarrow 2\,H_2O \quad \text{[acid solutions]}$$
$$O_2 + 2\,H_2O + 4\,e \longrightarrow 4\,OH^- \quad \text{[alkaline solutions]}$$

The presence of different species of electrolytes can lead to additional electronation reactions, such as one of the following:

$$Fe^{3+} + e \longrightarrow Fe^{2+} \uparrow$$
$$2\,H^+ + CO_3^{2-} + 2\,e \longrightarrow CO_2 \uparrow + H_2O$$
$$3\,H^+ + NO_3^- + 2\,e \longrightarrow HNO_2 + H_2O$$

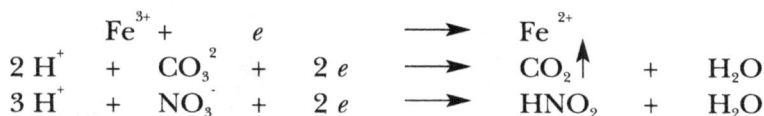

The occurrence of several possible electronation reactions increases the probability that metal-dissolution will occur, and the reaction yielding the highest corrosion current will control the rate of dissolution. Further, in accordance with the common ion effect (Le Chatelier's principle), the addition of ions present in a salt suppresses the solubility of the salt. As a corollary to this principle, then, the removal of ions present in a salt increases the solubility of the salt. Thus, if calcium carbonate is present in the production fluid, and the carbonate anion is reduced, as in the preceding electronation reaction, then the calcium salt concentration is increased.

The electronation of calcium carbonate can then be considered an electron sink, and the mild steel an electron source. Thus acid conversions of calcium carbonate scale, which occur at the scale metal interface, can act as electron acceptors. The steel acts as an electron source. Similar behavior is seen when the behavior of the Group 2A metals is involved. Thus, the low solubility salts of barium, calcium, magnesium, and strontium can exhibit combined electronation common ion effects. These effects aid in metal dissolution (corrosion) if their equilibrium potentials are positive with respect to metal dissolution equilibrium.

When the metal surface provides a site for the formation of scale, the deposition of the low K_{sp} salt/scale takes part in a series of reactions. These reactions have the potential for the corrosion of the metal. Figure 4–2 illustrates a possible reaction sequence for the corrosion of mild steel.

Protons attack the calcium carbonate scale, producing the highly soluble bicarbonate salt. This salt releases carbonate anions that take coordinate positions at the metal crystal. This process oxidizes the iron from metallic form, and water ligands occupy the remaining coordination sites. Continued oxidation of the ferrous ion can produce the ferric salts.

The attack of metal is most probable at the grain boundary, where interstitial spaces result from poor metal crystal packing. Thus,

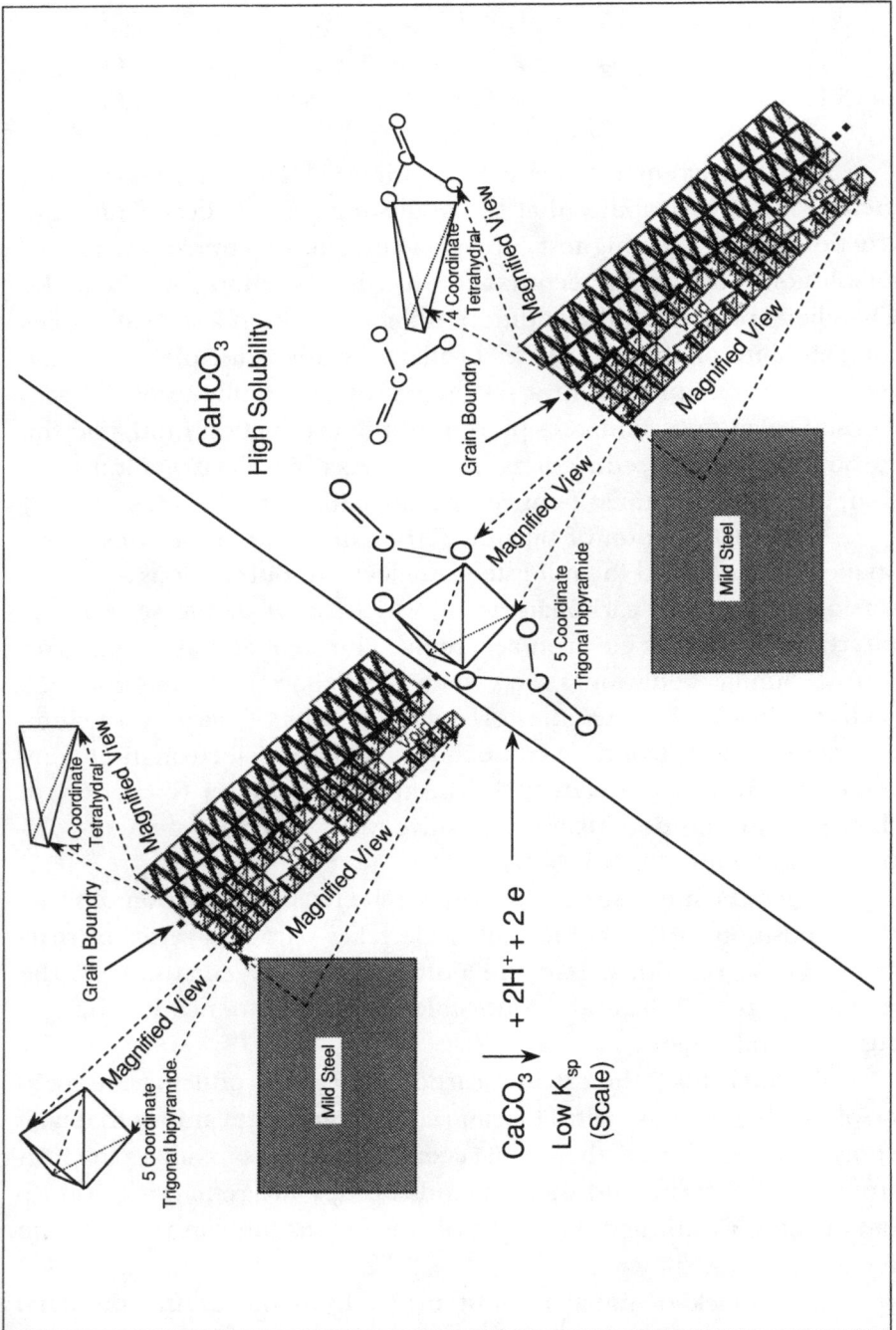

Fig. 4-2 Calcium carbonate assisted corrosion of mild steel

grain boundaries in the metal account for much of the corrosion patterns that are exhibited by metals as a result of the above reaction scheme. An interesting consequence of this behavior is exhibited by accelerated corrosion rates in some systems as a consequence of the addition of scale-control chemicals.

In general, the control of scale through the use of chemicals relies on the alteration of crystal morphologies involved in scale formation, rather than removal of the ion from reaction pathways. The economic viability of chemical treatments used by the petroleum industry demands that these treatments be effective at low levels. These treatment levels typically range from 1–500 parts per million (ppm) and seldom exceed 1,000 ppm. Scale treatment chemicals are generally expected to work from 1–50 ppm when continuous injections are performed.

However, squeeze treatments can involve concentrations of chemicals sufficient to disrupt the ionic salt concentrations in the system. (These treatments require that the scale compound be delivered to the formation in levels high enough to feed back for long periods of time.) The resulting disruptions can be significant enough to alter the corrosion rates in certain systems. Squeeze techniques are applied when the means for continuous injection are not available, and the only way to deliver the chemical is through back-flushing the well.

Sulfur and Iron: A Special Case

The high incidence of hydrogen sulfide in petroleum fluids containing water produces a special set of conditions that dramatically affect the processes of corrosion and scale in well tubing. Ferrous sulfide (FeS) has a very low solubility constant ($K_{sp} = 3.7 \times 10^{-18}$) in aqueous systems at neutral pH. It decomposes to Fe^{+2} and H_2SA as hydrogen ion concentration increases. The de-electronation of iron metal in aqueous systems containing H_2S can follow the classic anionic method:

$$H_2S + H_2O \longrightarrow \overset{s}{H} + H_3O^+$$
$$Fe + 2\,HS^- + 2\,H_3O^+ \longrightarrow (\,Fe^{2+} + S^{2-})\downarrow + 2\,H_2O + 3\,H_2\uparrow$$

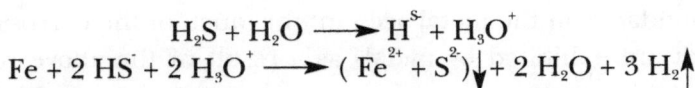

It also can occur with the polysulfide adducts arising from the radical mechanism described in chapter 3. Thus, pyrite (fool's gold—FeS_2) and other polysulfide salts can form as illustrated in the following sequence of reactions:

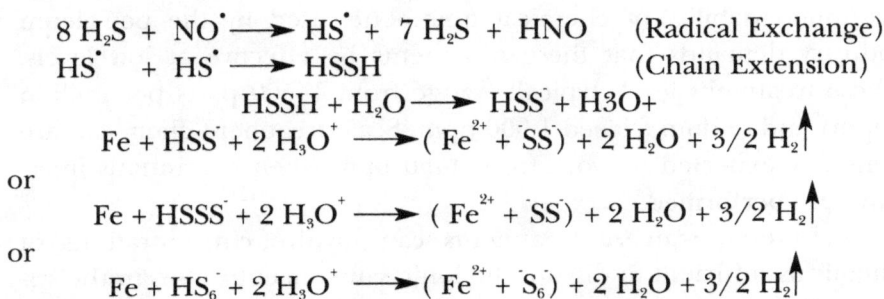

$$8\,H_2S + \overset{\bullet}{NO} \longrightarrow \overset{\bullet}{HS} + 7\,H_2S + HNO \qquad \text{(Radical Exchange)}$$
$$\overset{\bullet}{HS} + \overset{\bullet}{HS} \longrightarrow HSSH \qquad\qquad\qquad \text{(Chain Extension)}$$
$$HSSH + H_2O \longrightarrow HSS^- + H3O+$$
$$Fe + HSS^- + 2\,H_3O^+ \longrightarrow (\,Fe^{2+} + \,^-SS) + 2\,H_2O + 3/2\,H_2\uparrow$$

or

$$Fe + HSSS^- + 2\,H_3O^+ \longrightarrow (\,Fe^{2+} + \,^-SS) + 2\,H_2O + 3/2\,H_2\uparrow$$

or

$$Fe + HS_6^- + 2\,H_3O^+ \longrightarrow (\,Fe^{2+} + \,^-S_6) + 2\,H_2O + 3/2\,H_2\uparrow$$

The last equation in the preceding sequence represents the maximum number of repeating sulfurs obtainable without the formation of monoclinic sulfur. The low solubility of both the ferrous sulfide and the pyrite convey properties of corrosion resistance to iron (mild steel). This occurs by forming intercrystalline metallic salts within the metal crystal lattice and strongly resisting solvation. However, under strongly acidic conditions polysulfide salts (radical products) yield hydrogen sulfide and elemental sulfur, while the iron sulfide yields ferric hydroxide [$Fe\,(OH)_3$] and hydrogen sulfide.

The reactions of the polysulfide forms with iron metals can be examined more closely to determine a few possible orientations of the polysulfide metal salts. Thus Figure 4–3 shows some examples of the polysulfide metal salts of a four-coordinate square planar complex.

Several complex salt arrangements are possible from the presence of the polysulfide di-anions. Intra-ionic and interionic complex forms of S_2^{2-} through S_6^{2-} can form single and linked complex salts with four to eight coordinate iron forms. It is very important to note that the polysulfide salts fit the listed criteria for effectiveness as scale-control agents mentioned in chapter 2. These are:

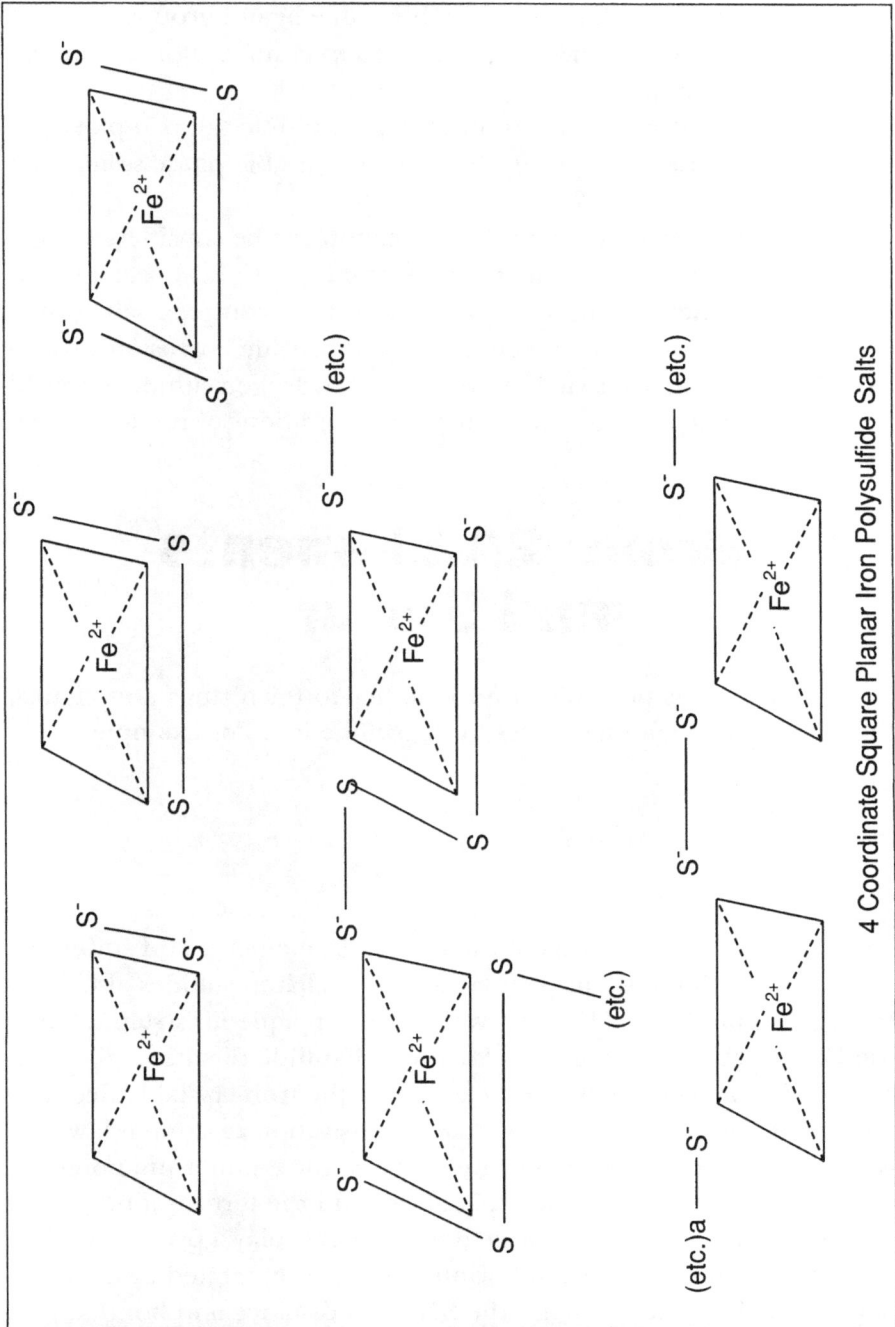

Fig. 4-3 Some possible orientations of square planar polysulfide iron salts

- Complexes that possess highly stable ligand groups
- Complexes formed from molecules containing multiple ligand groups
- Molecular complexes that bridge multiple salt complexes
- Molecular complexes that contain variable phase solubility

The polysulfide iron complex formation can be considered a natural form of scale control that results from the radical reactions of hydrogen sulfide. Additionally, polysulfide iron complex salt formation and its intermetallic crystal deposition provide corrosion protection. Thus, the radical reaction products of hydrogen sulfide, it would appear, are beneficial to mild steel under conditions of moderate pH.

Group 2A Elements and Sulfur

Sulfur atoms not only bond together to form rings and chains, they may also attach themselves to the sulfide ion. For example:

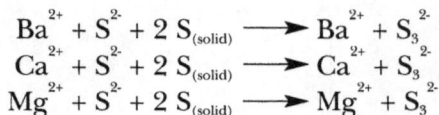

$$Ba^{2+} + S^{2-} + 2\ S_{(solid)} \longrightarrow Ba^{2+} + S_3^{2-}$$
$$Ca^{2+} + S^{2-} + 2\ S_{(solid)} \longrightarrow Ca^{2+} + S_3^{2-}$$
$$Mg^{2+} + S^{2-} + 2\ S_{(solid)} \longrightarrow Mg^{2+} + S_3^{2-}$$

The dehydrated forms of the Group 2A elements and sulfur are quite low in solubility, but the barium, and calcium sulfides also form hydration complexes. These are very soluble in aqueous systems. Since the Group 2A cations compete for the polysulfide di-anions, the incidence of beneficial passivation reactions at the iron crystal lattice sites is lowered. The extent to which the iron passivation reaction is lowered is, then, a function of the concentration and the equilibrium potential difference between the Group 2A cations and the ferrous ion.

Cationic ion sizes or charge density effects play a primary role in the determination of the equilibrium potentials exhibited by the competing reactions. Additionally, the S-S bond distance and bond angles will restrict the geometry of the resulting complex. The ferrous ion (Fe^{2+}) presents nearly perfect coordination site geometry for the formation of a cubic lattice structure. S_2^{2-} is analogous to sodium chloride and the S-S bonds assume the position of the chlorine. Figure 4–4 illustrates this structure.

The calcium cation (Ca^{2+}) is close in size to the ferrous cation (Fe^{2+}) and is capable of forming the same coordination geometry. Consequently, it competes most effectively with the iron polysulfide salt formation. The barium cation is larger than either the ferrous or calcium anion, and the magnesium cation is smaller. Thus the coordination geometry of these polysulfide salts involves deformation.

The deformation required shifts the equilibrium in favor of the ferrous complex. Consequently, the electronation reactions of barium and magnesium help shift the equilibrium in the favor of de-electronation (dissolution or corrosion) of the iron metal. Thus, the presence of calcium ions competitively interferes with polysulfide passivation of the iron surface, but in so doing, it interferes with the de-electronation (corrosion) of the metallic iron.

Summary

The preceding chapter briefly discusses the processes involved in the fabrication of mild steel products that are widely used by the petroleum industry. Some types of metallic crystal structures are discussed in terms of coordination states, and the reasons they arise in mild steel are briefly examined. Grain boundaries, the reason they arise, and how they affect and are affected by scale and corrosion processes are treated in terms of coordination geometry. The effects of electronation and de-electronation reactions are discussed. Ionic equilibria external to the metal can either promote or compete with these reactions.

The development of the forms of polysulfide anions is focused on to indicate the special place they hold in anionic metal complex formation. The interesting fact that the polysulfide di-anions possess some of the necessary criteria for effective scale inhibition is discussed.

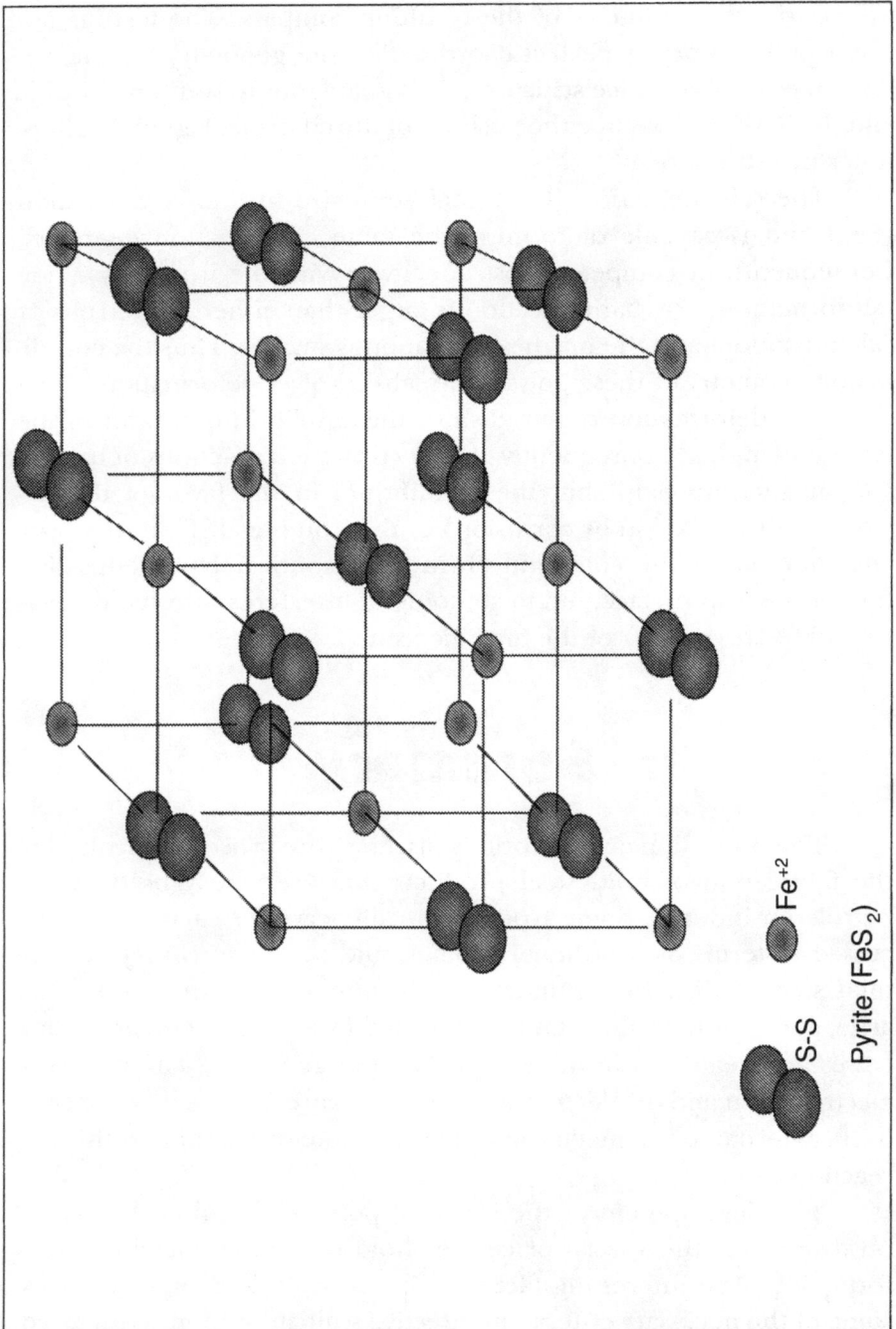

Fig. 4-4 Pyrite (FeS₂) crystal structure showing the cubic lattice arrangement

Finally, the reactions of the Group 2A elements with sulfur, and some of the roles they play as corrosion accelerators or inhibitors, is discussed.

References

Barrow, Gordon M. 1966. *Physical Chemistry.* 2d ed. New York, St. Louis, San Francisco, Toronto, London, and Sydney: McGraw Hill Book Company.

Becker, J. R. 1997. *Crude Oil Waxes, Emulsions, and Asphaltenes.* Tulsa: PennWell Publishing Company.

Bockris, J. O'M. and A. K. N. Reddy. 1973. *Modern Electrochemistry.* New York: Plenum Publishing Corporation.

Dickerson, R. F., H. B. Gray, and G. P. Haight, Jr. 1970. *Chemical Principles.* 1st ed. New York: W. A. Benjamin, Inc.

Hamill, William H. and Russell R. Williams, Jr. 1966. *Principles of Physical Chemistry.* 2d ed. Englewood Cliffs, New Jersey: Prentice-Hall.

Huheey, James E. 1978. *Inorganic Chemistry Principles of Structure and Reactivity.* 2d ed. New York, Hagerstown, San Francisco, and London: Harper & Row.

Noller, Carl R. 1966. *Textbook of Organic Chemistry.* 3d ed. Philadelphia and London: W. B. Saunders Company.

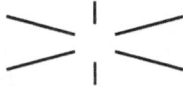

5
Petroleum Fluid Scale Growth and Inhibition

Crystal Growth

Water-containing petroleum fluid systems are particularly rich in calcium, barium, and magnesium cations, which occur in combinations with the anions carbonate, sulfate, and sulfide. Each of these combinations produces salts that possess variable solubility in aqueous systems. Depending on conditions of pH, temperature, and pressure, these salts, as a group, exhibit low solubility (10^{-16} to 10^{-3} mole/liter). Consequently, the aqueous solvent systems containing these salts tend to reach a point of saturation at low concentrations.

Saturation of a solvent (H_2O) by a solute (ionic salt) produces a metastable condition. This condition leads to phase separations (ionic salts change from dispersed ions to crystals) necessary to reestablish a satisfactory condition of equilibrium. This phase change requires that the salts in excess of those freely soluble in the solvent accumulate at points in the liquid phase. The salts then interact to exclude solvent and grow to a size that is great enough to rise or settle in the solvent.

The interaction of the salt forms and the solvent exclusion from these salts constitute the phenomenon of crystallization.

The forms that these solvent-excluded multiple salt aggregates take determine the morphology of the resulting scales. There are six fundamentally different crystal arrangements. These consist of *isometric, tetragonal, monoclinic, triclinic, orthorhombic,* and *hexagonal.* These structures are illustrated in Figure 5–1. Two major crystalline forms are most associated with the three cations (barium, calcium, and magnesium): the orthorhombic and the monoclinic. Thus, when discussing the pure salt crystal structures, these two forms will be the focus of attention.

The driving force for the formation of multiple aggregate metallic salt crystals is provided by the system's need to maintain an equilibrium concentration of solute in solvent. The forces of aggregation involve the efficiency of the ligand groups' shielding of the metal salts (cationic metal charge neutralization by ligand groups). Thus, metallic salt complexes possessing strongly shielding ligands tend to form macro-aggregate crystals less readily than those surrounded by less effective shielding ligands.

Petroleum fluid systems are largely composed of two ligand forms: ionic and nonionic. The ionic forms are the anions commonly associated with the barium, calcium, and magnesium salts. The nonionic ligands can range from water (the most common nonionic ligand) to complex cyclic alkyl, aryl, and alkyl aryl amines, alcohols, and ethers. The statistically higher incidence of water in these systems favors the water ligand group, unless the affinity of a polar or bipolar species for the coordination site is greatly favored.

Considering water as the predominant nonionic ligand group and the occupant of the available ligand sites of the metal salt, a qualitative picture of the shielding effect can be developed. The dipole nature of water is produced by the two pair of unshared electrons on the oxygen. This suggests that the orientation of water at the metallic salt's ligand site is situated with the oxygen pointed toward the metal cation and the hydrogens pointed away from it. This arrangement favors the solvation of the metallic complex by the forces of hydrogen bonding with the free solvent water.

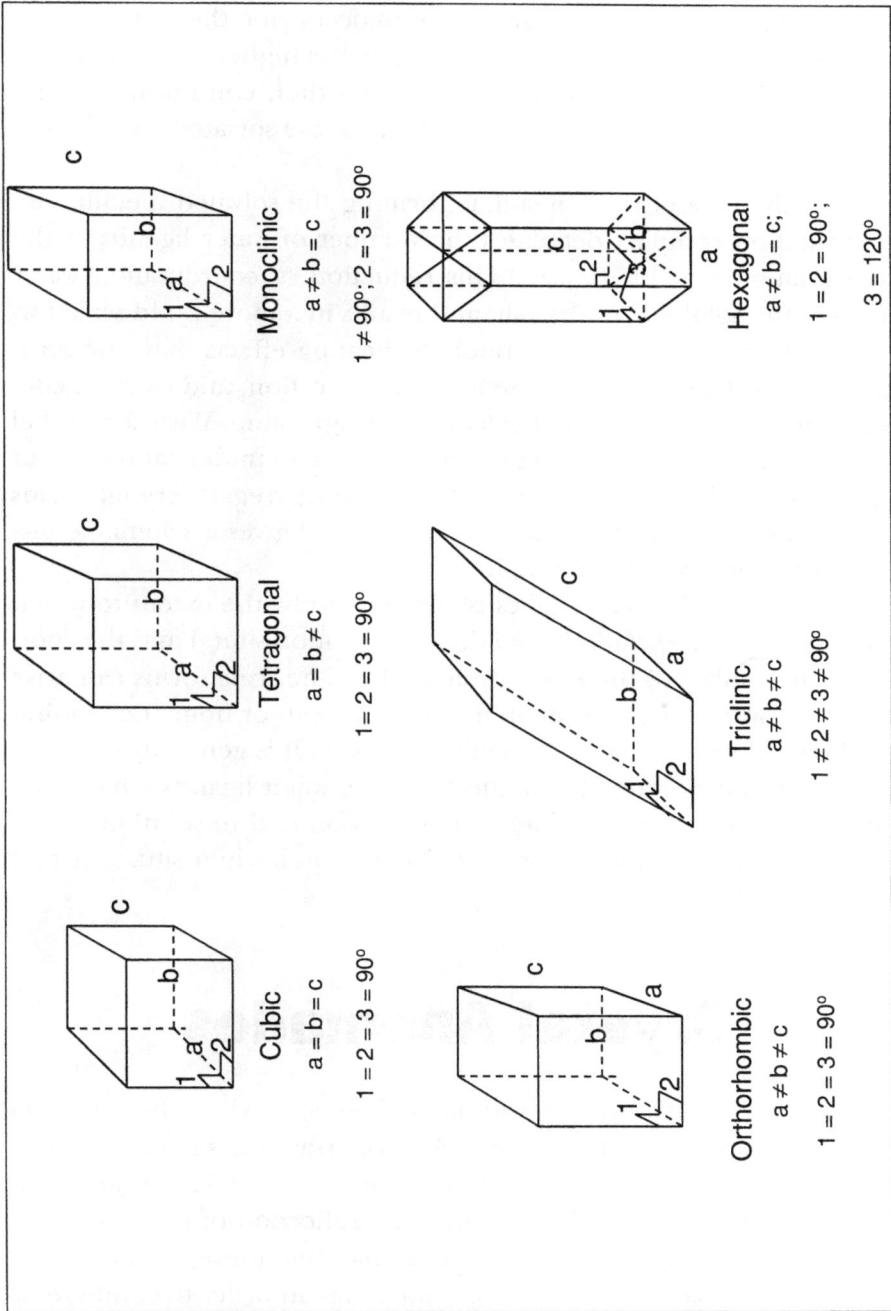

Fig. 5-1 Six forms of crystal arrangements (angles are denoted by numbers and lengths as a, b, and c)

The number of water ligand groups occupying the coordination sites of the metal salt remaining in solution is higher than those taking place in the crystallization process. Further, conditions of temperature and pressure affect the solubility of the solvated metallic salt complexes.

In the case of barium sulfate, heating the solvated metallic salt coordination complex decreases the number of water ligands in the coordination sites of the salt. This reduction of coordination water reduces the solubility of the salt, and results in a more rapid scale formation. Barium is more susceptible to heating effects than the magnesium or calcium salts because it is a larger cation, and its charge-to-mass ratio is less than that of calcium or magnesium. With this verbal picture of the aqueous complexes, it is possible to understand some of the factors influencing the types of macro-aggregate crystal forms encountered in petroleum systems. Figure 5–2 gives a schematic picture of the preceding paragraph.

The form the crystal takes is determined by the extent to which water ligands are excluded from the coordination site(s) and the coordination number of the salt complex. Thus, the scale forms can arise from salts with some coordination water present, or none, depending on the solubility of the coordination complex. It is generally true that the complex containing the highest number water ligands is more soluble in aqueous systems. Thus, the formation and dissolution of the aqueous ligand complexes prove to be an equilibrium situation that favors the dissolution.

Crystal Anomalies

Scale crystals form from ionic salt solutions when the temperature falls below a critical minimum for dispersal forces to prevent salt-salt combinations. This effect is best accounted for by kinetic and geometric arguments. Since temperature is a reflection of the molecular velocities of the molecules within a sample, the average velocities of the salt complexes must reach a specific range in order to combine. If the velocities are above this range, the salt complexes will be too ener-

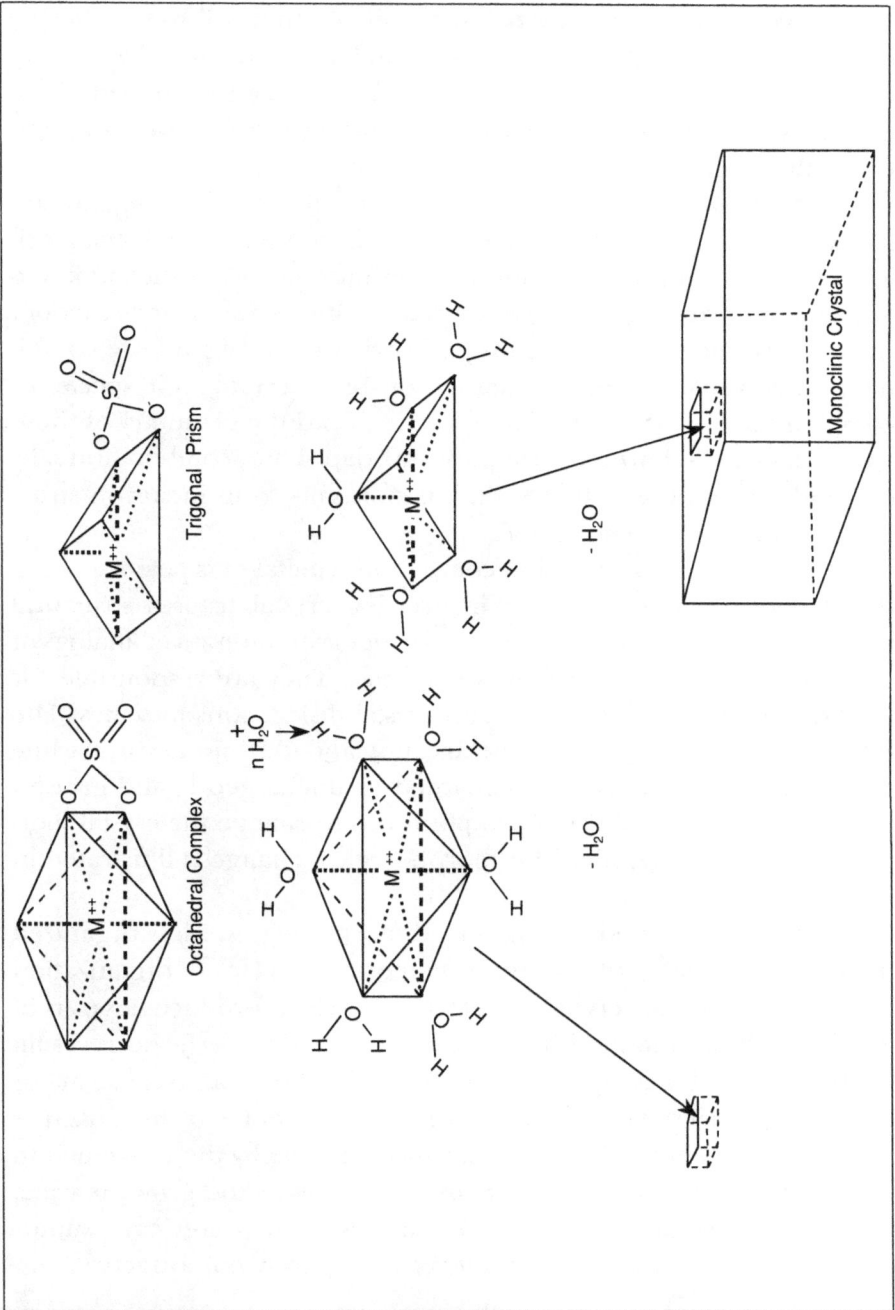

Fig. 5-2 Six and eight coordinate metal salts lose water ligands to form crystals

getic to combine. If they are below the range, they will lack sufficient energy to combine. As is the case with all kinetic systems, the velocity versus population curve for these salt systems has a bell-shaped distribution, with some salt molecules possessing velocities above and some below the optimum.

The geometric considerations include the need for ligand site interactions that are either facilitated or hindered by molecular orientations. Thus for salts to combine, they must possess sufficient kinetic energy to bring ligand sites within a minimum radius for reaction. In addition, the salts must be properly oriented for ligand site combinations. Petroleum fluids frequently consist of several ionic species in combination, and the critical temperatures and the geometry of these complex salts can have a considerable overlap. Thus, combination salts can and do result, and these combination salts form inclusion structures within the growing crystal.

The presence of inclusion compounds (metal salts possessing different coordination numbers) in growing crystals causes structural irregularities. These structural irregularities are cationic salt analogs of the uncharged metal coordination species. They are responsible for the grain boundaries in metals and crystal dislocations in scales. The scale forms always contain the metal cation and the anion ligand, while the metals seldom possess any charged or uncharged ligand groups. The effect of these inclusion compounds is to change the crystal morphology of forming scale. This morphological change is illustrated in Figure 5–3.

Crystal systems possessing mixed morphology are less organized than those consisting of a single crystal form. This lack of organization is manifested by scale crystal imperfections, which produce deposits of low durability and low stability. The quantities of imperfections resulting from inclusion compounds do not need to be great to cause major morphological crystal weaknesses. Thus, an avenue of approach is afforded for the intentional modification of scale by the introduction of chemicals capable of causing imperfections in the growing scale. Further, because only very small quantities of inclusion compounds are required, the economic viability of this approach is attractive.

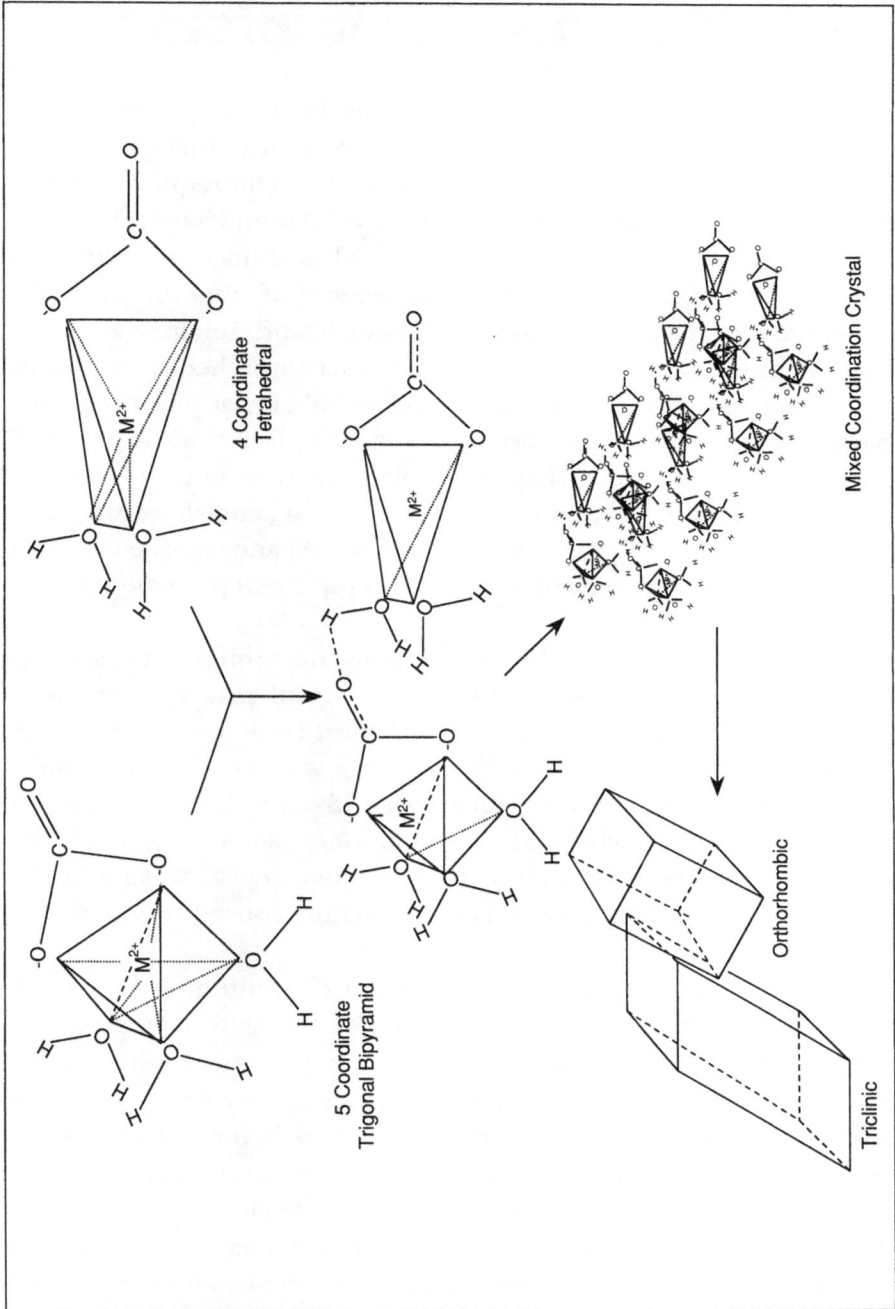

Fig. 5-3 Combined coordinated complex salts in process of forming crystal irregularities

Inhibition Mechanisms

All chemical inhibition mechanisms involving effective scale deposition control require the incorporation of an active ligand group at coordinate sites of the scaling metal complex. This requirement frequently, but not always, demands that the inhibition chemical replace the anionic ligand groups of the scaling salt with its own. Inhibition can also be accomplished by the replacement of some or all of the non-ionic ligands, while leaving the anionic ligands unchanged.

Anionic ligand replacement may be accomplished by introducing chemicals with stronger anions, such as sulfates or phosphates for carbonates. Replacements may also be accomplished with geometrically concerted dianions (e.g., diglycolates or diacetates for carbonates) and combined non-ionic anion effects (e.g., chelate compounds EDTA for carbonates and sulfates). Non-ionic ligand replacement can also alter the morphology of the scale without the replacement of the anionic groups.

The effectiveness of the non-ionic ligand groups is dependent upon its ability to displace other non-ionic ligand groups in the metal salt complex. As mentioned before, the most common non-ionic ligand group found in petroleum fluid systems is water. Thus, the ability of the non-ionic inhibitor to replace water from its ligand sites is key to the inhibitor's effectiveness. Two important factors determine the effectiveness of the inhibitor's success in replacing the water ligands: ligand group electronegativity and its relative concentration to the water present.

Obviously, the adjustment of inhibitor concentration is an inappropriate and costly approach; thus the electronegativity approach is most viable. Very few molecules possess the polarity of water or the electronegativity of oxygen. Therefore, advantage is taken of the combined effects of several electronegative groups being present in the same molecule. Thus, chemicals containing several electronegative groups with appropriate orientations can effectively displace water by concerted group interactions. Many of these chemical types are represented by polyamines (e.g., ethylenediameine, diethylenetriamine, triethylenetetraamine, and tetraethylenepentaamine).

The temperature and pressure at which scaling occurs is an important aspect of effective scale control. Consequently, these physical factors should be considered when applying a chemical treatment. If the scaling salt crystals form at temperatures and/or pressures above or below those of the inhibited salts, then the mechanism is mainly concentration dependent. However, if the scaling salts and the inhibited salts crystallize at or near the same conditions of temperature and pressure, the mechanism of inhibition favors crystal distortion effects. Crystal distortions, as pointed out above, are far more cost effective than concentration effects.

It is important to realize that the forms of scale that occur in petroleum fluids are strongly influenced by the partial pressure of water under prevailing conditions of temperature and pressure. When the temperature is sufficiently high to boil water at prevailing pressures, the equilibrium balance between the solvent water and the coordination water of the salt complexes is changed. When this equilibrium balance is changed, the rate of entrance and exit of water ligands is also changed, thus altering the solubility of the salt complex.

This behavior is important when considering the type of inhibitor to use, since vapor pressure can affect availability for ligand group complex site interactions. Further, the phase behavior of the inhibitor is very important. If, under prevailing conditions of temperature and pressure, the inhibitor exhibits a higher solubility in the continuous oil phase as opposed to the water phase, it will be unavailable to the forming scale. This phase partitioning is most pronounced in the non-ionic ligand group chemicals, in which the solubility of the polyamines in organic solvents is increased by temperature effects.

Crystal distortion effects can be attributed to two basic mechanisms: ligand-ligand group interactions and coordination number alterations. Ligand-ligand group interactions are dependent on the size and composition of the groups attached to the ligand sites of the metal. Coordination number alterations involve extremely strong interaction forces between ligand sites and ligand groups. These strong interaction forces generally require a combination of drastic physical conditions of temperature and pressure and strongly electronegative ligand group influences. Coordination site geometry and numbers are the result of the central metal's electron orbital arrangement. This geometry is locked into position by the overlap and combination of site and ligand group electron orbitals.

Ligand-ligand group interaction and complex coordination number alteration can occur as a combined effect, or they can operate independently. Multiple complex linkage is a subset of the ligand-ligand group interaction effect. It involves the occupation of ligand sites of multiple salt complexes through multifunctional anionic or nonionic electronegative groups. The relative scale inhibition effectiveness of these two basic mechanisms is determined by the magnitude and extent of the resulting crystal distortion. Figure 5–4 schematically illustrates ligand-ligand group effect. It also illustrates the coordination alteration of the metallic salts caused by strong interactive forces between the ligand and the coordination site of the metal.

Crystal Progression

Under fixed physical conditions of pressure and temperature, scale crystals should be nearly perfect (e.g., consist of a single coordination number salt with a uniform geometry). However, petroleum fluids are produced under highly variable physical conditions. This dynamic condition of temperature and pressure assures that crystal perfection is the exception rather than the rule. As temperature declines, those crystallizing salts that form at higher temperature proceed to form in a downward progression, and as they form, they release heat to the surroundings. This heat release facilitates the loss in energy attending the attainment of higher order by the crystal.

When the reservoir of the salt population has been reduced to a level below the K_{sp} value of the salt at a given temperature, the next salt form possessing a sufficiently high population will proceed to crystallize. The preceding salt may form a nucleation site for the newly forming crystals, or they may form separately. This process continues until the system temperature and pressure reach a condition of equilibrium. Thus, the scale that forms from petroleum fluid systems can be a collection of several salt crystals that have formed independently, or it may be a combination of several forms. If the crystals form as combi-

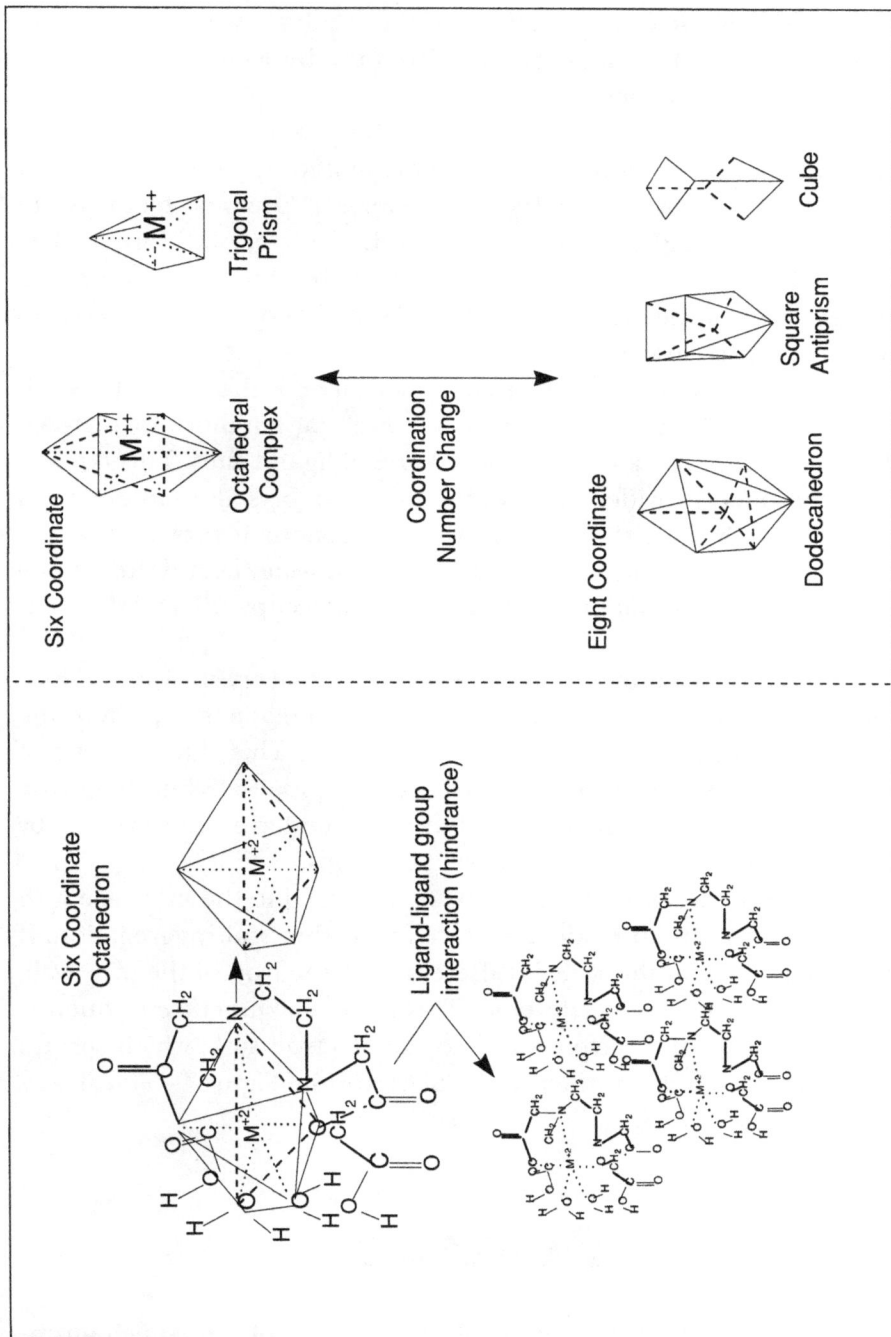

Fig. 5-4 Ligand-ligand group hindrance and coordination number changes

nations, morphological alterations of the resultant scale are least evident, whereas those scales resulting from the individual crystal forms exhibit greater changes.

The gross appearance of scales is the result of these secondary processes of aggregation (e.g., primary crystal formation followed by a second crystal form growing from the primary). Or it may be the result of mechanical aggregation of the individual crystal forms. The mechanical aggregation effects of scale result from the irregular geometry assumed by the individual crystals. As mixing occurs, these crystals are forced into close association.

It is important to realize that the surfaces of the formed crystals remain dynamic, and that there is a continual equilibrium between the solvating matrix and the exposed crystal ligand sites. This equilibrium condition provides a pathway for adjacent crystals to interact. An extreme example of these ligand site interactions is provided by the interactions of ligand sites in metal (ligand-site/ligand-site crystal packing). In this situation, no ligand groups occupy the metal's coordination sites.

Thus, aggregation of combined or individual crystal forms can be considered analogous to a weld between two metallic salts, where the interface possesses qualities of a metal or alloy. This characteristic of scales can be extended to interfaces between pipe walls and the growing scale, where coordination sites in the scale have been vacated by ligand groups and a metal-metal weld develops.

Once the weld has formed, the scale is held to the interface with a great tenacity and provides a surface for further scale aggregation. If the metal cation of the scale is different than the iron of the pipe wall, a bimetallic juncture can develop. This bimetallic juncture frequently provides either an electron source or an electron sink, which sets up conditions favorable for the process of corrosion. Figure 5–5 illustrates some of these effects.

Summary

The preceding chapter deals with some of the mechanisms involved in the process of scale formation from petroleum systems. A

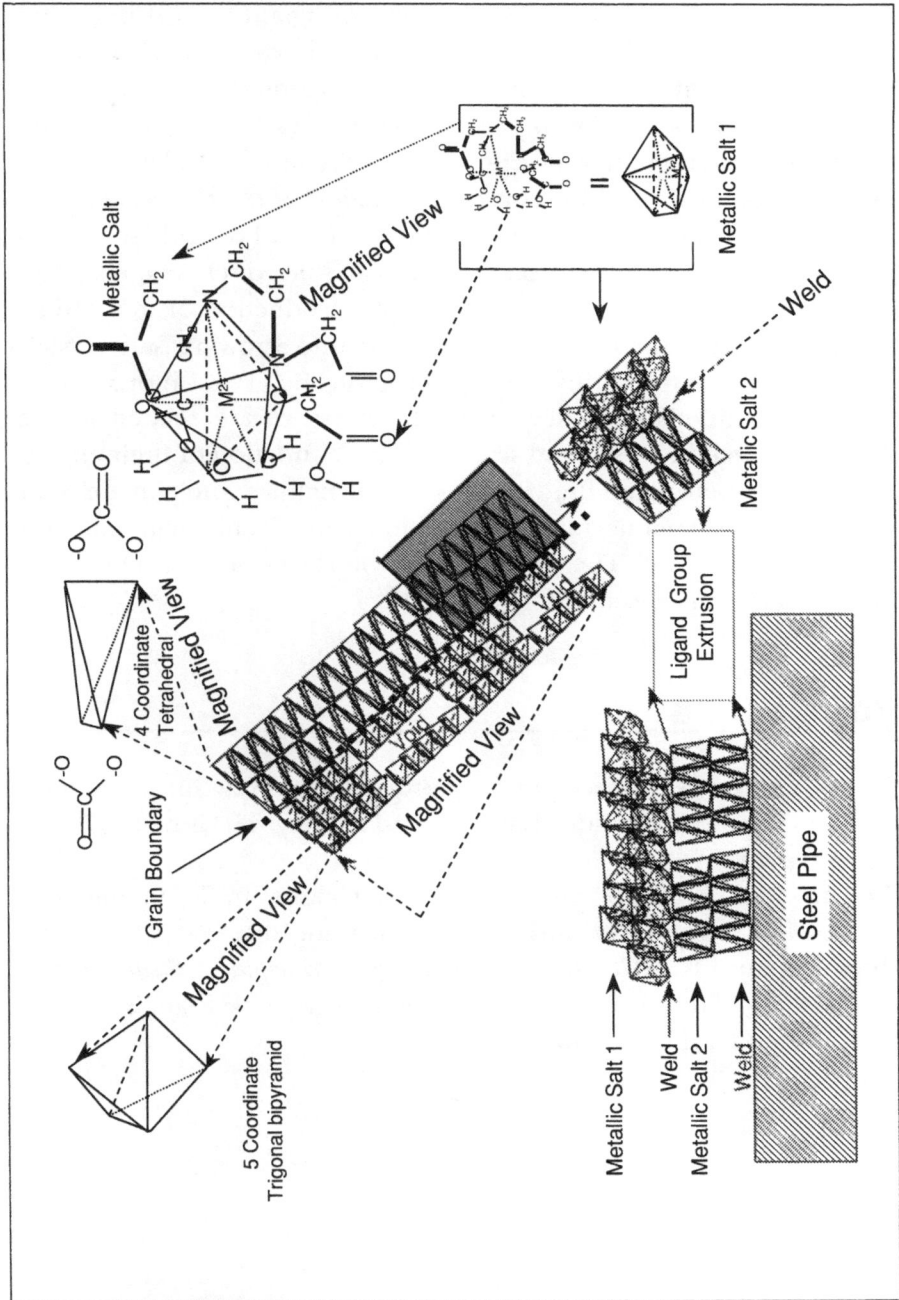

Fig. 5-5 Metal-salt weld, and metal-metal weld formed by vacant ligand site interactions

discussion of water as a ligand site occupant (ligand group) indicated how important the effect of solvent population statistics are in determining the coordination complex assumed. Additionally, the important role water ligands play in producing soluble salts is discussed. The nature of ionic and non-ionic interaction is discussed in relation to the configuration assumed by crystals as a function of their presence.

Crystal formation from variable salt forms and the role salt forms play in determining the morphology of scales are discussed. A discussion of the interference produced by the introduction of monofunctional and multifunctional ligand groups into growing salt crystals shows the effects the various groups have on the crystal morphology. Physical conditions of temperature and pressure are related to the effects observed in crystal and scale forms as a function of their changing nature in petroleum fluid systems. The dynamics, and concomitant effects of these dynamics, are discussed in terms of multi-aggregate formation (scale forms). The nature of the interfaces of the metallic salt and metal are also discussed.

References

Barrow, Gordon M. 1966. *Physical Chemistry*. 2d ed. New York, St. Louis, San Francisco, Toronto, London, and Sydney: McGraw Hill Book Company.

Dickerson, R. F., H. B. Gray, and G. P. Haight, Jr. 1970. *Chemical Principles*. 1st ed. New York: W. A. Benjamin, Inc.

Huheey, James E. 1978. *Inorganic Chemistry Principles of Structure and Reactivity*. 2d ed. New York, Hagerstown, San Francisco, and London: Harper & Row.

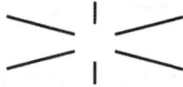

6
Solid-Solid Surfaces

Metal Surface
and Inorganic Scale

The effect of scale on metal surfaces is an extremely important aspect of damage caused by petroleum fluids. If a close examination of the interface between the metal and the metallic scale deposit is performed, several important features begin to stand out. Under acidic conditions, the iron of the mild steel undergoes proton attack by the hydronium ions present in the aqueous salt system. This reaction is represented by the following equilibrium equation:

$$2\,H_3O^+ + Fe + 2\,e^- \rightleftharpoons Fe^{2+} + 2\,OH^- + 2\,H_2\uparrow$$

The ferrous iron formed in this reaction is capable of further oxidation to form the ferric iron cation, as illustrated in the following equilibrium reaction:

$$Fe^{+2} + 3\,H_3O^+ + e^- \rightleftharpoons Fe^{3+} + 3\,OH^- + 3\,H_2\uparrow$$

The addition of the two preceding equilibrium equations results in the following:

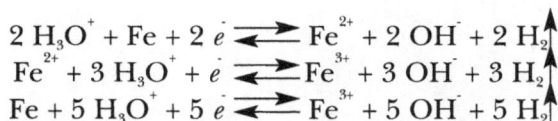

$$2\ H_3O^+ + Fe + 2\ e^- \rightleftharpoons Fe^{2+} + 2\ OH^- + 2\ H_2\uparrow$$
$$Fe^{2+} + 3\ H_3O^+ + e^- \rightleftharpoons Fe^{3+} + 3\ OH^- + 3\ H_2\uparrow$$
$$Fe + 5\ H_3O^+ + 5\ e^- \rightleftharpoons Fe^{3+} + 5\ OH^- + 5\ H_2\uparrow$$

Thus, the conversion of iron metal to the ferric ion is accomplished through the presence of five equivalents of hydronium ions. It results in the production of ferric hydroxide and molecular hydrogen. The molecular hydrogen generated by the oxidation of the iron can evolve freely from the oxidizing metal surface if the surface is free of scale. However, if metallic salt complexes are present, the metal can adsorb the molecular hydrogen as illustrated in Figure 6–1.

The disproportionation of molecular hydrogen between the metal and scale surface produces a proton and hydrogen radical. The proton attacks the more basic calcium hydroxide, while the hydrogen radical is adsorbed to the surface of the corroding metal. As the calcium hydroxide is converted back to calcium carbonate in the presence of carbon dioxide, water and molecular oxygen (O_2) are generated. The ferrous hydroxide [$Fe(OH)_2$] that results from the hydronium attack reacts with two hydrogen radicals to form two water molecules and the ferrous ion. The ferrous ion is then attacked by the molecular oxygen, which is produced in the conversion of calcium hydroxide to calcium carbonate, to form iron oxide. Thus, the corrosion cycle is completed.

However, there is an alternative path for the hydrogen radical, and this path involves the migration of these species into the metal. This is possible because of the extremely small size of the radical hydrogen. Hydrogen radical migration frequently occurs at metal grain boundaries, where poorly packed metal crystals provide channels between two metallic coordination species. When hydrogen radicals migrate along grain boundaries, they are often trapped for sufficient periods to build concentration adequate for radical recombination. When this occurs, the radical recombination produces molecular hydrogen that builds to high pressure within the channel. This causes blistering and pitting.

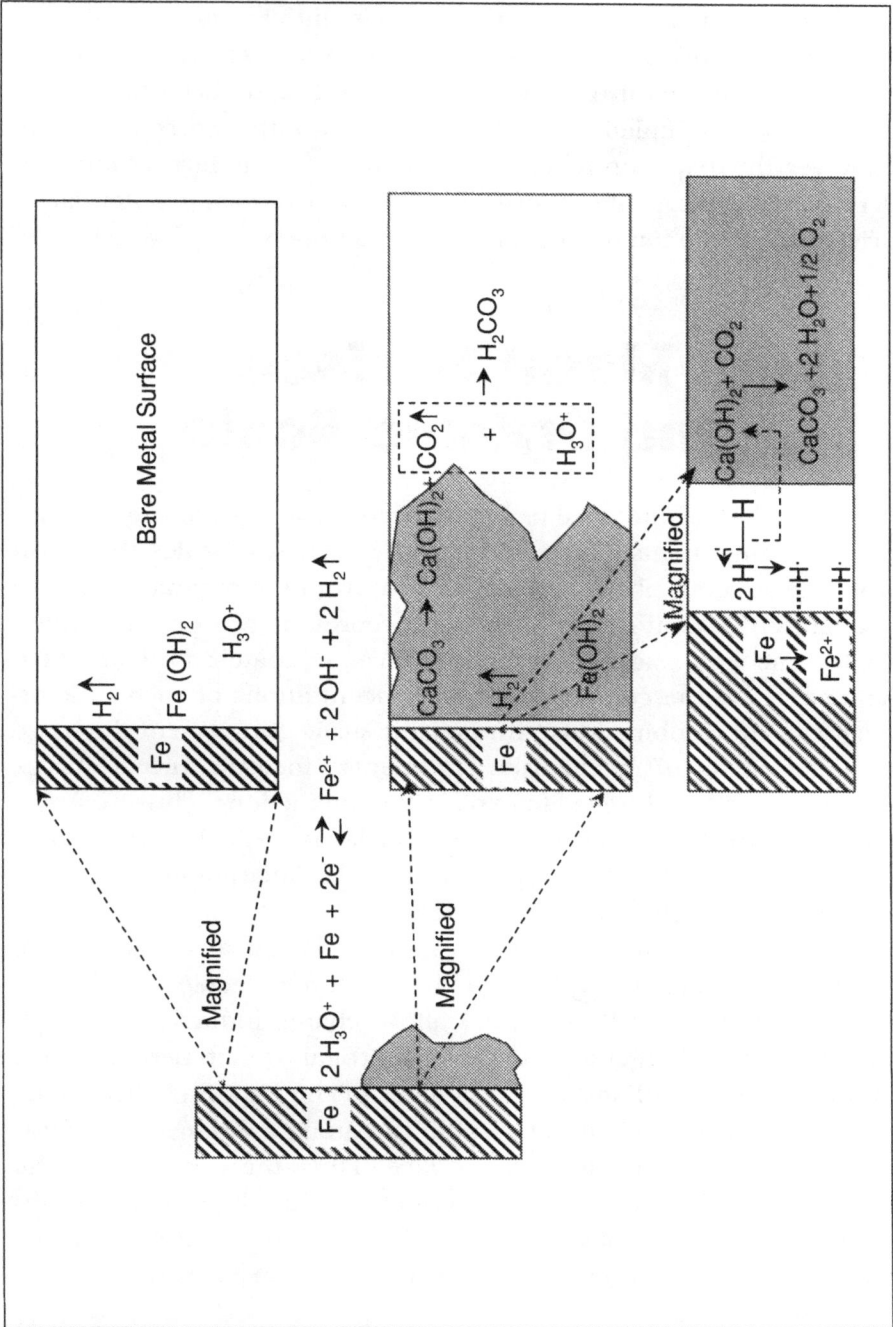

Fig. 6-1 Solid-solid interfacial reactions between metal and scale

This discussion follows a rather circuitous path in order to explain the nature of the combined corrosion scale effects. It would be far less complicated if the calcium ions could be reduced to metallic calcium by a coupled half-cell reaction with the corroding iron. However, the oxidation potential of the iron metal to ferrous ion ($\epsilon_0 = 0.44$ volts) is incapable of providing enough reducing potential to drive the calcium ion to calcium metal reaction ($\epsilon_0 = 2.87$ volts).

Metal Surface and Organic Scale

Petroleum fluid systems are complex systems of inorganic and organic components. In addition to the inorganic scales that form from the aqueous phase, scales that form from the organic phase are also present. Paraffin and asphaltene represent two organic scales. Asphaltene and paraffin scales can form separately or combined, depending on the composition and concentrations of each in a system. Paraffin combinations with metallic surfaces represent the more complicated set of mechanistic arguments, since no formal charge argument can be advanced to explain the interactions. Thus, complex quantum effects must be included in order to explain such interactions as London dispersion forces and van der Waals radii. Figure 6–2 schematically illustrates these interactions.

Although no formal charge exists on the metal surface or the paraffin molecule, the approach of the paraffin molecule to a distance equal to the van der Waals radius allows charge induction to occur. These induced charges arise because of orbital overlap between the d orbitals of the metal and the sp^3 orbitals of the paraffin. This orbital overlap is only possible if the paraffin can attain a distance of the van der Waals radii from the metal surface. Thus, the d orbitals of the metal surface and the paraffin form a dsp^3 hybrid orbital, which results in the inductively coupled charges. This mechanism is responsible for static charge interaction, leading to the attraction of paraffin to the metal surface.

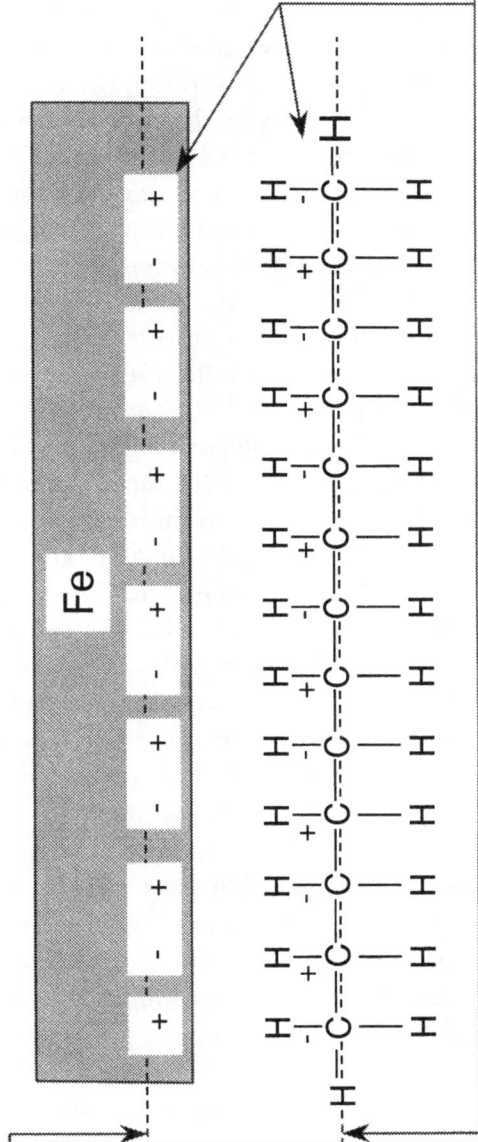

Fig. 6-2 Illustration of charge induction forces between paraffin and metal

Kinetic factors are extremely important in determining whether the paraffin will be deposited on the metal surface. A paraffin might approach the metal with a velocity or momentum greater than that required to bring it within range of the van der Waals radii. If this occurs, it will not be retained within this distance for a sufficient period to allow London dispersion forces to take over. If, on the other hand, the paraffin does not possess sufficient momentum to achieve a distance within the van der Waals radii, it will not be attracted. Thus, this situation can be thought of as a *quantum condition*, requiring that a specific energy level be possessed by the paraffin that is neither too great or too small to allow deposition. Quantum conditions required by metal-paraffin interactions then arise from kinetic effects.

Temperature and time are key elements in the attainment of orientations between the paraffin and metal affording induction forces. Additionally, temperature and time are critical to the aggregation of paraffin as it is crystallized from the organic solvent. Normal paraffins (C_nH_{n+2}) are neutral charge species, and as such they must form crystals (multiple molecular aggregates) by interactions similar to those previously described for metal-paraffin aggregates. In order to form a nucleate, two paraffin molecules also must meet these quantum conditions. This is the first stage of crystallization.

Only when these conditions are met can the London dispersion forces produce the inductively coupled molecular pair necessary for nucleation. Figure 6–3 illustrates this relationship for two interacting paraffin molecules. Note that neither the metal-paraffin nor the paraffin-paraffin aggregate possesses formal charges. The charges developed are a result of inductive effects.

The satisfaction of these quantum conditions is achieved by kinetic forces. Thus, momentum (mass x velocity) expressed for a specific time period [(mass x velocity x time) = (mass x distance)] provides the minimum mass/energy [(mass x velocity x time)/(van der Waals radii) = (mass)] required for interaction. According to Maxwell's speed distribution law, molecular velocity is determined solely by temperature.

$$N(v) = 4\pi N(m/2\pi kT)^{3/2} v^2 e^{-mvv/2kT}$$

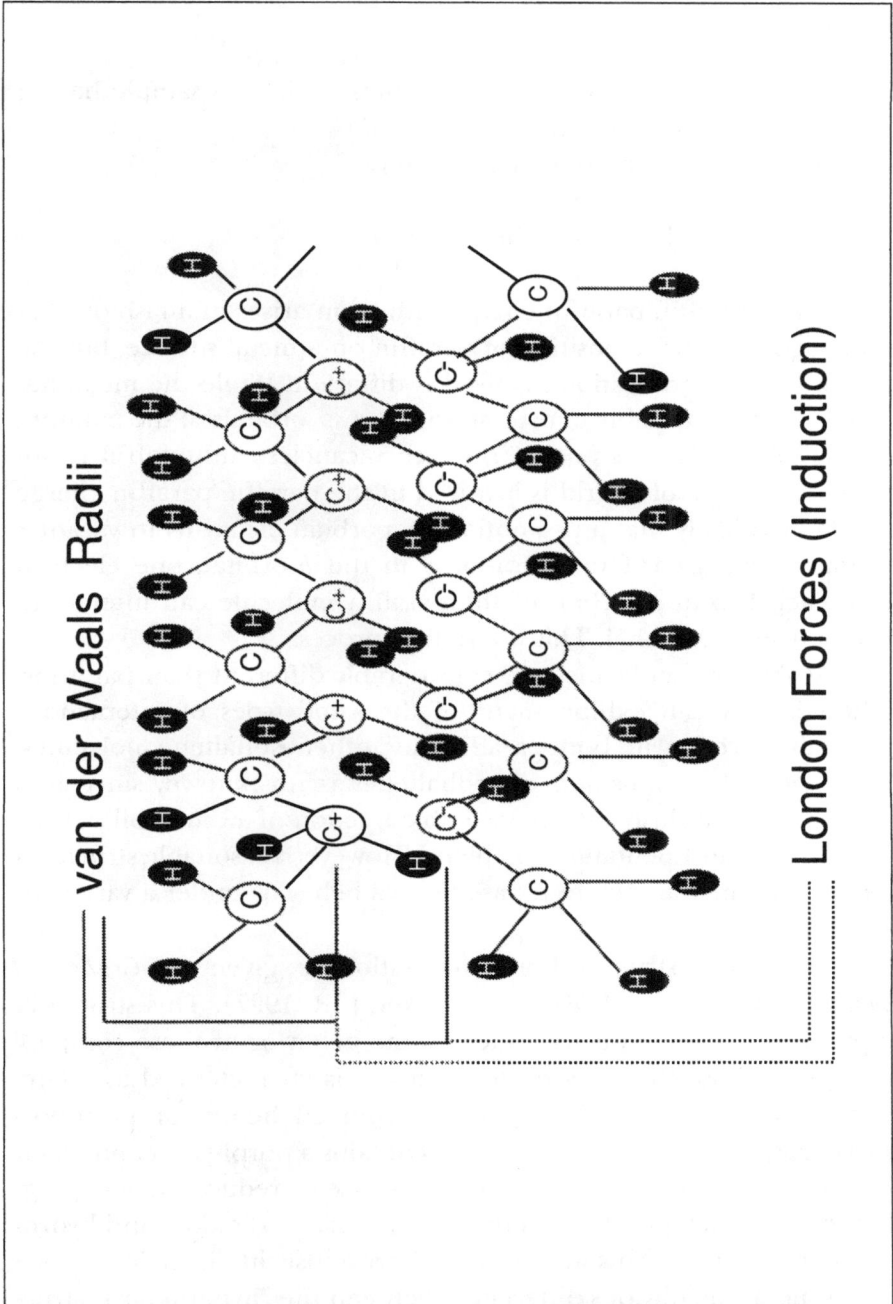

Fig. 6-3 Paraffin molecular nucleate arrangement showing van der Waals radii and charge arising from London dispersive forces of induction

where,

$N(v)dv$ is the number of molecules in the gas sample having speeds between v and $v + dv$

T is the absolute temperature

k is Boltzmann's constant

m is the mass of the molecule

Thus, paraffin-paraffin charge induction arises in a fashion that is analogous to the deposition of paraffin on a metal surface, but the type of bond hybridization involved is different. While the metal has vacant d orbitals capable of overlap with the sp^3 orbitals of the paraffin, the paraffin molecules possess no such vacancies. Although it is not known what type of hybrid is involved in the paraffin-paraffin charge couple, it is likely that a promotion of s orbital electrons to vacant p orbitals is required. Once promoted to the p orbital, one electron from each hybrid p carbon of the paraffin molecule can interact to form a p^2 hybrid orbital. This is a radical process.

Asphaltene molecules are considerably different than paraffins, although they can exhibit many of the same types of interactions (inductive forces) with both paraffins and other asphaltene molecules. The chemical composition of asphaltenes is not known, since it is defined in the industry as the insoluble fraction of a crude oil sample diluted by a 5 to 1 pentane to oil blend. However, reasonable structural approximations can be given based on its behavior under a variety of conditions.

One such structural approximation is given in *Crude Oil Emulsions, Waxes, and Asphaltenes* (Becker, J. R. 1997). This structural approximation accounts for much of the behavior of these complex amorphous crystalline aggregates. Thus, it is characterized as a fatty substituted pyrrole-sheathed polysulfide-linked hemin or protoporphyrin aggregate. The hemin version contains a porphyrin-complexed ferrous or ferric iron, which serves to oxidize or reduce incoming ligand groups. Such groups include oxygen, carbon dioxide, and hydrosulfide molecules. (This was illustrated previously in Fig. 3–5.)

One of the obvious differences between the "hypothetical" structure of the asphaltene and paraffin arises because the asphaltene is suspected to contain oxidized iron, making it an organometallic compound. As a result, the asphaltene aggregate can accept two anionic ligands to form the organometallic salt. 'The paraffin possesses no metal cations or any other group capable of taking on a formal charge. The charge character of the asphaltene then can be thought of as formal in nature. If the asphaltene is dispersed in a solvent capable of conducting charge (e.g., a polar solvent), it can act on and be acted upon by other charges through the intervening conducting solvent. Given this facility, asphaltenes would then be considered electrolytes, but electrolytes possessing very unique solubility characteristics.

From the hypothetical structure, some solubility characteristics of asphaltenes can be rationalized. The outer aliphatic coat of the asphaltene can be viewed as a structure capable of interacting with other nonpolar species through the London forces of dispersion previously discussed for paraffin. These dispersion forces then can be looked upon as those giving the asphaltene aggregate the capability of interacting with paraffin molecules contained in the system.

However, as the core of the asphaltene is penetrated from the exterior, the second set of forces becomes more important. These forces are characterized as pi bonding in nature and assist in keeping the asphaltene aggregates dispersed in an increasingly aromatic solvent environment. The pi bonding character possessed by the secondary layer is also involved with pi bonding interactions with the heteroaromatic character of the hemin and protoporphyrin complex. Thus, asphaltenes can never be thought of as soluble. (This is an artificial distinction indicating a minimal-sized particle surrounded by maximal number of solvent entities resulting in a continuous mixture.)

After penetrating the primary aliphatic sheath, the secondary pi bonding region, the core hemin, is encountered. This hemin portion is the salt of the organometallic complex. At this point the question of whether or not water and/or electrolytes can reach this salt is important. If so, then the diffusion of a conductive media (water) can act to transfer ions into and out of the core. This then can change the corrosion and inorganic scale processes. These alterations of scale and corrosion processes can be illustrated schematically as in Figure 6–4.

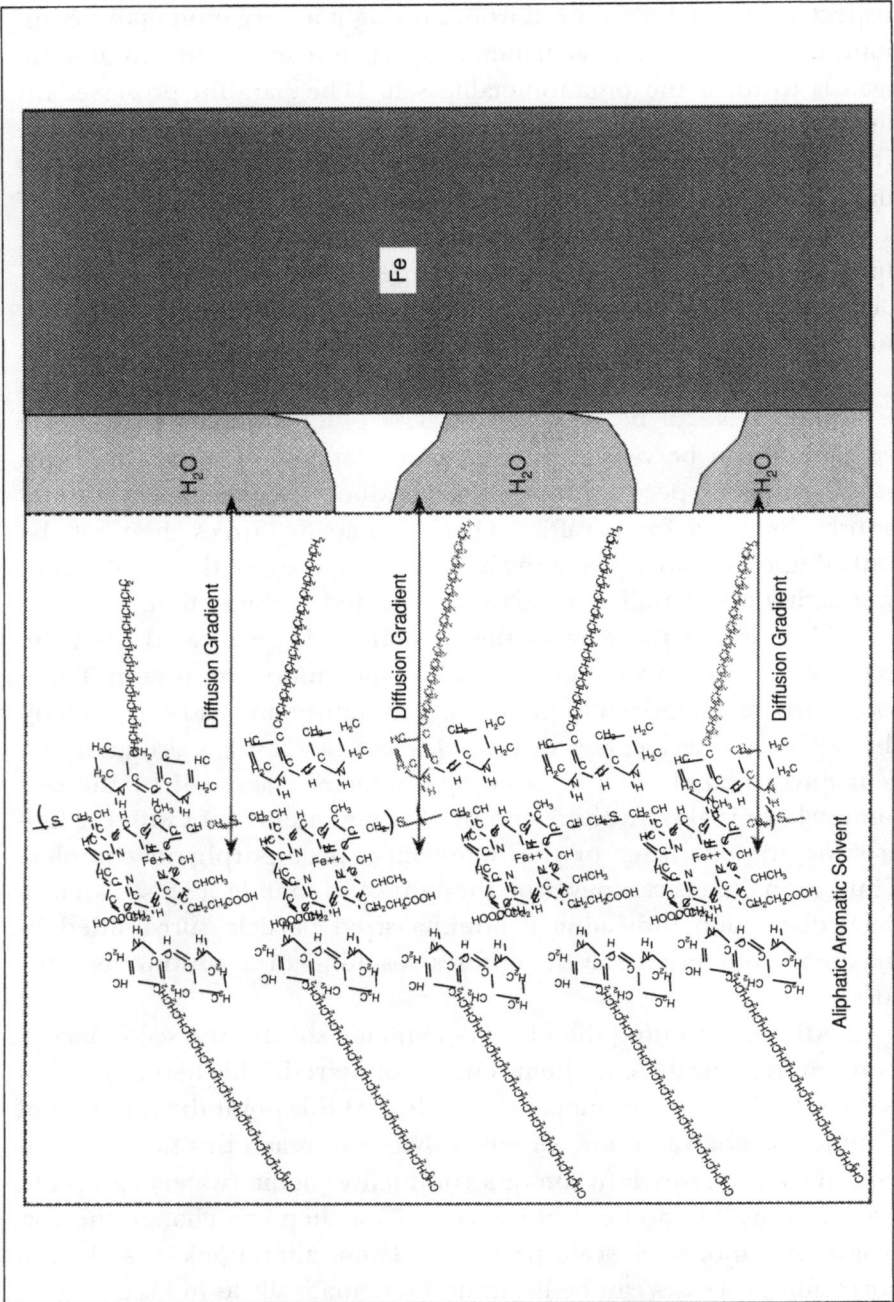

Fig. 6-4 Metal asphaltene diffusion gradient arising from solubility characteristics peculiar to asphaltene structures

Combined Inorganic and Organic Scale

Inorganic and organic scales can combine, and asphaltene participation in combined scale formation is mechanistically favored. This mechanistically favored combination does not preclude the inclusion of paraffin, since paraffin groups tend to form on the aliphatic tails of the asphaltene's outer shell. The highly acidic conditions of the aqueous fraction of a petroleum fluid favor corrosion reactions at the metal surface. The combined effect of metallic corrosion and the presence of asphaltene diffusion corridors is the establishment of an equilibrium ion exchange between the porphyrin complex and the corroding metal.

Thus, the intimate contact achieved by asphaltene deposition on the metal surface often develops an environment that assists in the corrosion process. Further, the diffusion gradient developed between the metal and the asphaltene tends to exacerbate the process of inorganic scale formation by providing local areas of high ionic concentrations. An interaction between asphaltenes and inorganic scales that is particularly noteworthy involves the exchange of anionic ligand groups. When asphaltene corridors are available between the aqueous phase and the hemin core, a pathway for the exchange of anion ligand groups is provided. This architecture is illustrated in Figure 6–5.

The hemin complex strongly associates with oxygen, carbon monoxide, hydrogen sulfide, and hydrogen cyanide. Thus, the anionic forms of these species combine with the ferrous and ferric forms of the porphyrin complex. If these forms are either free in the aqueous phase or occur as ligand groups in an external complex, then an exchange of anions occurs. The high specificity of the hemin group for these anions does not preclude the possibility of their coordination by other external anionic forms. However, it would appear that the hemin has a preference for monovalent molecular anions.

One additional structural feature of asphaltenes, or more particularly the hemin complex, is that the central ferrous cation is coordinated by two anions and four nitrogen bases. Recalling the chelate effect discussed previously, it would seem that asphaltenes should be

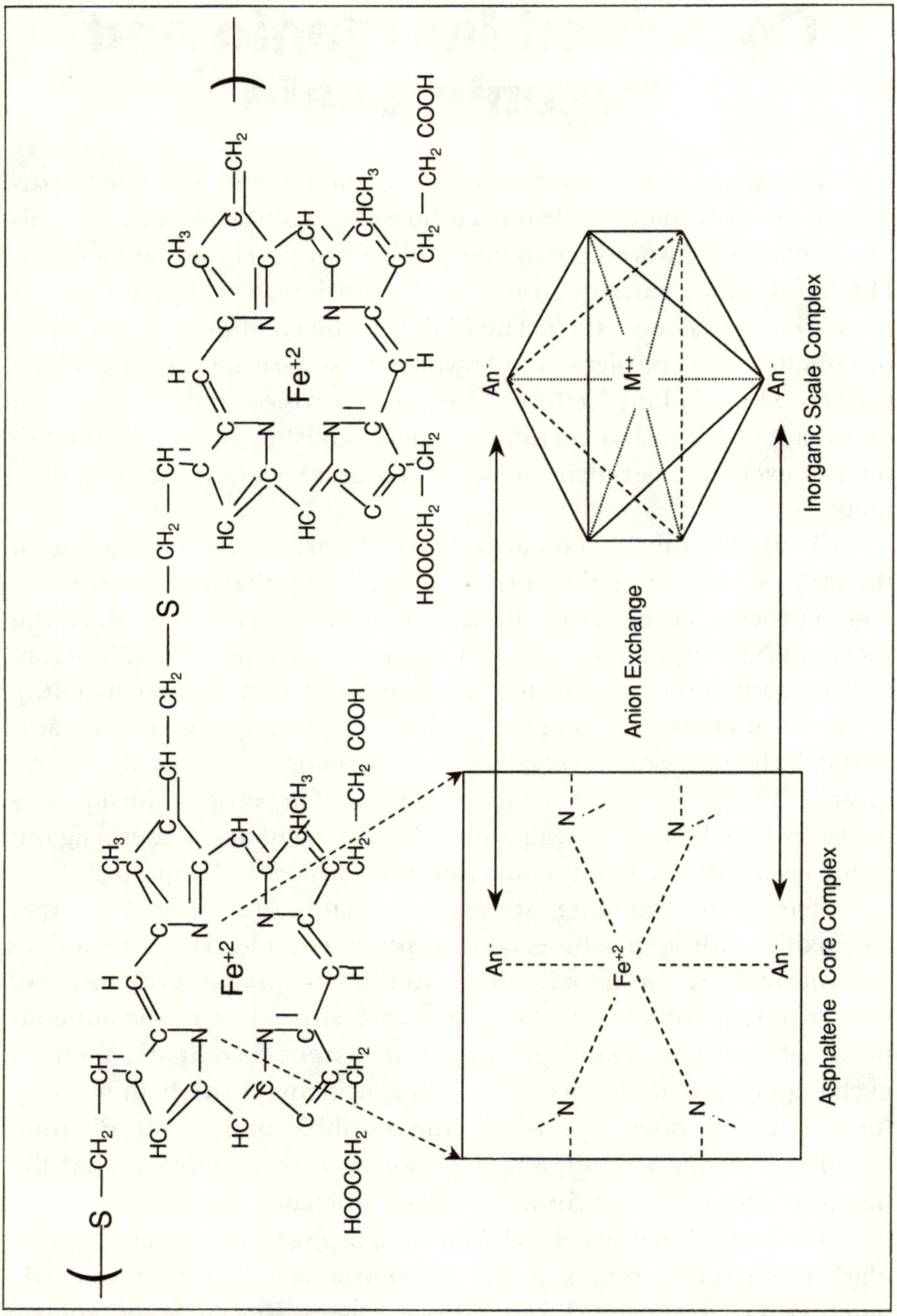

Fig. 6-5 Anion exchange between an inorganic scale and an asphaltene core complex

considered chelated. (Multiple ligand groups possessed by a single molecule give the resultant complex increased stability over those composed of individual ligand groups.) Thus the displacement of the iron cation by another chelating agent would require the resulting chelate be more stable than the hemin complex.

Bipolar Surfactant and Solid-Solid Scales

The presence of bipolar surfactants in petroleum fluid systems can have dramatic effects on the nature of solid-solid deposits. Because bipolar surfactants have both a polar group and a nonpolar group, they can be involved in the phase transfer of polar salts to nonpolar solvent fractions. The nonpolar group of the surfactant undergoes the same forces of aggregation as those of paraffin groups (London dispersion forces). Consequently, these species are frequently involved in the incorporation of aqueous solvent/solute mixtures in paraffin and asphaltene deposits.

This incorporation of the water-solvated polar salt within paraffin and asphaltene deposits produces discontinuous solvent environments at the solid-solid interface. The discontinuous interface (organic and water wetted metals or crude solids) causes random patterns of corrosion and scale formation at the metal surface. Figure 6–6 illustrates these effects.

Thus, the discontinuous coverage of metal surfaces by asphaltenes, paraffins, or combination paraffin-asphaltene deposits produces scale and corrosion effects that exhibit variations in surface corrosion and scale deposition.

Summary

The preceding chapter discussed the charge induction effects between metal and paraffin resulting from the London dispersion

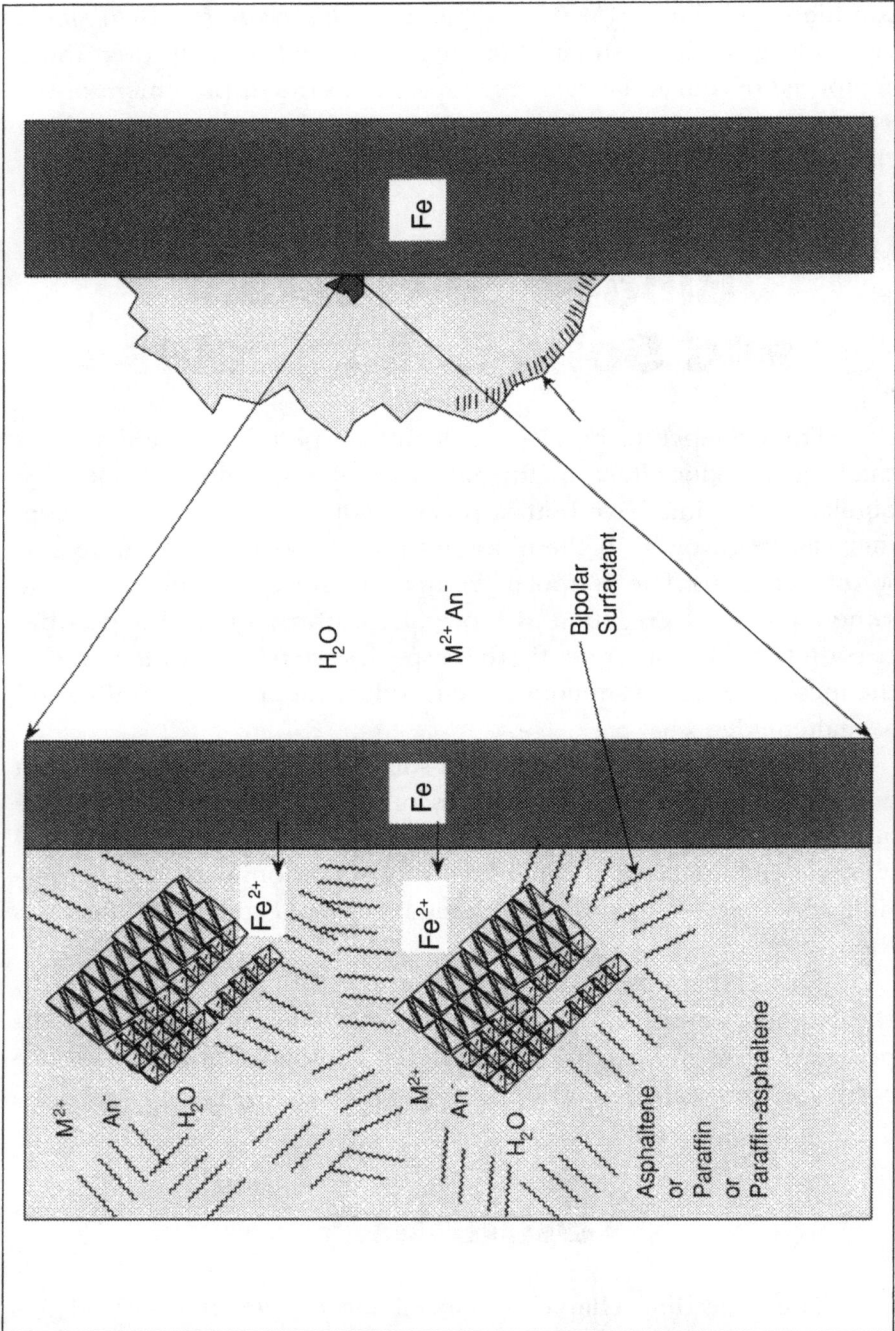

Fig. 6-6 Discontinuous solid-solid interface caused by bipolar surfactants

forces. Quantum effects were introduced to explain the requirements demanded by these inductive effects. Additionally, temperature and time were related to the velocity and momentum of the system to indicate their importance in the formation of hybrid orbitals. The asphaltene structure and its anion exchange possibilities were developed from a model that suggests that aliphatic sheath molecules will selectively pass certain ions through to the asphaltene core (hemin complex). The importance of bipolar molecules in the incorporation of scales within paraffin, asphaltene, or combination asphaltene-paraffin depositions was also discussed.

References

Barrow, Gordon M. 1966. *Physical Chemistry*. 2d ed. New York, St. Louis, San Francisco, Toronto, London, and Sydney: McGraw Hill Book Company.

Becker, J. R. 1997. *Crude Oil Waxes, Paraffins, and Asphaltenes*. Tulsa: PennWell Publishing Company.

Dickerson, R. F., H. B. Gray, and G. P. Haight, Jr. 1970. *Chemical Principles*. 1st ed. New York: W. A. Benjamin, Inc.

Hamill, William H. and Russell R. Williams, Jr. 1966. *Principles of Physical Chemistry*. 2d ed. Englewood Cliffs, New Jersey: Prentice Hall.

Huheey, James E. 1978. *Inorganic Chemistry Principles of Structure and Reactivity*. 2d ed. New York, Hagerstown, San Francisco, and London: Harper & Row.

7
A Closer Look at Petroleum Fluids and Scale

Scale and Metal Complexes

Complexes of the divalent Group 2A metal cations can exist in several coordinate forms from six through eight ligand sites. Additionally, the extent to which these ligand sites are occupied can range from two anions and no non-ionic ligands to four, five, and six non-ionic ligand group occupants. When groups of these complex salts aggregate to form crystals, the aggregate geometry depends on the number and composition of ligand group occupants.

Groups attached to the occupant ligand groups also have the capability of intercomplex group interactions involving adjacent complexed cations. These interactions can be attractive or repulsive in nature. Attractive intercomplex ligand group forces will tend to reduce the radius separating the cationic centers, while repulsive forces will increase this radius.

The interaction between intercomplex ligand group occupants can involve charge dispersal effects or stearic hindrance effects. In either case, the constituent producing the attraction, repulsion, or hindrance determines the packing efficiency of the component complexes. The packing efficiency can be assigned a value based on the tendency of the complexes to undergo subsequent aggregation leading to higher ordered crystalline structures.

In order to set a baseline value, certain assumptions must be made based on the behavior of the central metallic complex. Thus, the following list of assumptions can be used to describe the packing efficiency of metallic complexes:

1. Atomic metals represent the most efficient packing
2. Charge-satisfied or metallic salts are second in order of packing efficiency
3. Metallic salts decrease in packing efficiency with increasing numbers of nonionic ligand groups
4. Metallic salts with nonionic ligand groups attached decrease in packing efficiency as stearic hindrance and/or repulsive forces between complexes increase

Figure 7–1 illustrates the trends expected from ligand group interactions involved in crystal packing efficiency (excluding heteroatomic group repulsion).

Ligand Groups

Petroleum fluids are elaborate mixtures consisting of an extremely large array of possible ligand group candidates. Figure 7–2 illustrates the complexity of these fluids. The concentration, electronegativity, and aqueous solubility of the types of chemicals listed in Figure 7–2 determine their occurrence as ligand group occupants of the metallic salt complexes. The fact that metallic salts that exhibit inorganic scaling tendencies occur almost exclusively in the aqueous fraction of petroleum fluids makes water the most prevalent candidate for ligand site occupation.

Molecules possessing greater electronegativity, either through concerted group interaction or individually, can overcome the statisti-

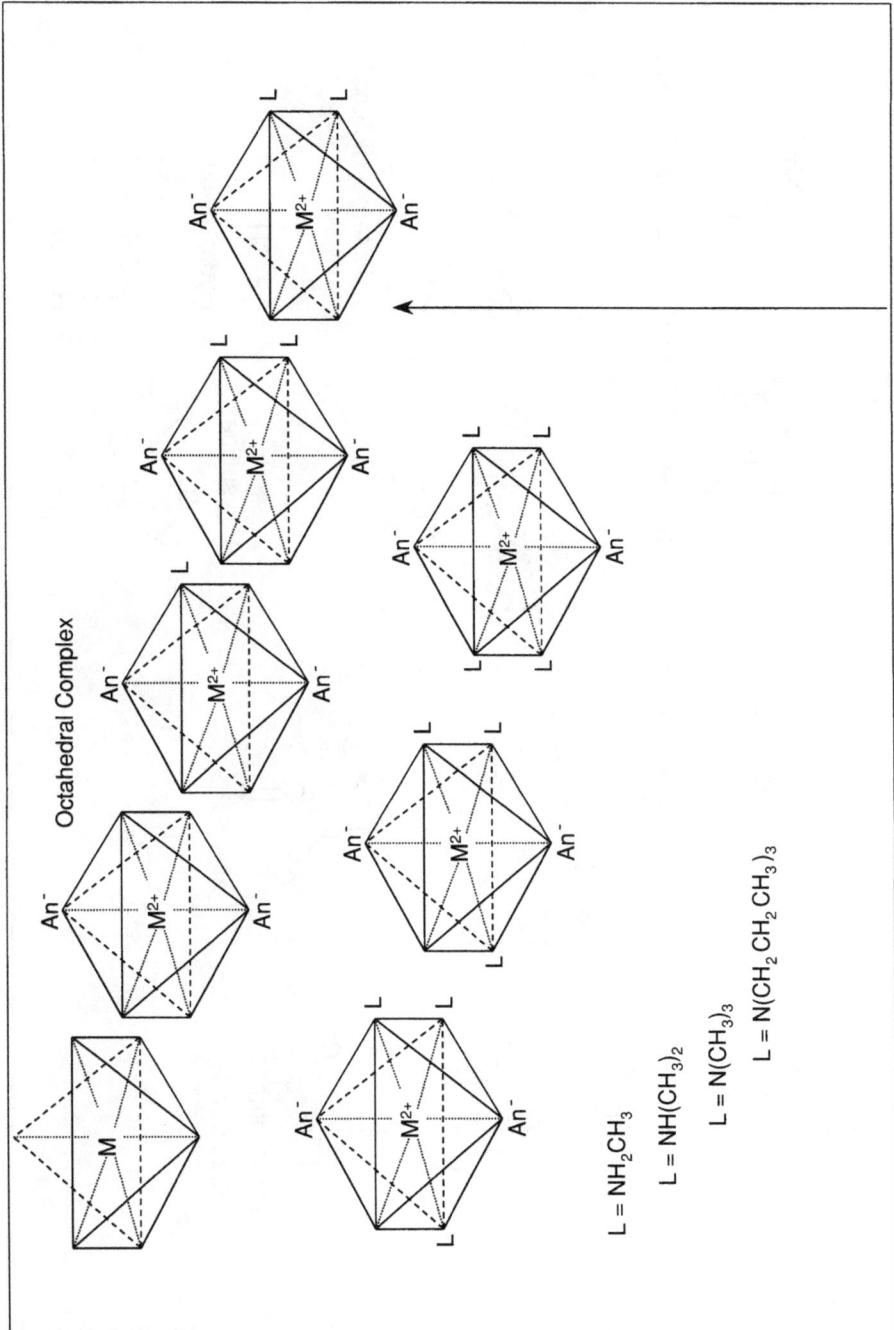

Fig. 7-1 Metallic complex packing efficiency

105

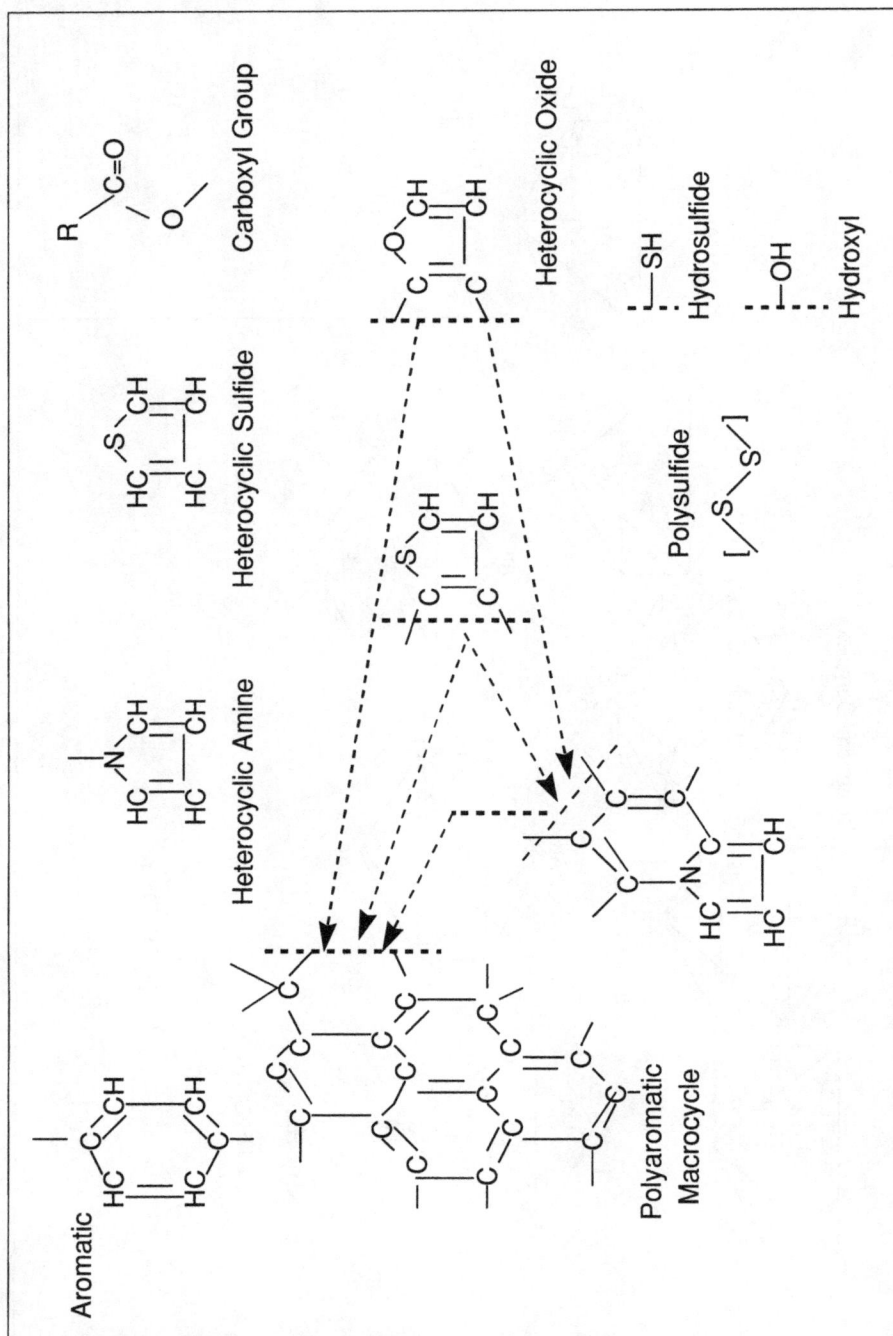

Fig. 7-2 Types of chemicals found in petroleum fluids

cally favored dominance of water. In reality, though, molecules possessing electronegativities greater than that of water are rare or nearly nonexistent in petroleum fluids. However, molecules possessing multiple electronegative substituents are frequently present. Some examples of these molecules include polysulfides, polyamines, polyhydroxyalkyls, and polyethers.

The reason that the polycomponent nonionic ligand candidates can preferentially occupy metallic salt ligand sites is due to the chelate effect that was mentioned earlier in connection with EDTA. Figure 7–3 illustrates this chelate effect.

Considering packing efficiency factors and ligand group reactivity and stability, it is possible to use a logical design approach to build molecules that interfere with crystal (scale) formation.

Scale Inhibitor Design

The previous discussion has pointed to some factors that are pertinent to the design of molecules with potential scale-inhibition qualities. However, additional factors should be known about the prospective chemical. These factors include the chemical's effect on the phase solubility behavior, anion ligand group exchange rate and affinity, and complex distortion effects on the resultant complexes. Phase solubility behavior, anion exchange affinity, and anion exchange rate can be altered by more than one method. These methods include the addition of non-ionic fatty polyligands and/or the replacement of the natural anionic ligands by fatty anions.

Complex distortion forces can be addressed by the inclusion of multiple anionic groups, structurally locked into a molecular configuration that maintains a specific distance between each anionic functional group. Thus, a certain set of ligand group structural effects can be engineered into a prospective chemical scale inhibitor. Keeping the above listed factors in mind, a first approximation to an idealized chemical scale inhibitor structure can be engineered. The criteria are as follows:

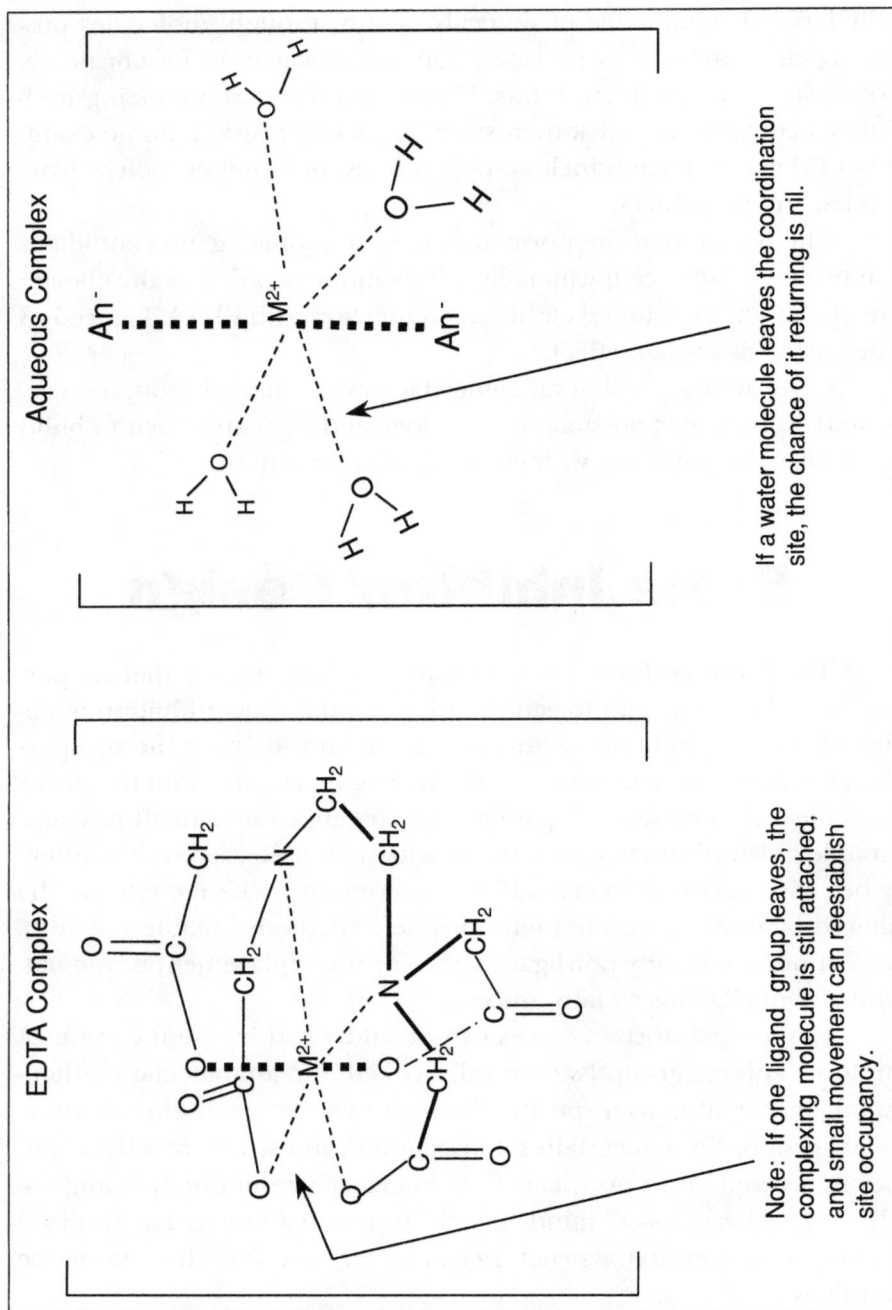

Aqueous Complex

If a water molecule leaves the coordination site, the chance of it returning is nil.

EDTA Complex

Note: If one ligand group leaves, the complexing molecule is still attached, and small movement can reestablish site occupancy.

Fig. 7-3 The chelate effect, showing the reason for the additional stability of a molecule possessing multiple ligands

1. The base structure should include from two to three anionic groups separated by no more than four carbons
2. The base structure should contain from two to six electronegative heteroatoms with spacing approximating the distance from one ligand site to another within the same metallic complex. The base structure should contain a bulky substituent group that can cause stearic hindrance between adjacent metallic complexes

Figure 7–4 shows how this chemical species could progress to the idealized first approximation of a structure. In Figure 7–4 each step follows the constraints imposed by the preceding list. The R groups listed are only intended to be illustrative of stearic hindrance, and the subscript n can range in whole number multiples from 0 to **n** (e.g., 0,1,2,3,...**n**). Selecting a structure derived from this procedure, a proposed arrangement of the complex can be visualized.

Figure 7–5 gives two ligand group arrangements of two of the idealized structures when complexed with six and seven coordinate divalent metal cations. The six-coordinate ligand group metal salt complex in Figure 7–5 contains bulky tertiary butyl [$-C(CH_3)_3$] groups that produce a substantial degree of stearic hindrance. The effect of this hindrance is to reduce the crystal packing efficiency of the resulting ligand complex. The presence of the tertiary butyl groups also reduces the likelihood of hydrogen bonding between the secondary amine groups [$(CH_2)_2NH$] of adjacent complex salts. Thus, the eight-coordinate ligand structure should exhibit greater crystal packing efficiency than the six-coordinate structure.

The preceding discussion on scale inhibitor design is intended to illustrate some of the principles involved in the development of an effective scale crystal modifier. The alkyl polyamine acetate structure represents a particularly simple example of structural effects that can be engineered into scale compounds. Disulfates, trisulfates, phosphates, and sulfides are also acceptable anionic ligand components, and polyalkyl sulfides and polyalkyl ethers may be substituted for the polyamine. Overall, the aim is to produce a chemical capable of preferentially reacting with potential scale forms, altering their structure, and reducing their crystal-packing capacity.

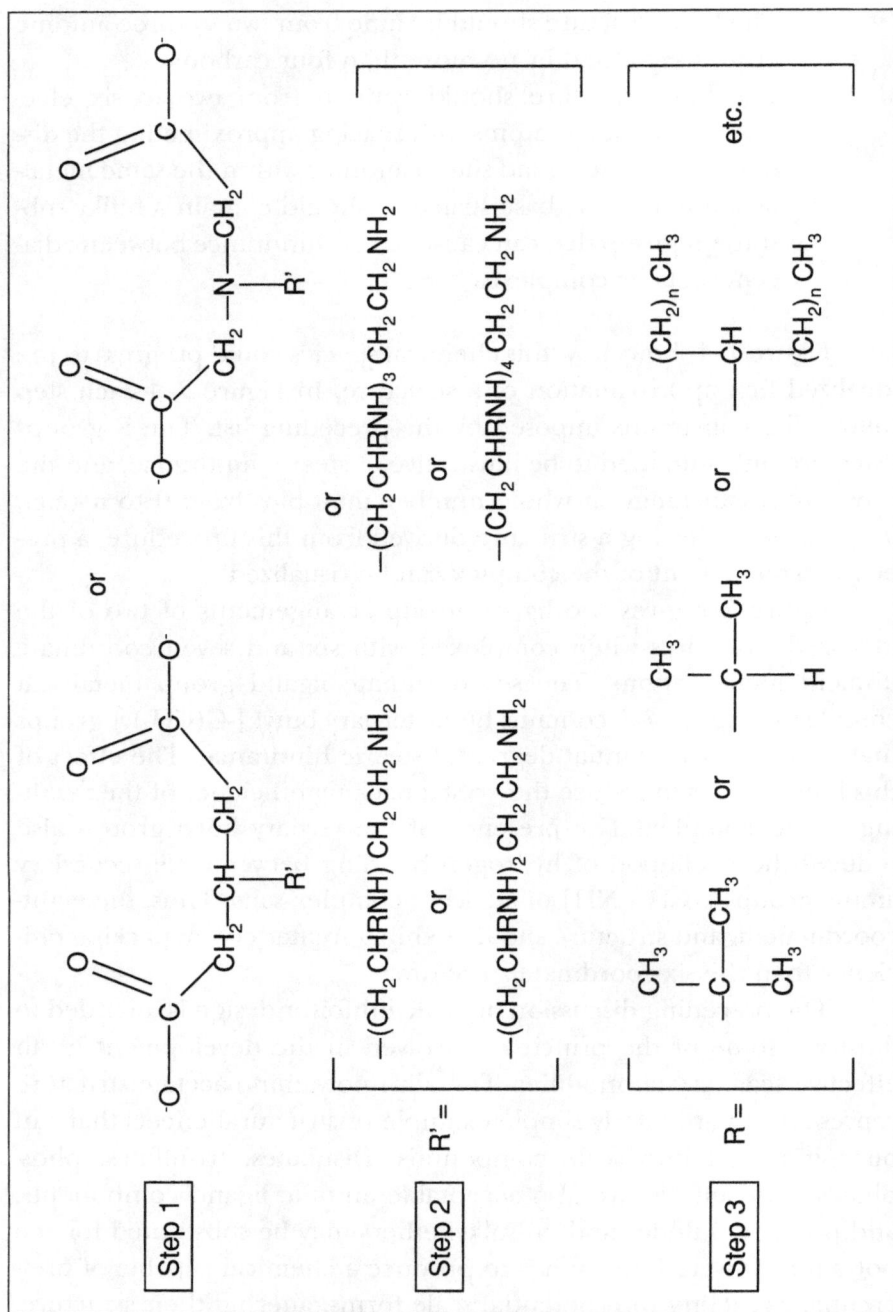

Fig. 7-4 Logical steps to a first approximation to an idealized scale inhibitor chemical

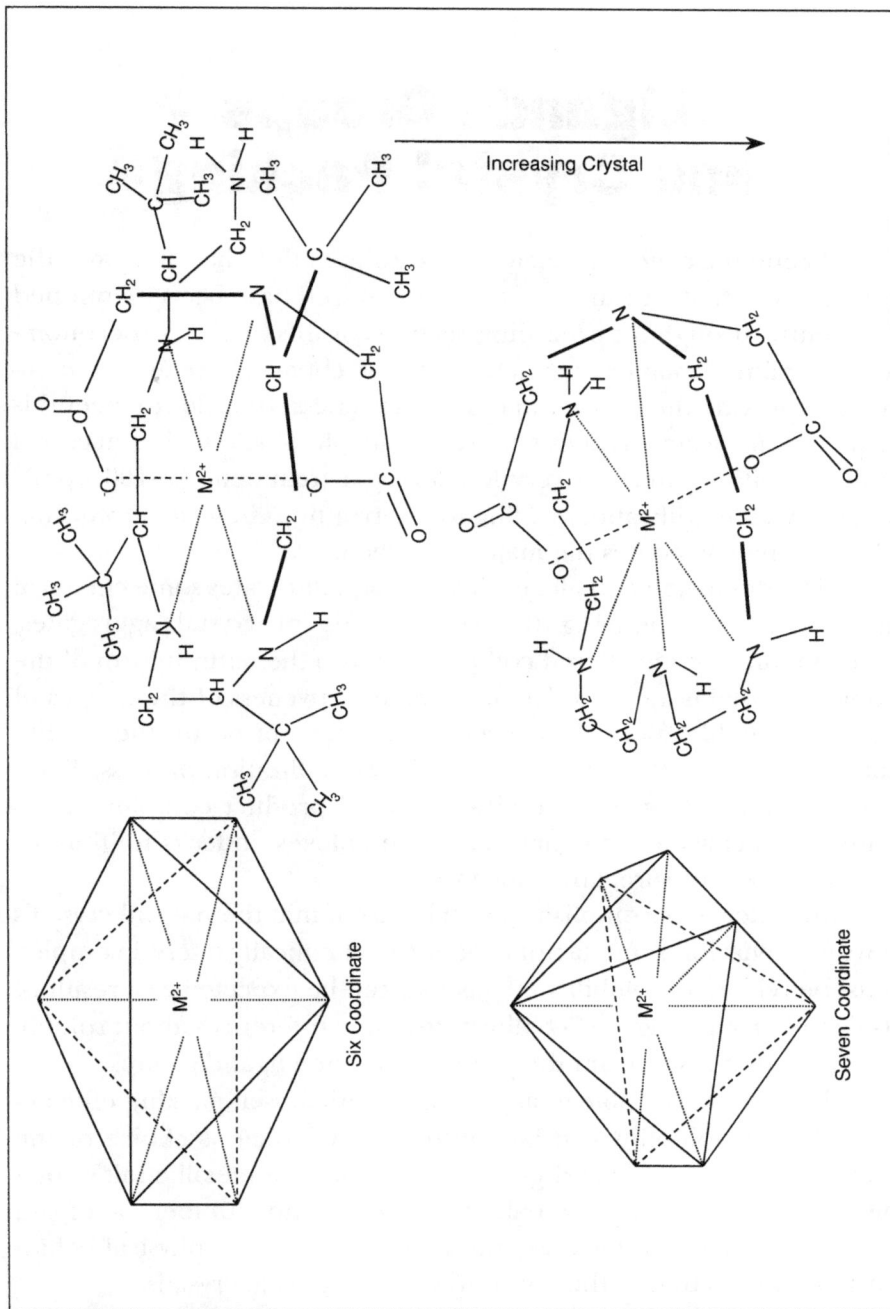

Fig. 7-5 Six and seven coordinate complexes with two idealized structures

Ligands Groups and Crystal Packing

Economic concerns related to petroleum fluid handling and the requirement that treatment costs be minimized have been mentioned frequently throughout preceding sections of this book. This economic constraint demands that low levels of chemical treatment compounds provide the maximum cost effectiveness. In order to meet this requirement, scale chemicals must be capable of producing maximal scale inhibition effects at very low levels of treatment (1–100 ppm). Thus, crystal modification leading to interruptions in scale networking phenomena constitutes the major approach.

The discussion of scale inhibitor design illustrates some chemical approaches at modifying the components of crystal aggregates. Incorporation of the altered complex salt into the natural form of the growing crystal is a critical factor in the effectiveness of these types of chemical modifications. If incorporation does not occur, the modification will have little or no effect on the crystallization process. Thus, it is imperative that the chemically altered product complex act to mimic the behavior of the natural salt complexes under conditions of temperature, pressure, and concentration.

In order to accomplish this ability to mimic the natural crystal's physical behavior, some factors about the chemically altered complex must be related to solubility effects that can be expected as a result of its ligand groups. Figure 7–6 illustrates some inferences about solubility effects that result from the presence of some ligand groups.

It is now reasonable to assume that both crystal packing efficiency and water solubility can be controlled by judicious choice of the number and type of ligand groups occupying the metallic salt's complex ligand sites. Thus, a balance between the number of ligand groups and their complex salt solubility must produce physical behavior that mimics that of the naturally occurring scale crystals.

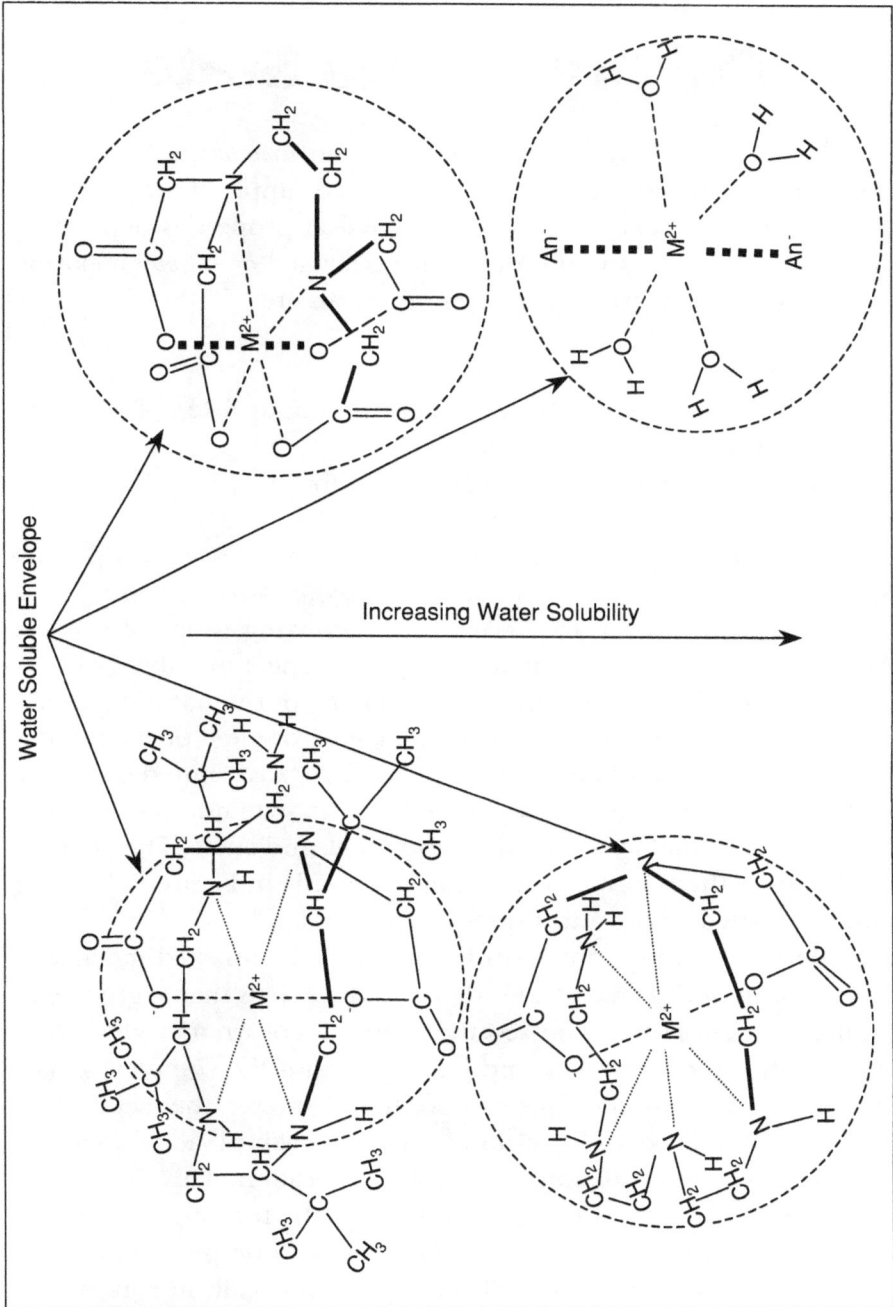

Fig. 7-6 Solubility effects due to ligand group type and occupancy

The Crystalline State

Solids occur in two forms: *crystalline* and *amorphous*. Most solids occur as crystals, and scales represent a good example of these types of solids. Many crystals exhibit different physical properties depending on their orientation, and this difference is termed *anisotropic* behavior. Examples of the external properties of crystals are:

1. Symmetry
2. Constant angles between corresponding faces of different crystals of the same substance
3. Production of plane faces on growth

The macroscopic properties of crystals, as indicated above, arise from the fact that atoms, molecules, or ions that compose crystals pack together to form a regular and ordered array. In the naturally occurring forms of crystals, these arrays consist of units that periodically repeat in a three-dimensional pattern. The three-dimensional pattern of the naturally occurring crystals repeats in such a way that the environment of each unit is identical to any other unit within the crystal. As was pointed out previously, all perfect crystals can be ordered into six systems: triclinic, monoclinic, orthorhombic, tetragonal, hexagonal, and cubic. The geometric shapes of these forms were shown previously in Figure 5–1, along with the related angles and symmetry elements.

Because of the properties of crystals, the complexed metal salts that make up scales can be considered as molecules that aggregate to form arrays. These arrays possess the properties of symmetry, constant angles between crystal faces, and plane growth if the complexes making up the crystal are the same composition. However, the inclusion of a different complex configuration into the growth plane of a natural crystal causes the formation of a crystal dislocation. This dislocation results in an anomaly in the growth plane of the forming crystal, and succeeding complex addition to the plane continue propagating the anomaly. This phenomenon is illustrated schematically in Figure 7–7.

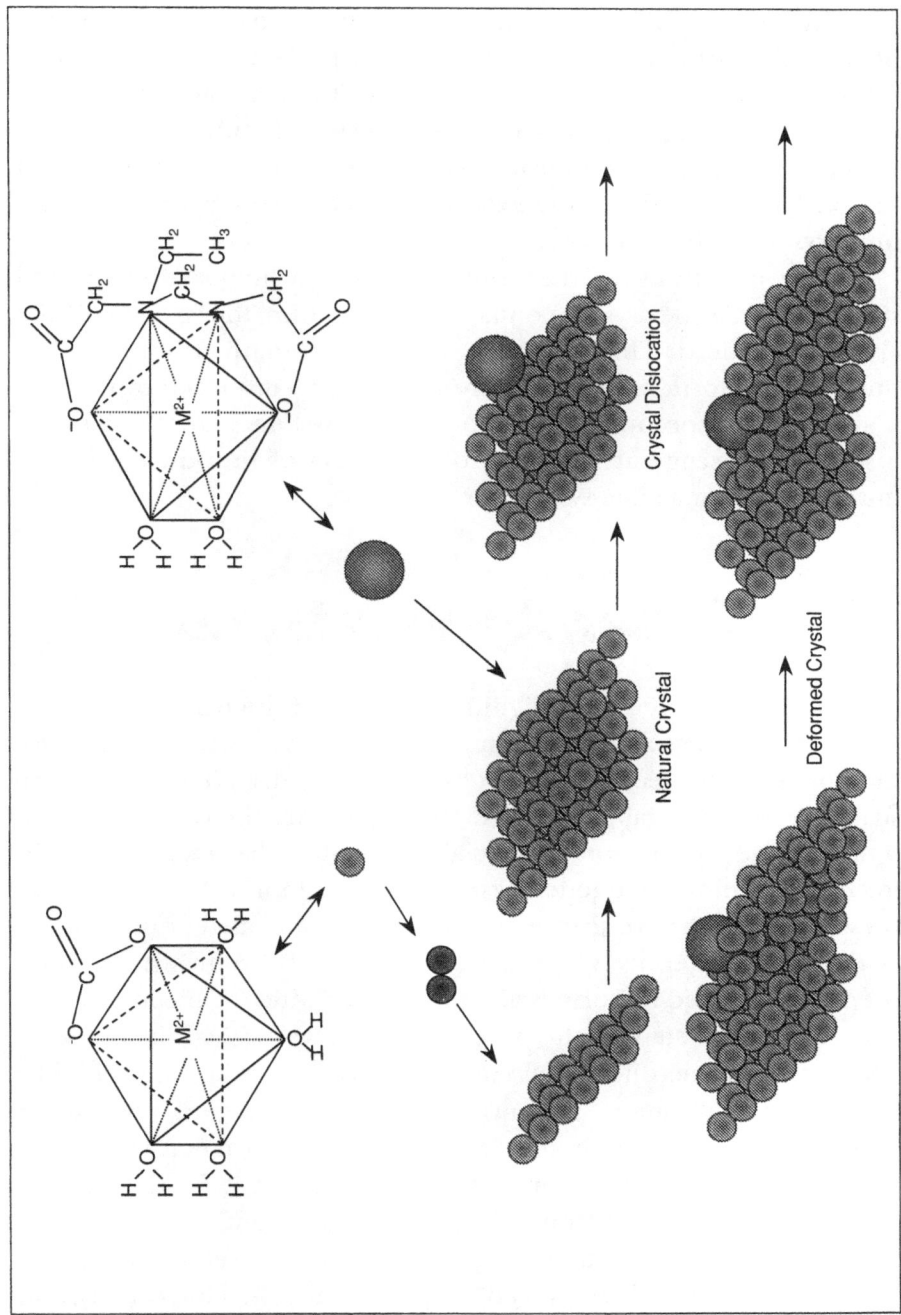

Fig. 7-7 The inclusion of a crystal anomaly and how it is propagated by continued addition of natural complex salts

Inclusion of differing complex salt morphology into the growth plane of the forming crystal causes crystal imperfections. These imperfections change the overall morphology of the resultant crystal. The continued packing (secondary aggregation) of the unit crystal forms (monoclinic, triclinic, orthorhombic, etc.) is disturbed. 'The nature of the packing depends on the extent to which the morphology of the unit crystal has been altered.

The continuity of the crystal forms determines their overall strength, and if the crystals contain discontinuities, the strength of the crystal is reduced. This reduction in crystal strength produces a less durable scale form, which, in many cases, can be displaced by the normal turbulence occurring within a dynamic system. Figure 7–8 shows a somewhat exaggerated picture of the effect of introducing crystal imperfections in a scale system.

Other Applications

Perhaps an even more significant aspect of the type of chemistry just described is its application in other disciplines. Much of the work in semiconductor technology involves engineered solid-solid interface structures. Considerable time and effort goes into the research, development, and production of specialized crystals that can be used in microelectronics. As a general rule, the crystalline materials used in these applications are grown under tightly controlled conditions to assure crystal integrity. Once the crystals have been grown, cut, and prepared, diffused impurities are added to produce semiconductors.

One of the steps in the preparation of these crystals involves the application of micron-sized circuit diagrams. This is accomplished by photoresist application, gaseous metal deposition, and removal of excess metal from the circuit. Once the patterned metal circuit has been prepared, gaseous impurities are added. It is in the pattern aspect of this microelectronic design that the intentional incorporation of crystal imperfections might find its most useful application. Thus, if crystal imperfections can be incorporated in a logical way, and patterns derived from their inclusion, the circuit design process might be possible at the crystal level.

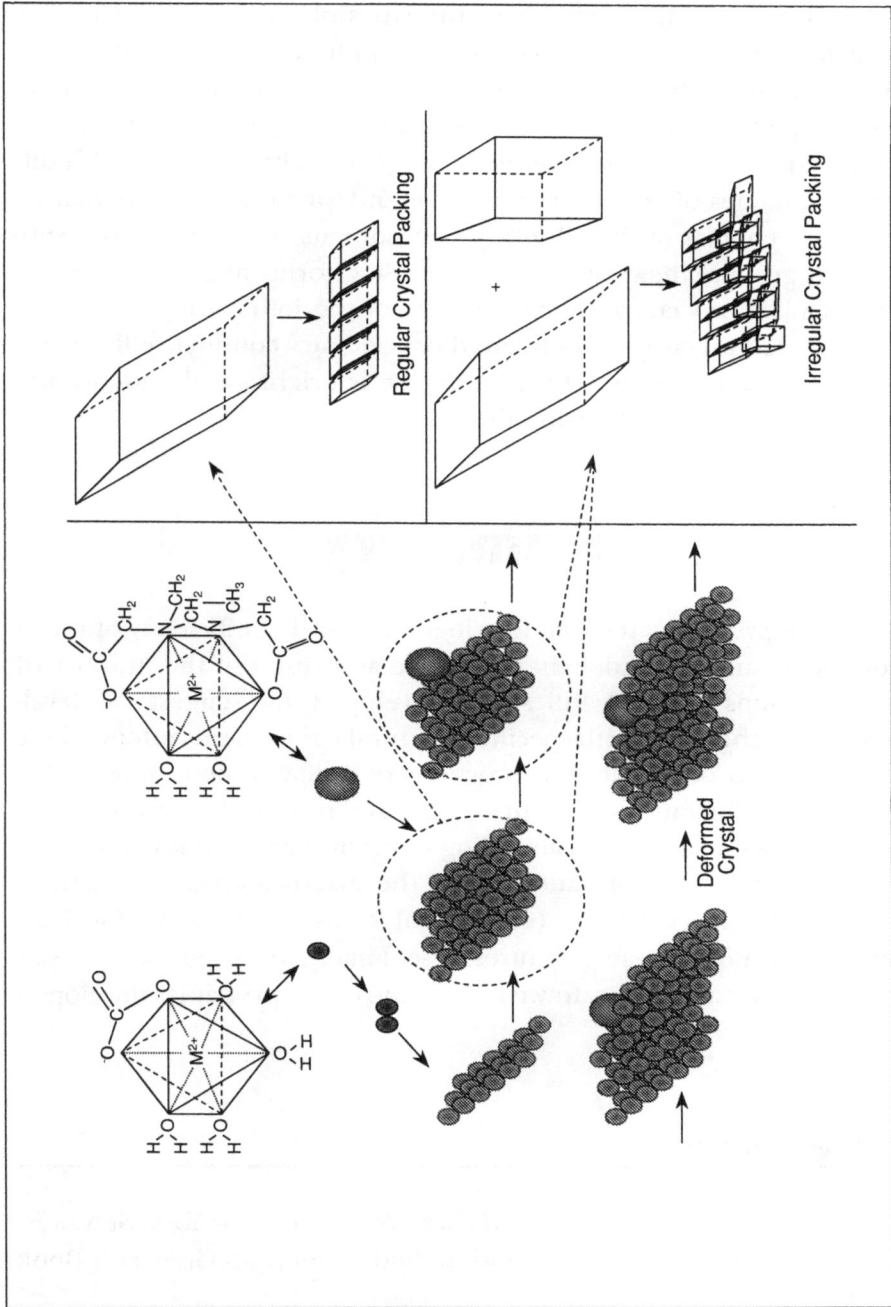

Fig. 7-8 The distortion effects on scale crystal aggregation by the inclusion of a different salt complex

The electronic character of the types of salts produced by this scale modification process might also be applied to the creation of new surface coatings for metal and other conductive surfaces. By producing adequately characterized metal complex salts with specific ligand group components, new types of coatings could be produced with differing qualities of corrosion resistance and/or durability. Further, by judicious choices of ligand group components (e.g., bulky alkyl substituent groups), new metal-plastic bonding forms might be possible. By a similar procedure, phase transfer of the ionic complex to normally nonconductive solvents could provide new conductive fluid and solid organic materials. Thus, the future is bright for the incorporation of these modified forms of crystals.

Summary

The preceding section described the packing efficiency of metal complexes and the order that they take as a result of the number of ligand groups attached. Rules were developed that suggest the development of chemicals with specific structural components intended for the alteration of the normal processes of crystallization. Several diagrams and illustrations were presented to illustrate the nature of the physical chemical changes attending many of the processes described. A fair amount of information about the macro-aggregate structures taken by the salt complexes (e.g., crystal forms) was presented with an emphasis on how these structures arise. Finally, an extension to possible new applications was drawn from some of the principles developed in the chapter.

References

Barrow, Gordon M. 1966. *Physical Chemistry.* 2d ed. New York, St. Louis, San Francisco, Toronto, London, and Sydney: McGraw Hill Book Company.

Bockris, J. O'M., and A. K. N. Reddy. 1973. *Modern Electrochemistry*. New York: Plenum Publishing Corporation.

Dickerson, R. F., H. B. Gray, and G. P. Haight, Jr. 1970. *Chemical Principles*. 1st ed. New York: W. A. Benjamin, Inc.

Hamill, William H., and Russell R. Williams, Jr. 1966. *Principles of Physical Chemistry*. 2d ed. Englewood Cliffs, New Jersey: Prentice-Hall.

Huheey, James E. 1978. *Inorganic Chemistry Principles of Structure and Reactivity*. 2d ed. New York, Hagerstown, San Francisco, and London: Harper & Row.

Noller, Carl R. 1966. *Textbook of Organic Chemistry*. 3d ed. Philadelphia and London: W. B. Saunders Company.

$$>\!\!\stackrel{|}{-}\!\!<$$

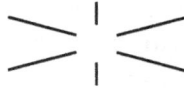

8
A Closer Look at Petroleum Fluids and Corrosion

Corrosion and Scale Analogies

To this point, much of the chemistry involved in scale and corrosion phenomena has been handled as if they were two completely separate processes. However, some inferences have been drawn over the course of the preceding material that tend to blur this distinction. The earlier treatments of the crystal natures of both steel and scale provide a means of drawing these two processes together.

If the formation of scale is looked upon as a process that ties into the process of corrosion, then the metallic salt is an element common to both. (The corrosion process consists of metallic salts that form crystals and then corrode to form metallic salts again.) Further, if the metallic salt is considered as an intermediate form of a metal, then it is reasonable to assume that some form of energy is necessary to change the salt to metal. The smelting process supplies sufficient heat energy to vaporize the majority of anionic materials present in the iron salts, leaving behind the atomic metal. Electrolytic methods are alternatives to the smelting process, but require the supply of electrons to the cationic metal ion in order that it may be released from its anionic counter ion(s).

In the case of corrosion, the metal atoms must release electrons to the environment in order to produce the metal cation that can combine with nearby anions to form the metallic salt. Figure 8–1 illustrates the conversion of iron metal to the ferric ion assuming no participation by the solvent water. In this simple system, iron and ferric ions are in equilibrium as long as electrons are supplied to the salt or removed from the metal. If it is assumed that the anions released to the solvent attack the metal and remove it as the ferric salt, then the electrons for the equilibrium reaction are supplied by the anions. This then accounts for the equilibrium, and no other explanation is required.

However, by omitting the solvent water from participation in the reaction, and neglecting the ferrous ion in this scheme, some errors are introduced. Since water does participate in the reaction process, and ferrous ions arise as intermediates of the ferric ion oxidation states, some accounting must take place to correct for these errors. The following equations are included as an attempt to correct this oversight:

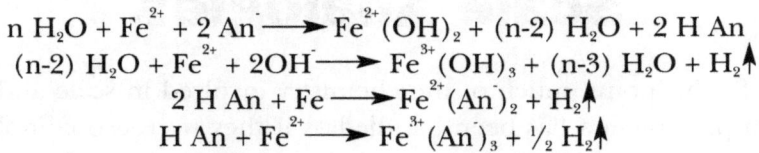

$$n\ H_2O + Fe^{2+} + 2\ An^- \longrightarrow Fe^{2+}(OH)_2 + (n\text{-}2)\ H_2O + 2\ H\ An$$
$$(n\text{-}2)\ H_2O + Fe^{2+} + 2OH^- \longrightarrow Fe^{3+}(OH)_3 + (n\text{-}3)\ H_2O + H_2\uparrow$$
$$2\ H\ An + Fe \longrightarrow Fe^{2+}(An^-)_2 + H_2\uparrow$$
$$H\ An + Fe^{2+} \longrightarrow Fe^{3+}(An^-)_3 + \tfrac{1}{2}\ H_2\uparrow$$

If the anion (An^-) in the above equations is replaced by hydroxyl (OH^-), then these equations can be written as follows:

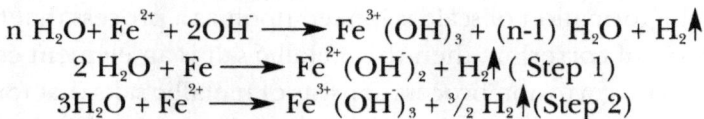

$$n\ H_2O + Fe^{2+} + 2OH^- \longrightarrow Fe^{3+}(OH)_3 + (n\text{-}1)\ H_2O + H_2\uparrow$$
$$2\ H_2O + Fe \longrightarrow Fe^{2+}(OH)_2 + H_2\uparrow\ (\text{Step 1})$$
$$3H_2O + Fe^{2+} \longrightarrow Fe^{3+}(OH)_3 + \tfrac{3}{2}\ H_2\uparrow(\text{Step 2})$$

Choosing the hydroxyl anion simplifies the system of equations, and it can be written as follows:

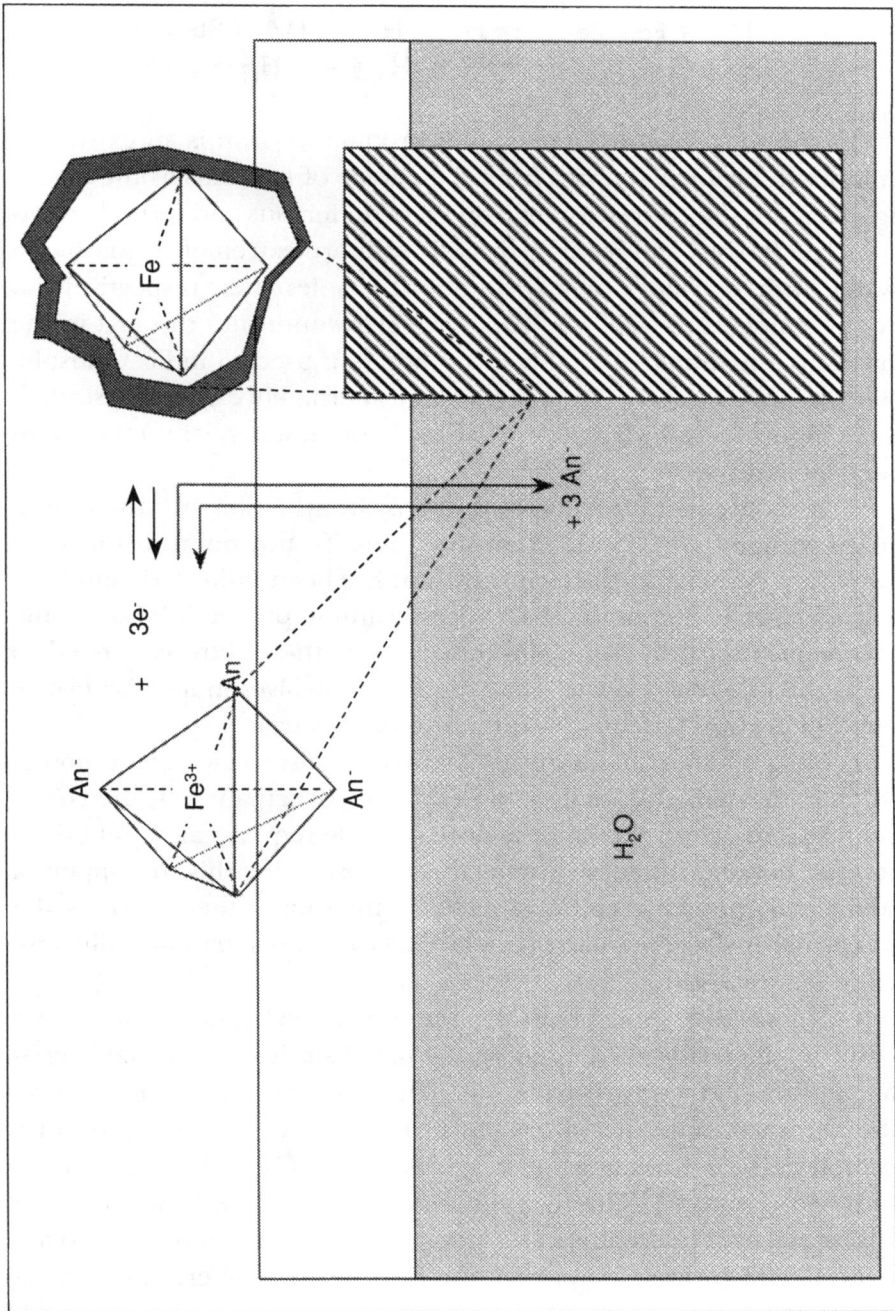

Fig. 8-1 Iron metal in equilibrium with ferric ion in a nonparticipating water solvent

$$2\,H_2O + Fe + 2e^- \longrightarrow Fe^{2+}(OH^-)_2\!\!\downarrow + H_2\!\!\uparrow \quad (\text{Step 1})$$
$$3H_2O + Fe^{2+} + e^- \longrightarrow Fe^{3+}(OH^-)_3\!\!\downarrow + \tfrac{3}{2}\,H_2\!\!\uparrow \quad (\text{Step 2})$$

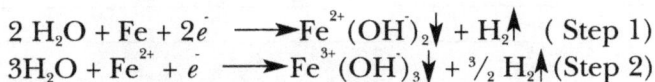

The solubility of the oxidation products of iron is an extremely important factor that arises from the choice of hydroxyl as the anionic species in the conversion of iron metal to ferrous and ferric hydroxide. The ferrous and ferric hydroxide salts are extremely insoluble in water, with K_{sp} values of $\approx 10^{-14}$ and 10^{-36} moles/liter, respectively, at room temperature (25° C). The down arrow indicates this fact in the previous equations, since this is the notation used to indicate insolubility or precipitation. Thus Figure 8–1 alternately can be illustrated, as in Figure 8–2, to depict the hydroxyl anion, water, iron metal, and metallic salt system.

It should be noted that both the hydroxyl anion and the molecular hydrogen are derived from the water. To this point in the scale corrosion couple, no electron provision has been indicated, and lacking electron generation, the process cannot proceed. Thus, some mechanism must be rationalized to provide the electrons needed to drive the process. A key to this mechanism involves molecular hydrogen, the hydroxyl anions, and the iron metal surface.

Since the ferrous and ferric hydroxide salts (scale) are so poorly soluble, they are essentially removed from participation in the corrosion process. Thus, the water hydroxide cycle requires an equilibrium to exist between these two molecules. In order for this to happen, a source of protons must be provided. This source then involves the molecular hydrogen and iron metal surface. This situation is illustrated in Figure 8–3.

Figure 8–3 shows basically three hydroxyl attack points: the adsorbed molecular hydrogen on the metal surface, the metallic crystal, and the ferrous hydroxide salt. The attack by a hydroxide ion on the adsorbed molecular hydrogen forms water from abstraction of a proton and the formation of a hydride anion (H$^-$ or hydrogen atom with two electrons). The hydride anion is very unstable and loses its added electron to the metal, resulting in the formation of a hydrogen radical. This process allows electron flow to a site where it can react with water.

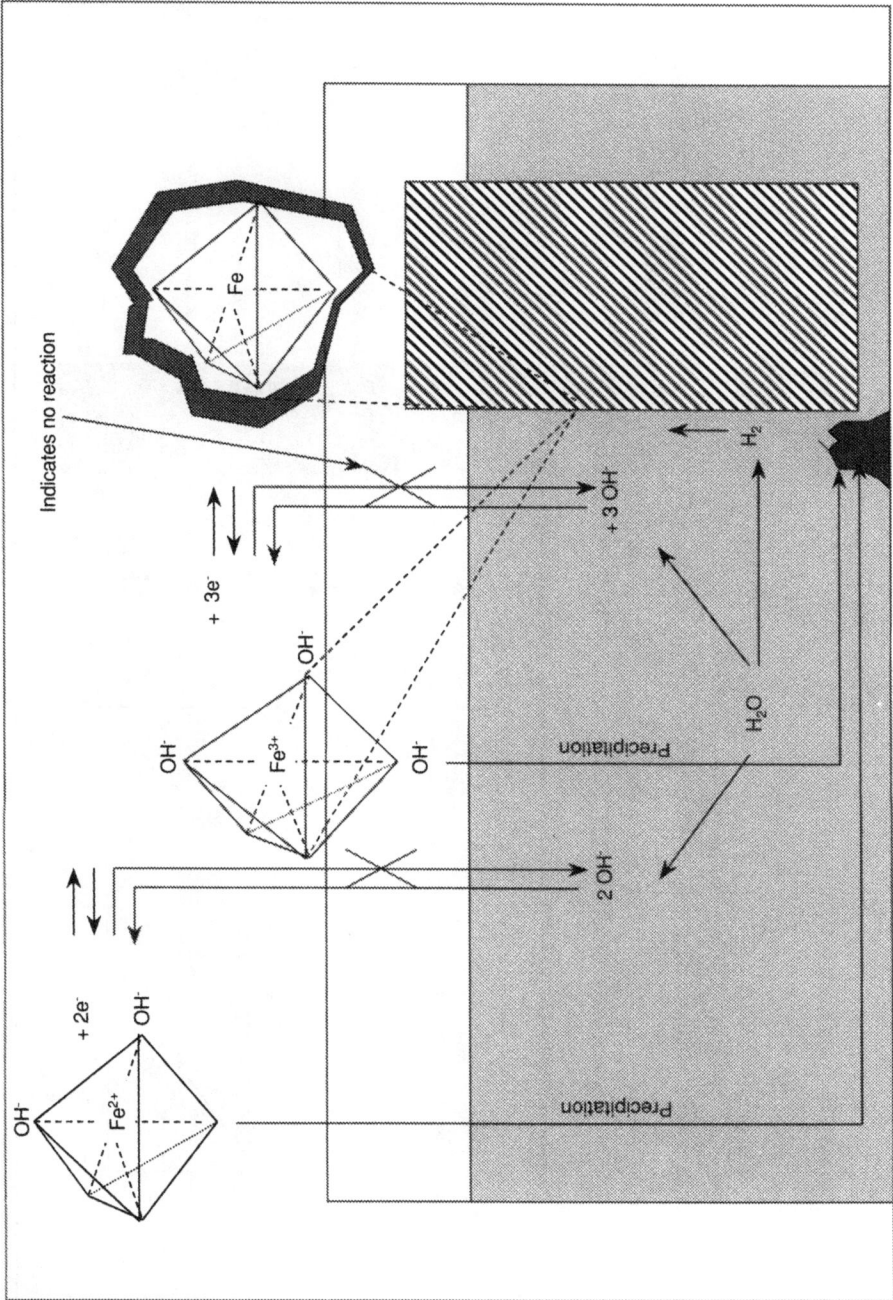

Fig. 8-2 Iron corrosion to ferrous and ferric ions assisted by the participation of water

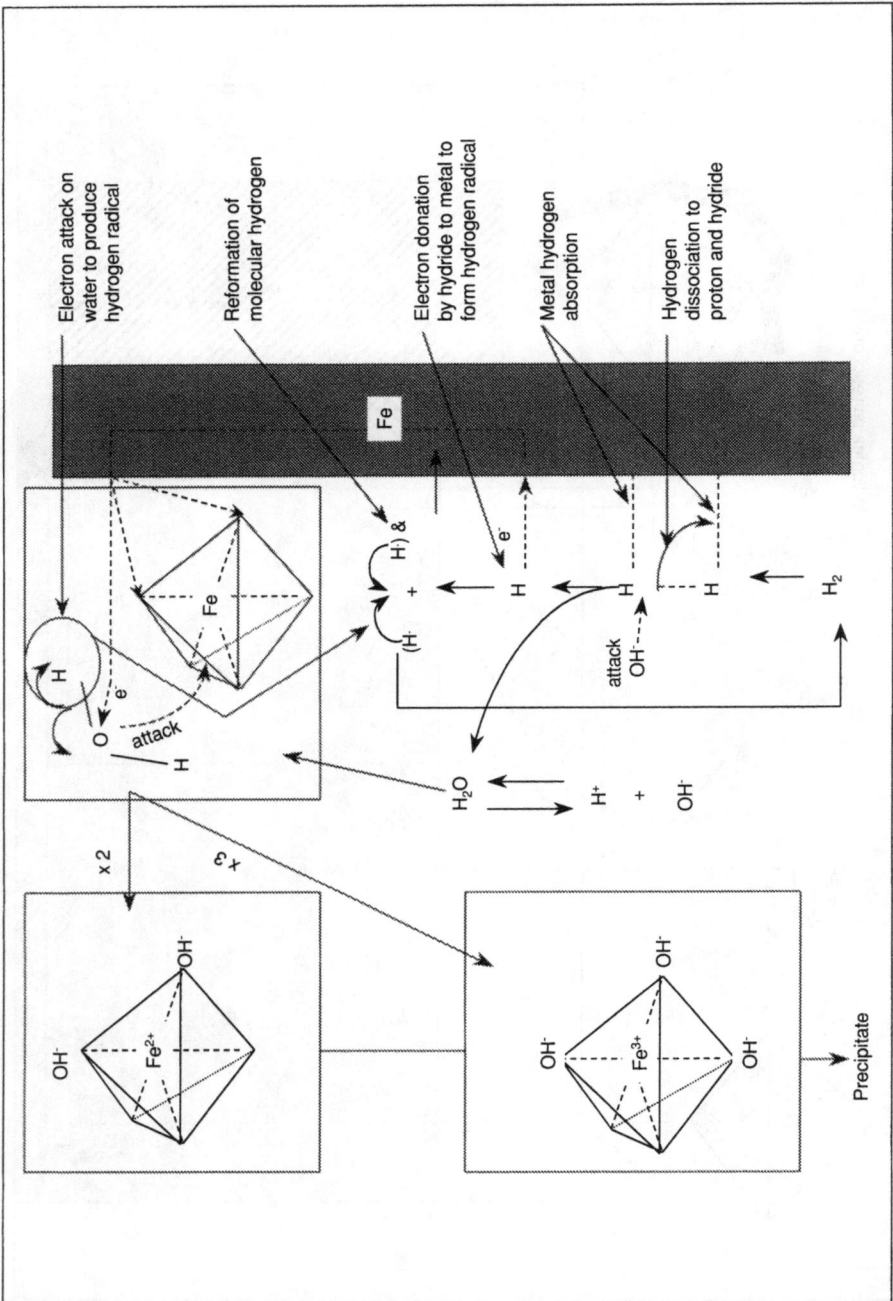

Fig. 8-3 The participation of water in the corrosion scale couple of ferrous and ferric hydroxide

When this free electron attacks a water molecule, one electron from the oxygen combines with it, and the other leaves with the hydrogen. The departing hydrogen is a radical (monomolecular hydrogen). The monomolecular hydrogen or hydrogen radical can now either recombine to form molecular hydrogen (H_2) or, because of its extremely small size, it may enter the metal lattice. This reaction scheme explains how corrosion could occur in a pure water system without invoking the equilibrium between water and its dissociated forms of hydronium and hydroxyl ions (H_3O^+ and OH^-). The inclusion of this equilibrium condition leads to interesting conclusions. Although the hydronium ion represents an additional species in the mechanism illustrated in Figure 8–3, there are no apparent significant differences in the overall mechanism or the products of the reaction. In fact, the presence of the equilibrium equation clarifies how the hydroxide anion is formed.

Thus, from this mechanism, it appears that the hydroxyl anion is most responsible for iron corrosion, and ferrous and ferric hydroxide scale formation. Since the product of the two ions is water, the determining factor of corrosion/scale reactions is the degree of dissociation of the water producing the two ions. Figure 8–4 illustrates the effect of the dissociation reaction of water on the formation of scale and the corrosion of iron.

Electron Source and Electron Sink

In the preceding reaction mechanisms, the hydride ion is the source of electrons, water serves as the sink, and the metal surface is the electron conductor. When the hydride ion releases its excess electron to the metal lattice, the sea of electrons within the metal propagates the charge. This process facilitates the transmission of charge to another point on the metal surface that is near a properly oriented water molecule.

When the electron encounters a properly oriented water molecule, the electron tunnels its way to the oxygen of the water. The

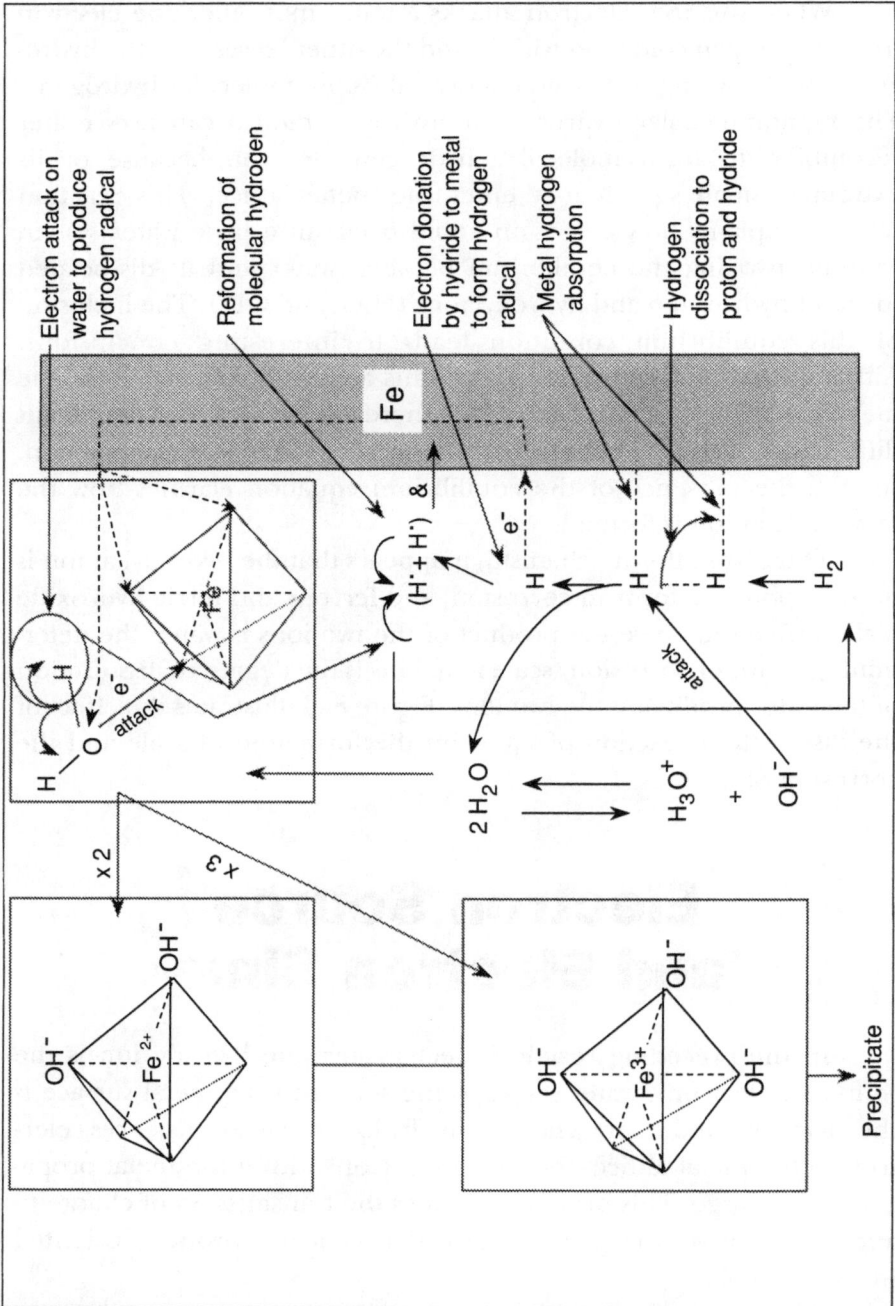

Fig. 8-4 The participation of water in the corrosion scale couple of ferrous and ferric hydroxide, including the dissociation of water

additional negativity imparted to the oxygen by the tunneling electron produces an unstable condition that requires two things. These requirements are 1) the loss of monoatomic hydrogen ($H^{\bullet}$ or $H^+ + \dot{e}$) and 2) the formation of a hydroxyl ion to return to a stable energy condition. Thus, the electronation and de-electronation reactions can be written as follows:

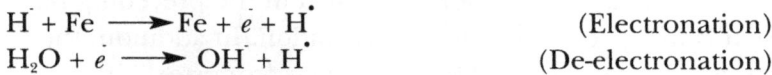

$$H^- + Fe \longrightarrow Fe + \dot{e} + H^{\bullet} \qquad \text{(Electronation)}$$
$$H_2O + \dot{e} \longrightarrow OH^- + H^{\bullet} \qquad \text{(De-electronation)}$$

Note that the iron metal does not take part in the electronation reaction, except to act as a charge transfer medium. The oxidation of the metallic iron takes place by the action of the hydroxyl on the metallic crystal and the metallic salt, as indicated in the following reactions:

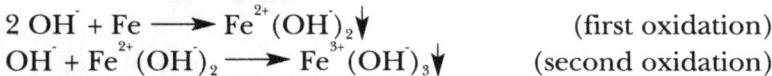

$$2\,OH^- + Fe \longrightarrow Fe^{2+}(OH^-)_2 \downarrow \qquad \text{(first oxidation)}$$
$$OH^- + Fe^{2+}(OH^-)_2 \longrightarrow Fe^{3+}(OH^-)_3 \downarrow \qquad \text{(second oxidation)}$$

Acidic Conditions

The acid corrosion of iron frequently produces ferrous and ferric ions. These combine with anions in aqueous solution to produce largely insoluble corrosion products. Additionally, Lewis acids such as ferric chloride can produce acid conditions capable of converting the insoluble ferrous hydroxide salts to insoluble oxide derivatives. The ferric hydroxide is converted to ferric oxide (commonly known as rust) by the addition of strong acids. Iron oxides appear slimy on the metal surface and gradually build to a mass that falls under the influence of gravity. Agitation can also dislodge the oxide after a sufficient quantity has accumulated on the surface of the corroding metal.

The following reactions and their combined reaction sum illustrate some of the steps involved:

$$5\ H_2O + FeCl_3 \longrightarrow 3\ H_3O^+ + 3\ Cl^- + Fe^{3+} + O_2\uparrow + \tfrac{1}{2}\ H_2\uparrow + 3\ \dot e^-$$
$$2\ H_3O^+ + Fe(OH)_2\downarrow + 2\ \dot e^- \longrightarrow 3\ H_2O + FeO + H_2\uparrow$$
$$3\ H_3O^+ + 2\ Fe(OH)_3\downarrow + 3\ \dot e^- \longrightarrow 6\ H_2O + Fe_2O_3 + \tfrac{3}{2}\ H_2\uparrow$$
$$2\ H_3O^+ + FeCl_3 + Fe(OH)_2 + 2\ Fe(OH)_3 + 2\ \dot e^- \longrightarrow \uparrow$$
$$4\ H_2O + FeO\uparrow + Fe_2O_2\uparrow + 3\ Cl^- + Fe^{3+} + O_2\uparrow + 3\ H_2\uparrow$$

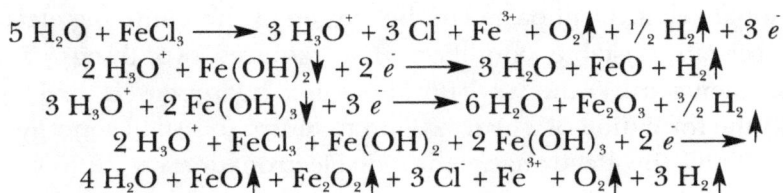

Ferric chloride was chosen as the acid in the preceding reaction because it is freely soluble in aqueous solution. In addition, the ferric cation is common to the corrosion products from steel (iron). It also helps to point out the exchange of scale forms ferrous and ferric hydroxide for ferrous and ferric oxide, respectively, under conditions of low pH.

Basic Conditions

The basic corrosion of steel (iron) follows much the same process as that illustrated for aqueous assisted corrosion. However, the hydronium-hydroxyl equilibrium is shifted in favor of the hydroxyl. Thus, the addition of an alkali base will force the equilibrium to a high pH value, favoring the increased concentration of the hydroxyl anion. The increased hydroxyl concentration then increases the rate of attack of the metal crystal and its incompletely oxidized metallic salts (ferrous hydroxide).

Calcium hydroxide is a particularly good choice for adding base (hydroxyl ions) to the system of water and iron metal, since it is freely soluble in water. In addition, the calcium cation is very difficult to reduce. (It requires high electromotive force to reduce Ca^{2+}.) This high reduction potential makes the Ca^{2+} a poor candidate for interference with the electronation/de-electronation couple discussed previously. Figure 8–5 illustrates the effect of the addition of calcium hydroxide to the iron water system.

Based on the acid and base reactions, the mechanisms for hydride metal electronation, and the electron de-electronation of water, we can further develop this concept. It is apparent that the ferrous and ferric hydroxide scales occur as intermediates in the processes of aqueous acid and base iron corrosion.

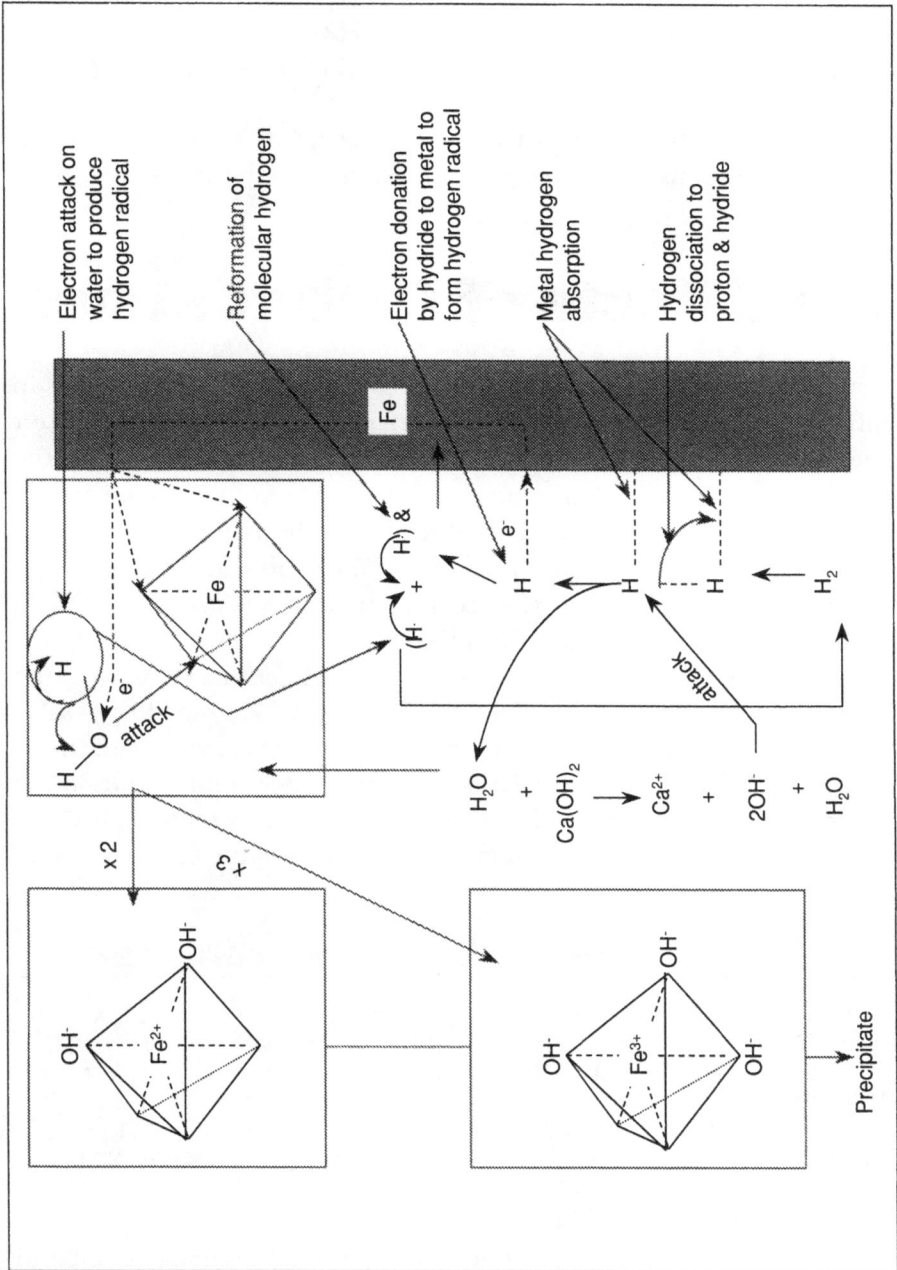

Fig. 8-5 The effect of adding calcium hydroxide to the iron metal in water system (Note: The increase in hydroxyl concentration increases the hydroxyl attack, and the calcium ion does not interfere with oxidation/reduction.)

$$Fe + 4\,H_3O^+ + 4\dot{e} \longrightarrow \left| Fe(OH)_2 + 2\,H_2 \uparrow \right. \left[+ 2\,H_3O^+ + 2\underline{e} \longrightarrow FeO + 4H_2O + 2H_2 \uparrow \right.$$

$$2Fe + 8\,H_3O^+ + 8e \longrightarrow \left| 2Fe(OH)_2 + 4\,H_2 \uparrow \right. \left[+ 4\,H_3O^+ + 4\underline{e} \longrightarrow Fe_2O_3 + 5H_2O + 7H_2 \uparrow \right.$$

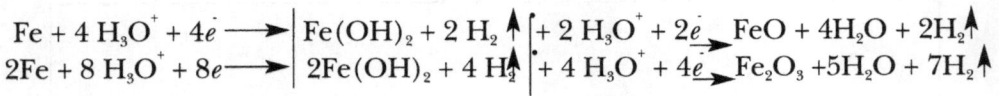

This situation provides a unifying concept that can be extended to various aqueous systems containing different anionic species.

The Effects of Other Cations

The identification of the hydroxy metal salts as intermediates in either the aqueous acid or basic corrosion process simplifies our understanding of the process. It also makes it easier to rationalize the effects of other anions. In aqueous systems, the anions added by the acidifying chemical, or the cations added by the basic salts, produce soluble salts of either the corroding metal or the cation of the basic salt. In these cases the metallic corrosion products are either metal oxides (acid) or metal hydroxides (base).

This situation changes in the presence of cations or anions that form aqueous insoluble or low solubility (low K_{sp} metallic) salts. (These include silver, copper, mercury, aluminum, magnesium, manganese, nickel, zinc, or lead.) Thus, the presence of these metals in the aqueous corrosion of iron can produce low solubility metallic hydroxyl and chloride salts of these metals. Table 8–1 gives the corresponding K_{sp} values at 286° K.

Table 8–1 Low Solubility Hydroxyl and Chloride Metallic Salts (at 286° K)

$[Ag^+][Cl^-]$	=	1.0×10^{-10}	$[Cu^+][Cl^+]$	=	1.0×10^{-6}
$[Hg_2^{2+}][Cl^-]_2$	=	2.0×10^{-18}	$[Pb^{2+}][Cl^-]_2$	=	1.7×10^{-5}
$[Cu^+][OH]_2$	=	1.0×10^{-19}	$[Ag^+][OH^-]$	=	1.0×10^{-8}
$[Al^{3+}][OH]_3$	=	2.0×10^{-33}	$[Mg^{2+}][OH^+]_2$	=	1.2×10^{-11}
$[Fe^{3+}][OH]_3$	=	1.1×10^{-36}	$[Fe^+][OH]_2$	=	1.1×10^{-14}
$[Mn^{2+}][OH]_2$	=	4×10^{-14}	$[Ni2+][OH+]_2$	=	4×10^{-14}
$[Zn^+][OH]_2$	=	1.8×10^{-14}			

It can be seen that in the acid system, silver, mercury, copper, and lead can interfere with the corrosion of metallic iron, since their cations remove chloride ions from the system. The interference caused by the chloride anion removal involves an *increase* in the rate of iron corrosion. This is because the hydroxyl ion must increase to counter the imbalance between it and the hydronium ion. Thus, aqueous acid iron corrosion rates increase with increasing concentrations of these other cationic species in the following order:

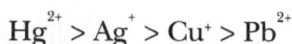

$$Hg^{2+} > Ag^+ > Cu^+ > Pb^{2+}$$

The basic aqueous corrosion of iron is also interfered with by the presence of other metallic cations. However, only those cationic salts that possess a lower K_{sp} value than the ferrous (Fe^{2+}) cationic salt can appreciably *decrease* the corrosion rate. Thus, the aqueous basic corrosion rate of iron decreases as the concentrations of these other cations increase in the following order:

$$Al^{3+} >> Cu^{2+}$$

Conversely, the basic aqueous corrosion rate is *increased* by the presence of concentrations of those ions with K_{sp} values greater than the ferrous (Fe^{2+}) cationic salt. It is increased as the concentrations of these other ions increase in the following order:

$$Ag^+ > Mg^{2+} > Mn^{2+} = Ni^{2+} > Zn^{2+}$$

The reason for the corrosion rate increase is that the ferrous hydroxide salt is less soluble than the other salts. This lower solubility favors the combination of the hydroxyl anions and the ferrous cations over the formation of the other cations.

The Effects of Other Anions

The presence of different anions in iron-containing aqueous systems produces a closely analogous effect to that which occurs in the presence of different cations. Solubility effects are again involved. However, the solubility of the anionic iron salts is now considered for only those anions commonly present in petroleum fluid systems. These anions (excluding the overwhelmingly prevalent hydroxyl ions) are comprised of carbonate, sulfides, sulfates, and chlorides. Thus, the same arguments presented for the solubility products of the iron hydroxide versus those of the alternate cations can be advanced for alternate anions.

The solubilities of the iron salts of the carbonates, sulfides, sulfates, and chlorides are all greater than those of the hydroxides in basic aqueous corrosion systems. Thus, the increased anionic character of these systems increases the corrosion rate as the metallic hydroxide intermediate is precipitated from the aqueous phase.

Combined Metal Salt Effects

For the most part, the cations mentioned previously (except Al^{2+}, Mn^{2+}, Mg^{2+}, and Cu^{+}) are not very common in aqueous petroleum systems. The Group 2A elements (barium, calcium, magnesium, and strontium) are most often found in these fluid systems. These salts, and those of aluminum and manganese, must be considered when the corrosion of iron takes place. The solubilities of these salts in combination with some common anions (at 298° K) are presented in Table 8–2.

Table 8-2 Solubilities of Some Metal Salts**

M^{n+}	Radius (10^{-8} cm)	Chlorides	Hydroxides	Carbonates	Sulfates
Cu^{+}	0.96	1.0×10^{-6}	1.0×10^{-19}	Insoluble	
Decomposes					
Mn^{2+}	0.80	72.3	4×10^{-14}	6.5×10^{-3}	52
Al^{3+}	0.50	69.9	2.0×10^{-33}	Not found	31.3
Mg^{2+}	0.31	5.6	1.2×10^{-11}	1.2×10^{-3}	2.4
Ca^{2+}	0.65	5.4	2.1×10^{-2}	1.5×10^{-4}	1.5×10^{-2}
Sr^{2+}	1.13	3.0	6.5×10^{-2}	7.0×10^{-4}	5.0×10^{-3}
Ba^{2+}	1.35	1.5	2.8×10^{-1}	1.0×10^{-4}	1.0×10^{-5}

**Moles per 1000 Grams of water

We can apply the same logic to this situation as that employed in the case in which chloride removal from the acid system increases the rate of corrosion of the iron. Doing so, it can be seen that only under acidic conditions will the copper increase corrosion rates. Copper and aluminum act to form the only metal hydroxides in the table that will lower the corrosion rate of the iron. Thus, the sulfate, chloride, and carbonate salts of calcium, manganese, magnesium, strontium, and barium will increase the aqueous basic corrosion of iron.

The Common Ion Effect

An interesting correlation between scale formation and corrosion emerges from these derivations. This correlation is based on the common ion effect. According to the common ion effect, if an ion present in a salt is added to the solution, it suppresses the solubility of the salt in accordance with Le Chatelier's principle. Simply stated, *when a system at equilibrium is subjected to a stress of any kind, the system shifts toward a new equilibrium condition in such a way as to relieve that stress.*

In the case of corrosion reactions, the hydroxyl ion is the one specific to the process. In either the aqueous acid or base system, it is the concentration of this common ion that determines the corrosion rate. Thus, those processes that act to suppress hydroxyl concentration lower corrosion rates. Conversely, those processes that act to increase hydroxyl concentration increase corrosion rates.

Summary

The preceding chapter developed a picture of the electrolytic nature of corrosion and suggests that aqueous corrosion of iron proceeds through an iron hydroxyl intermediate. It is pointed out that the adsorption, heterolytic cleavage, proton transfer, and hydride donation of an electron to the metal from molecular hydrogen drive the aqueous iron corrosion process. It is also pointed out that the formation of an insoluble scale (removal from solution of the corrosion product) shifts the equilibrium reaction in favor of further metal oxidation.

The effects of cations, anions, and combined anion/cation mixtures in the aqueous acid or base corrosion process were discussed. Those processes that act to suppress hydroxyl concentration lower corrosion rates; conversely, those processes that act to increase hydroxyl concentration increase corrosion rates. Finally, the application of the common ion effect specifically to the intermediate metal hydroxide salts ties corrosion and scale processes together.

References

Barrow, Gordon M. 1966. *Physical Chemistry*. 2d ed. New York, St. Louis, San Francisco, Toronto, London, and Sydney: McGraw Hill Book Company.

Dickerson, R. F., H. B. Gray, and G. P. Haight, Jr. 1970. *Chemical Principles*. 1st ed. New York: W. A. Benjamin, Inc.

Huheey, James E. 1978. *Inorganic Chemistry Principles of Structure and Reactivity*. 2d ed. New York, Hagerstown, San Francisco, and London: Harper & Row.

Jones, Loyd W. 1988. *Corrosion and Water Technology for Petroleum Producers*. 1st ed. Tulsa: OGCI Publications, Oil & Gas Consultants International, Inc.

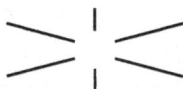

9
Electrochemical Effects on Scale

Scale and Electronegativity

In the preceding chapter, the analogous properties and interdependence of scale and corrosion processes were discussed. A considerably more detailed understanding of this interdependence is obtainable by considering some of the electrochemical processes involved. The electronegativities of the various species involved in these processes can be used as a starting point, and the reactions driven by this property can be rationalized.

While it is not our intent to delve into the details of electronegative theory, a brief discussion of a few of the concepts involved should be included. First, it might be instructive to point out the meaning of electronegativity. Linus Pauling (1960, p. 88) defined electronegativity as "the power of an atom in a molecule to attract electrons to itself." After Pauling offered his definition, R. S. Mulliken (1934, 1935) and W. Moffitt (1950) determined that "the attraction of an atom for electrons should be an average of the ionization energy and electron affinity of the atom." Finally, A. L. Allred and E. G. Rochow (1958) defined electronegativity as "the electrostatic force exerted by the nucleus on the valence electrons."

These definitions all share a common thread. According to James E. Huheey (1978), "the ability and capacity of an atom (or more specifically an orbital on an atom) to attract electrons will be rigorously defined by the environment of that atom." He determined that "two factors will determine the attraction of atoms for electrons, the average charge on the atom and the hybridization of the atom."

All of these definitions seek to draw together the idea of charge interactions of atoms or molecules with other molecules or atoms. These definitions are used to explain why some form bonds, some do not, and what the relative strengths of the bonds formed might be. To this end Pauling used thermochemical data, and Mulliken used effective nuclear charge based on calculations of the covalent radii to assign values to the elements. These calculations give a picture of the atom, its nuclear charge, and its tendency to form molecules.

Since in this book, we are mainly concerned with those atoms, ions, and molecules that commonly occur in petroleum fluids, the present discussion will be confined to these species. Therefore, the calculations of electronegativity can be used to gain an appreciation of the results of the above definitions. These calculations will give us a background toward understanding the factors involved in the bonding properties of those elements commonly found in petroleum fluids.

Table 9–1 summarizes the results of these various methodologies. It also gives information about the electronic charge and the types of orbital hybrids involved with certain atoms. It is hoped that by studying these factors, a more complete understanding of the processes of scale formation will be gained. It should also be useful in understanding the correlation between the scale and corrosion processes.

Table 9-1 Electronegativities of Some Petroleum Fluid Elements

Elements Orbital	Pauling Pauling	Allred-Rochow	Volts Hybrid	Mulliken-Jaffe Scale	Volts	electron
H	2.2	2.20	s	2.21	7.17	12.85
C	2.55	2.50	p	1.75	5.80	10.93
			sp^3	2.48	7.98	13.27
			sp^2	2.75	8.79	13.67
			sp	3.29	10.39	14.08
N	3.04	3.07	p	2.28	7.39	13.10
			23% s	3.56	11.21	14.64
			sp^3	3.68	11.54	14.78
			sp^2	4.13	12.87	15.46
			sp	5.07	15.68	16.46
O	3.44	3.50	p	3.04	9.65	15.27
			20% s	4.63	14.39	17.65
			sp^3	4.93	15.25	18.28
			sp^2	5.54	17.07	19.16
Na	0.93	1.01	s	0.74	2.80	4.67
Mg	1.31	1.23	sp	1.17	4.09	6.02
Al	1.61	1.47	sp^2	1.64	5.47	6.72
P	2.19	2.06	p	1.84	6.08	9.31
			sp^3	2.79	8.90	11.33
S	2.58	2.44	p	2.28	7.39	10.01
			sp^3	3.21	10.14	10.73
Cl	3.16	2.83	p	2.95	9.38	11.30
K	0.82	0.91	s	0.77	2.90	2.88
Ca	1.00	1.04	sp	0.99	3.30	4.74
Mn^{2+}	1.55	1.60				
Fe^{2+}	1.83	1.64				
Fe^{3+}	1.96					
Sr	0.95	0.99	s	0.50	2.09	4.18
Ba	0.89	0.97				

From Table 9–1, it can be seen that both the ferric and ferrous cations exhibit higher electronegativities than the other listed metals (Al, Ba, Ca, Mg, Mn^{2+}, K, and Na). Thus, the order of increasing electronegativities of the metals listed is as follows:

$$Fe^{3+} > Fe^{2+} > Al > Mn^{2+} > Mg > Na > Ca > Sr > Ba > K$$

This order of electronegativity lists only those metals commonly found in petroleum fluid systems. Likewise, the electronegativity of interactive anionic groups found in petroleum fluids follows an increasing order:

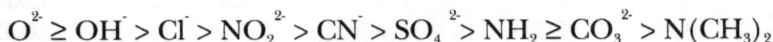

$$O^{2-} \geq OH^- > Cl^- > NO_2^{2-} > CN^- > SO_4^{2-} > NH_2 \geq CO_3^{2-} > N(CH_3)_2$$

Thus, if one were to base the affinity of reaction solely on electronegativity factors, some of the reaction product salts might be expected to show the following order of reactivity:

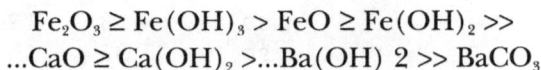

$$Fe_2O_3 \geq Fe(OH)_3 > FeO \geq Fe(OH)_2 \gg$$
$$...CaO \geq Ca(OH)_2 > ...Ba(OH)\,2 \gg BaCO_3$$

These results are in line with the experimental data. They indicate that a rough approximation to the expected products can be made by employing the metal and group electronegativities. This predictability is very valuable when complex mixtures of various metal cations and ions are examined.

Electromotive Force and Scale

Scale formation occurs under both acid and basic conditions. Furthermore, electromotive force across a range of pHs plays a significant role in the ease and speed with which scales form. Thus, an examination of the half-cell potentials of some of the metallic salts involved in petroleum fluid scales is instructive. It reveals trends that can be applied to an understanding of the scale formation process. (See Table 9–2 for standard aqueous reduction EMFs.)

Table 9–2 Standard Aqueous Reduction EMFs at 25°C

Acid Solutions

Electrode	ϵ^o
$Ba^{2+} + 2\ e^- = Ba$	-2.906
$Sr^{2+} + 2\ e^- = Sr$	-2.888
$Ca^{2+} + 2\ e^- = Ca$	-2.886
$Mg^{2+} + 2\ e^- = Mg$	-2.363
$\frac{1}{2}\ H_2 + e^- = H-$	-2.25*
$Al^{3+} + 3\ e^- = Al$	-1.662
$Mn^{2+} + 2e^- = Mn$	-1.180
$Fe^{2+} + 2e^- = Fe$	-0.4402
$Fe^{3+} + 3e^- = Fe$	+0.771
$O2 + 4\ H^+ + 4e^- = 2\ H_2O$ (g)	+1.185

Base Solutions

Electrode	ϵ^o
$Ca(OH)_2 + 2e^- = Ca + 2\ OH^-$	-3.02
$Ba(OH)_2 \bullet 8H_2O + 2e^- = Ba + 2\ OH^- + 8H_2O$	-2.99
$H_2O + e^- = H(g) + OH^-$	-2.9345
$Sr(OH)_2 + 2e^- = Sr + 2\ OH^-$	-2.88
$Ba(OH)_2 + 2e^- = Ba + 2\ OH^-$	-2.81
$Mg(OH)_2 + 2e^- = Mg + 2\ OH^-$	-2.69
$Mn(OH)_2 + 2e^- = Mn + 2\ OH^-$	-1.55
$FeS + 2e^- = Fe + 2\ S_2^-$	-0.95
$Fe(OH)_2 + 2e^- = Fe + 2\ OH^-$	-0.877

*Note: The reduction potential of the hydride formation from molecular hydrogen is greater than that of the ferrous and ferric ions.

Many of the Group 2A elements (Ba, Ca, Sr, and Mg) are prepared from molten salts. This is due to the difficulty of finding anything with a higher oxidation potential with which to reduce them chemically. It is interesting to note that the acid reduction potentials of Ca^{2+} and Mg^{2+} yield values a few hundred millivolts lower than the base reduction potential. In contrast, the Ba^{2+} and Sr^{2+} base reduction potentials are lower or equal respectively to the acid (see Table 9–3).

Table 9–3 Reduction Potentials

Electrode	ϵ^{0}	
$Ca^{2+} + 2e^{-} = Ca$	-2.886	(acid)
$Ca(OH)_2 + 2e^{-} = Ca + 2\ OH^{-}$	-3.02	(base)
$Ba^{2+} + 2e^{-} = Ba$	-2.906*	(acid)
$Ba(OH)_2 + 2e^{-} = Ba + 2\ OH^{-}$	-2.81	(base)
$Sr^{2+} + 2e^{-} = Sr$	-2.888*	(acid)
$Sr(OH)_2 + 2e^{-} = Sr + 2\ OH^{-}$	-2.88	(base)
$Mg^{2+} + 2e^{-} = Mg$	-2.363	(acid)
$Mg(OH)_2 + 2e^{-} = Mg + 2\ OH^{-}$	-2.69	(base)

*Base values less than or equal to the acid values.

Additionally, the acid reduction of molecular hydrogen to form the hydride anion is lower than the acid reduction potential of all the Group 2A elements. The acid reduction of molecular hydrogen to the hydride anion is greater than those of Al, Mn, Fe, and O_2. However, the base reduction potential of water to hydroxyl is greater than all the Group 2A elements except calcium. This information can then be summarized for aqueous petroleum fluid systems:

1. Calcium and magnesium hydroxides are readily converted to acid salts
2. Barium and strontium hydroxides are difficult to convert to acid salts
3. Acid reduction of all Group 2A elements is difficult
4. All Group 2A elements except calcium are readily converted to hydroxides
5. No Group 2A elements are reduced by hydride ions
6. Divalent aluminum, manganese, iron, and oxygen are reduced by hydride ions in acid solutions

Using the oxidation/reduction potentials, these six statements, and the preceding electronegativity discussion, it is possible to summarize. *The aqueous acid or base oxidation of iron to ferrous and ferric salts is favored over the oxidation of the Group 2A elements. However, the presence of aluminum and manganese compete with this oxidation under aqueous acid conditions.*

This statement then suggests that the acid or base corrosion of iron and the formation of its insoluble salts (scale) are favored over the formation of all the Group 2A element salts. However, in the presence of aluminum or manganese ions, a competition is set up with the oxidation of iron. The positive oxidation potential of oxygen in aqueous acid systems is sufficient to oxidize iron to either the ferrous or the ferric ion. However, it is not sufficient to oxidize any of the Group 2A elements or manganese.

Hard and Soft Acids and Bases

For a given ligand, the stability of complexes with dipositive metal ions follows the order:

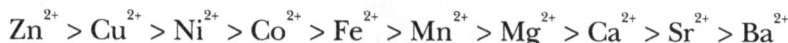

$$Zn^{2+} > Cu^{2+} > Ni^{2+} > Co^{2+} > Fe^{2+} > Mn^{2+} > Mg^{2+} > Ca^{2+} > Sr^{2+} > Ba^{2+}$$

This order arises from a decrease in size across the series and from the ligand field effects. Thus, the hard acid metal cations include those of the alkali metals (Group 1A), alkaline earth metals (Group 2A), and lighter transition metals in higher oxidation states such as Ti^{4+}, Cr^{3+}, Fe^{3+}, Co^{3+}, and the hydrogen ion, H^+. The soft acid metals include those of the heavier transition metals and those in lower oxidation states, such as Cu^+, Ag^+, Hg^+, Hg^{2+}, Pd^{2+}, and Pt^{2+}.

According to their preferences toward either the hard or soft metal acid ions, ligands may be classified as having tendencies of forming in order given in Table 9–4. Thus Ca^{2+}, Sr^{2+}, Ba^{2+}, Fe^{2+} are all classified as hard acids. Further, soft acids prefer combining with soft bases, and hard acids with hard bases. The hard and soft bases commonly encountered in petroleum fluids are listed in Table 9–5. It is interesting to note that the order of increasing electronegativity and oxidation potential follows the increase in complex stability (see Table 9–6).

Table 9–4 Hard and Soft Acids

Hard (acid) metal	Soft (acid) metal
N >> P > As > Sb	N << P > As > Sb
O >> S > Se > Te	O << S < Se < Te
F > Cl > Br > I	F < Cl < Br < I

Table 9–5 Hard and Soft Bases Common in Petroleum Fluids

Hard Base	Soft Base
H_2O, OH^-, O^{2-}, RO^-, CO_3^{2-}, PO_4^{3-}, SO_4^{2-}, NH_2^-, Cl^-	H^-, CO, RSH, RS^-, I^-

Table 9–6 Relative Electronegativity, Oxidation Potential, and Complex Stability

Fe^{3+} > Fe^{2+} > Al > Mn^{2+} > Mg > Na > Ca > Sr > Ba > K   Electronegativity

Fe^{3+} > Fe^{2+} > Al > Mn^{2+} > Mg > Na > Ca > Sr > Ba   Oxidation Potential

Cu^{2+} > Ni^{2+} > Co^{2+} > Fe^{2+} > Mn^{2+} > Mg^{2+} > Ca^{2+} > Sr^{2+} > Ba^{2+} Complex Stability

None of this is surprising, since charge dispersed over a given atomic or ionic radius affects all of these factors. If the anionic radius is known, then a relative rate of complex formation can also be suggested. The following order of anionic radii frequently found in petroleum liquid systems is determined from thermochemical data:

$$SO_4^{2-} > HS^- > NO_3^- > O_2^{2-} > CO_3^{2-} > HCO_3^- > NO_2^- > CH_3COO^- > OH^- > NH_2^-$$

We can rearrange this order to reflect expected reaction rates and combine it with the hard base data, the electronegativity, oxidation potentials, and complex stability. The results indicate the scale formation order expected in a system containing all or some of the species listed.

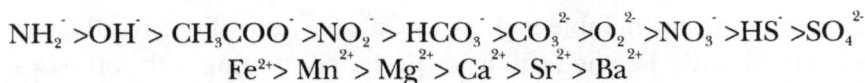

$$NH_2^- > OH^- > CH_3COO^- > NO_2^- > HCO_3^- > CO_3^{2-} > O_2^{2-} > NO_3^- > HS^- > SO_4^{2-}$$
$$Fe^{2+} > Mn^{2+} > Mg^{2+} > Ca^{2+} > Sr^{2+} > Ba^{2+}$$

Thus, ferrous hydroxide would be expected to form before ferrous sulfide, calcium hydroxide before calcium carbonate, and barium hydroxide before barium carbonate. Gathered together in the form of these reaction rate expectations, this information gives valuable information about the qualitative nature of the expected products. However, concentration effects must be considered before any quantitative predictions can be made regarding the end products of these molecular combinations.

Solubility Products and Scale Prediction

The presence of aluminum, barium, calcium, magnesium, manganese, or strontium in the aqueous corrosion of iron can produce low solubility metallic hydroxyl salts of these metals. The K_{sp} values at 286°K are given in Table 9–7. Thus to a first approximation the solubility products of the hydroxyl salt forms can show the reactivity order as follows:

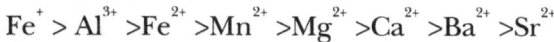

$$Fe^+ > Al^{3+} > Fe^{2+} > Mn^{2+} > Mg^{2+} > Ca^{2+} > Ba^{2+} > Sr^{2+}$$

Table 9–7 Low Solubility Metallic Hydroxyl Salts

$[Fe^{3+}] [OH^-]_3 = 1.1 \times 10^{-36}$	$[Al^{3+}] [OH^-]_3 = 2.0 \times 10^{-33}$
$[Fe^{2+}] [OH^-]_2 = 1.1 \times 10^{-14}$	$[Mn^{2+}] [OH^-]_2 = 4 \times 10^{-14}$
$[Mg^{2+}] [OH^-]_2 = 1.2 \times 10^{-11}$	$[Ca^{2+}] [OH^-]_2 = 2.1 \times 10^{-2}$

This follows (with the exception of barium and strontium) the same order as the electronegativity, complex stability, oxidation potential, and hard and soft acid trends. At this point, it should be recognized that a very strong case for the reaction order of scale products has been developed. Consequently, the application of these principles can be an effective means of predicting scale formation tendencies.

Scale Formation Predictions

The qualitative development of scale prediction methods follows a logical progression. Futhermore, electronegativity, electromotive force, hard and soft acid complex stability, and solubility constant data are all consistent. Knowing this, EMF data can be used to determine quantitative relationships. This is possible because of the Nernst equation relating stoichiometry of the reacting species to the product and the electromotive force involved.

$$\epsilon = \epsilon^{0} - (RT/n\tau)\ ln\ (\pi_{products}/\pi_{reactants})$$

where

ϵ^{0} represents the EMF with all species at unit activity

π represents the usual products of activities raised to the power corresponding to stoichiometry of the reaction (as K_{sp} and K_{eq})

τ = 96,500 coulombs/equivalent

Once this is done for the products and the reactants, the spontaneity of the reaction can be determined by the relation between Gibb's free energy and the resultant ϵ values. This is calculated through the following relationship:

$$\Delta G = -n\ \tau\ \epsilon$$

A negative value of ΔG indicates a spontaneous reaction, and conversely a positive value indicates the reaction does not take place spontaneously. An example is given for the formation of ferric hydroxide from ferric chloride in aqueous acid:

$$\epsilon = (+0.771) - (.059/n)\ log\ (1.1\ x\ 10^{-36}/74.2) = +3.003/n$$

and

$$\Delta G = (+5.702/n) \ [(-n)] = -2.699$$

So the free energy is negative, and the reaction is expected to proceed spontaneously. Conducting the same calculations on the Group 2A elements can produce values for the free energy of each of the various species. It reveals whether or not they will occur spontaneously, and if they will form more readily than the iron scale. In this way, the concentration effects can be included in a model to predict the scale formation tendencies of mixtures of metallic salts. It should be noted the term ($\pi_{products}/\pi_{reactants}$) assures that the formation of ferric hydroxide will overshadow the formation of any of the Group 2A elements. It also assures that the formation of ferrous hydroxide will be faster than all Group 2A elements.

Summary

The preceding chapter attempts to draw together the concepts of electronegativity, electromotive force, hard and soft acids/bases, and solubility products to form a qualitative view of the scale process. It points out the relationship between ionic size and charge, and how these factors combine to indicate the reaction direction and progression. By unifying the concepts around the ionic species present in petroleum fluid systems, it is possible to make predictions about the formation several scale forms. Finally, the Nernst equation is brought into play to produce some quantitative information about the scales formed, and which forms should be expected from a mixture of ions.

References

Allred, A. L., and E. G. Rochow. 1958. *Journal of Inorganic Nuclear Chemistry.* 5:264.

Barrow, Gordon M. 1966. *Physical Chemistry.* 2d ed. New York, St. Louis, San Francisco, Toronto, London, and Sydney: McGraw Hill Book Company.

Bockris, J. O'M., and A. K. N. Reddy. 1973. *Modern Electrochemistry.* New York: Plenum Publishing Corporation.

Dickerson, R. F., H. B. Gray, and G. P. Haight, Jr. 1970. *Chemical Principles.* 1st ed. New York: W. A. Benjamin, Inc.

Hamill, William H., and Russell R. Williams, Jr. 1966. *Principles of Physical Chemistry.* 2d ed. Englewood Cliffs, New Jersey: Prentice Hall.

Huheey, James E. 1978. *Inorganic Chemistry: Principles of Structure and Reactivity.* 2d ed. New York, Hagerstown, San Francisco, and London: Harper & Row.

Moffitt, W. 1950. *Proceedings of the Royal Society.* (London) A202:548.

Mulliken, R. S. 1934. *Journal of Chemical Physics.* 2:782.

Mulliken, R. S. 1935. *Journal of Chemical Physics.* 3:573.

Noller, Carl R. 1966. *Textbook of Organic Chemistry.* 3d ed. Philadelphia and London: W. B. Saunders Company.

Pauling, Linus. 1960. *The Nature of the Chemical Bond.* 3d ed. Ithica, New York: Cornell University Press.

10
The Special Case of Hydrogen

Hydrogen Evolution and Corrosion

In order for a metal to undergo oxidation, electrons must be stripped away from the metal atom. Corrosion of iron involves a change in the oxidation state of the metal from an uncharged atom to a charged ion. The charged ion is either in the 2+ or 3+ state in the case of iron, which indicates that the ion results from the loss of two or three valence electrons. If iron is placed in de-ionized water with a neutral pH, the iron is pure, and oxygen is excluded from the system, will corrosion occur? (Pure iron contains no interstitial heterometal atoms of differing ionization potential.)

Paraphrasing from a landmark paper by C. Wagner and W. Trand (1938), spatially separated electron-sink and source areas on the corroding metal and impurities on the surface are not essential for corrosion. The necessary and sufficient condition for corrosion is that the metal dissolution reaction and some electronation reaction proceed simultaneously. Thus, for these two processes to occur simultaneously, there are two requirements. The potential difference across the interface must be more positive than the equilibrium potential reaction:

$$M^{n+} + n\overset{.}{e} = M$$

It must also be more negative than the equilibrium potential of the electronation reaction:

$$A + n\overset{.}{e} = D$$

The electronation reaction involves electron acceptors contained in the electrolyte.

If these are the necessary and sufficient conditions for corrosion, then a test of the hypothetical mechanism for hydride involvement in the corrosion process is possible. It should satisfy the first equilibrium requirement:

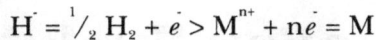

$$\overset{.}{H} = {}^1/_2\,H_2 + \overset{.}{e} > M^{n+} + n\overset{.}{e} = M$$

According to Table 10–1, the first requirement is satisfied. The electronation reactions are shown in Table 10–2.

Table 10–1 Equilibrium Potential Reactions

Electrode	$\overset{o}{\epsilon}$
$Fe^{+2} + 2\overset{.}{e} = Fe$	-0.4402
$Fe^{+3} + 3\overset{.}{e} = Fe$	+0.771
$\overset{.}{H} = {}^1/_2\,H_2 + \overset{.}{e}$	+2.25

Table 10–2 Electronation Reactions

Electrode	$\overset{o}{\epsilon}$	
$H_3O+ + \overset{.}{e} = H_2O + 1/2\,H_2$	0.00	(acid)
$O_2 + 4\,H^+ + 4\,\overset{.}{e} = 2\,H_2O$	+1.229	(acid)
$O_2 + 2\,H_2O + 4\,\overset{.}{e} = 4\,OH^-$	+0.401	(base)
${}^1/_2\,H_2 + \overset{.}{e} = \overset{.}{H}$	-2.25	(acid)

In aqueous environments the electron acceptor is always H_3O^+. Consequently, both necessary and sufficient criteria for corrosion are met by the hydride anion conjecture. This result leads to the conclusion that the hydride ion takes place in the iron metal corrosion process. In a previous section of this book, a mechanism was suggested by which the iron adsorbed molecular hydrogen is cleaved by a hydroxyl anion to form water and a hydride ion. It would appear that from the above justification that the previously suggested mechanism is reasonable.

J. O'M. Bockris and A. K. N. Reddy (1973) have described the adsorption of hydrogen on metal surfaces. This adsorption occurs as proton abstraction from hydronium ions in aqueous acid and proton abstraction from water in aqueous base by an electrified metal. Thus the adsorbed hydrogen is considered monomolecular, and as such, must recombine at the surface to allow hydrogen evolution to take place. This mechanism is entirely reasonable if one is considering an electrified interface (externally supplied current).

This mechanism does not, however, explain hydrogen evolution in a system possessing no externally applied current (corrosion). Consequently, one is left trying to rationalize the adsorption of hydrogen on a nonelectrified interface, since molecular hydrogen evolution takes place in the bulk aqueous acid or base metal systems.

The adsorption of hydrogen by an electrified or nonelectrified interface is a quantum process. Thus it requires the existence of specific energy levels for the establishment of bonds between the metal and the hydrogen. In the case of the electrified interface, the quanta are supplied by the external current. However, in the case of the nonelectrified interface, the quanta must be supplied by a different mechanism.

Charge induction or London dispersion forces can be called upon to provide insight into the nature of metallic hydrogen adsorption. These forces are said to arise from the inductive forces of noncharged species with other noncharged species. Since no externally applied current is provided in the corrosion process, it follows that neither the hydrogen nor the metal surface possesses an initial charge. We can invoke the quantum condition, which consists of occupation by molecular hydrogen of a minimum distance from the metal surface at a specific energy. It produces conditions conducive to the establishment of London dispersion forces.

The London dispersion force model requires that there be a possibility for charge dispersal on both the adsorbing and adsorbed species. Thus diatomic hydrogen (molecular hydrogen) must be the adsorbed species. If monatomic hydrogen (H) were the adsorbed species, there would be no possibility for charge separation, since this species is completely symmetrical (e.g., $1s^1$ orbital). Figure 10–1 illustrates some of the points raised in this paragraph.

The inductive charges thus produced by the adsorption of molecular hydrogen and the metallic crystal distortion effects comprise opposing forces. These opposing forces seek to regain equilibrium by destabilizing the hydrogen-hydrogen bond. The destabilized hydrogen-hydrogen bond seeks equilibrium by heterolytic cleavage of the $2s^2$ bond to form a proton (H^+) and hydride anion (H^-). The proton is then abstracted by hydroxyl ions that, along with molecular hydrogen, are in equilibrium with water in the external electrolyte. The hydride anion donates electrons to the metal. The electrons donated by the hydride ion are conducted through the metal. They emerge at some distant point and abstract protons from water to form the hydroxyl anion, which reestablishes the water molecular hydrogen and hydroxyl equilibrium.

Metallic De-electronation

The loss of electrons by the metal to form metal cations has still not been explained. The exit point of the electron produced by the hydride transfer can be considered the electron source. The external water molecules can be considered the electron sink. The conductance of the electron through the metal from the hydride anion does not involve the entrance and exit of a specific electron. Instead, the repetitive charge transfer results from alternate electron excitation and relaxation.

The excess charge on the metal at the exit point must be neutralized. This neutralization occurs by the exit of an electron from the sink, which occurs with its attack on a water molecule. This water molecule must be close enough in proximity and properly oriented to

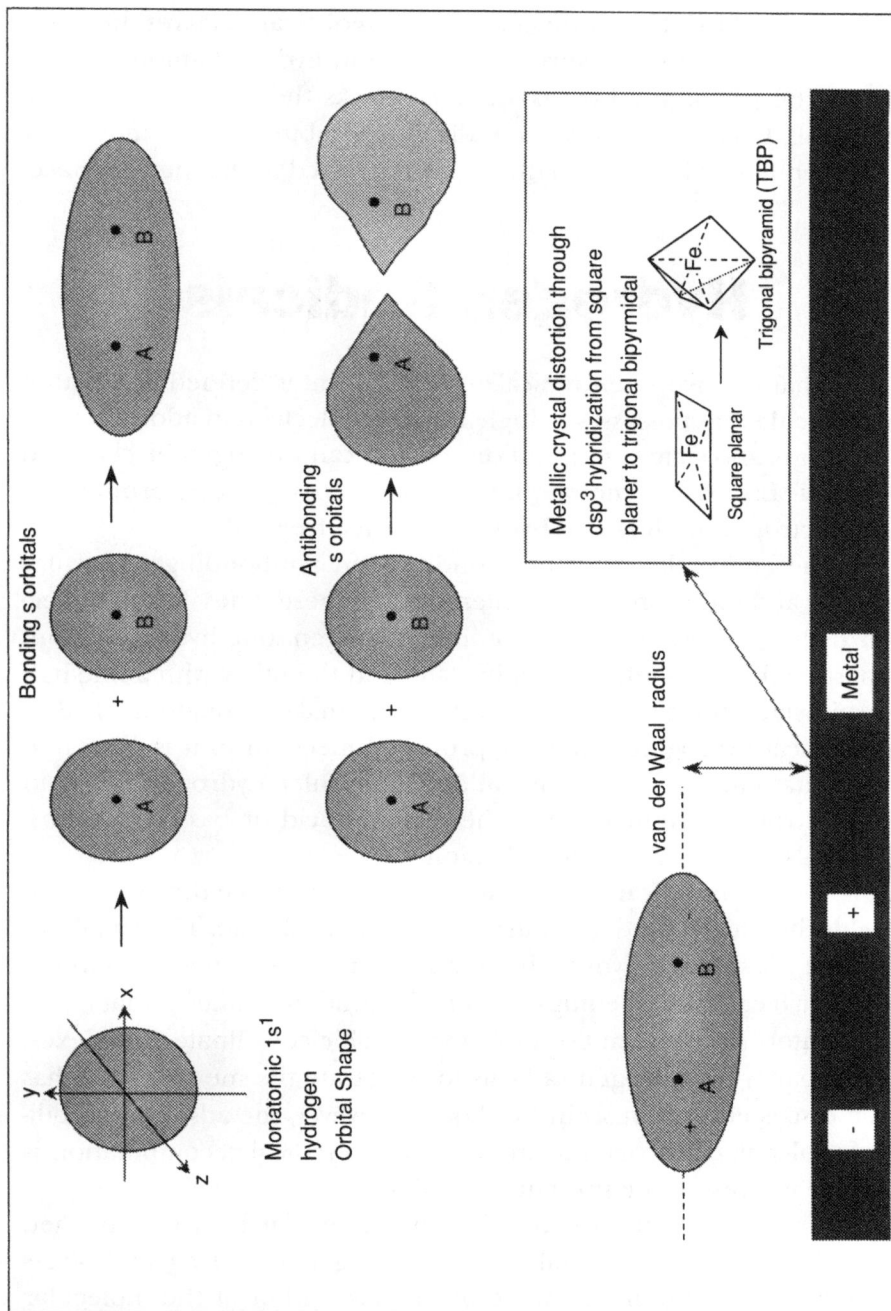

Fig. 10-1 The London dispersion forces of induction providing energy levels for molecular hydrogen's adsorption to the metallic surface

the metal surface. The exiting electron homolytically cleaves the water molecule to generate a hydrogen radical and hydroxyl anion ($H^{\bullet}$ and OH^{-}). The proximate hydroxyl anion attacks the metal to form the metal salt. Finally, the hydrogen radicals recombine to form molecular hydrogen (e.g., $H^{\bullet} + H^{\bullet} \longrightarrow H_2$), which is recycled to the metal surface.

Hydrogen Radicals

What is a hydrogen radical? A free radical is defined as an atom or molecule that possesses a single unpaired electron in addition to its normal ground state configuration. So how can hydrogen be classified a radical? Likewise, if monatomic hydrogen possesses one proton and one electron, then how can hydrogen form a free radical?

The answer lies with the bonding and antibonding molecular orbitals, and the symmetry considerations of these states. When molecular hydrogen is split into monatomic hydrogen, one hydrogen atom departs with a one-half down spin state, and the other with a one-half up spin state. It is this paired spin state that makes monatomic hydrogen radical hydrogen. From the proposed mechanism it is clear that the monatomic hydrogen radical and molecular hydrogen occur in the electrolyte as under either the aqueous acid or base conditions. This is also true of the hydroxyl anion.

Hydrogen radicals are smaller than protons since the 1s electron shields the nucleus and neutralizes the proton's charge. This small size makes it possible for hydrogen radicals to enter small openings in the metal surface. Such openings include the grain boundaries that occur at the interface between two different metallic coordinate complexes. This capacity of hydrogen radicals to penetrate the metal surface has been discussed previously in this book. However, the adsorption ability of molecular hydrogen formed by radical-radical recombination is of more interest in the present discussion.

Earlier, it was pointed out that the molecular hydrogen formed in pockets within the metal caused hydrogen blistering and stress cracking. No mention was made of the adsorption of the molecular hydrogen formed. The surface area covered by the molecular hydrogen in these pockets is limited. The increasing concentration of the molecular hydrogen formed in the radical-radical recombinations will quickly occupy this available surface area.

Molecular hydrogen adsorption takes place only to the extent that the free bare metal is available. It is at this point that the similarity between the interstitial surfaces and the electrolyte metal interface breaks down. These interstitial spaces only accommodate hydrogen radicals and their product molecular hydrogen. Thus the unavailability of hydroxide anions makes the heterolytic cleavage of the adsorbed molecular hydrogen impossible.

However, as the concentration of the molecular hydrogen builds, stress fractures and blistering occur. As the stress fractures and blisters increase in size, the electrolyte is able to enter these structures. Thus the processes of heterolytic molecular hydrogen cleavage, hydride and proton formation, hydride to metal electron transfer, homolytic hydroxyl and hydrogen radical formation, and hydroxyl attack of the metal begin. This is illustrated in Figure 10–2.

One very important result emerges from the information developed in the hydride mechanism presented in the previous several sections. *In aqueous acid or base electrolyte systems, hydrogen radicals are only formed at the metal surface when radical generation is instigated by another free radical.* An example of free radical generation that occurs away from the metal surface involves the hydrogen sulfide oxidation process. The following reaction shows an example of how hydrogen radicals can form within the electrolyte, and away from the metal surface:

$$HSH + R^\bullet \longrightarrow HSR + H^\bullet$$

where

$R^\bullet$ is any radical species with energy high enough to homolytically cleave the hydrogen sulfur bond (e.g., $NO^\bullet$, $O^\bullet$, $HS^\bullet$, $OH^\bullet$, $SO_4^{\bullet 2-}$, etc.).

There are also cases involving mercaptans [e.g., RSR, where $R = (C_nH_{2n+2})$], polysulfides, metallohydroxides (e.g., $M^{n+}(OH^-)_n + R^\bullet \longrightarrow M^{(+n-1)}[(OH)_{(n-1)}O^\bullet] + RH$). All of these radicals are short lived

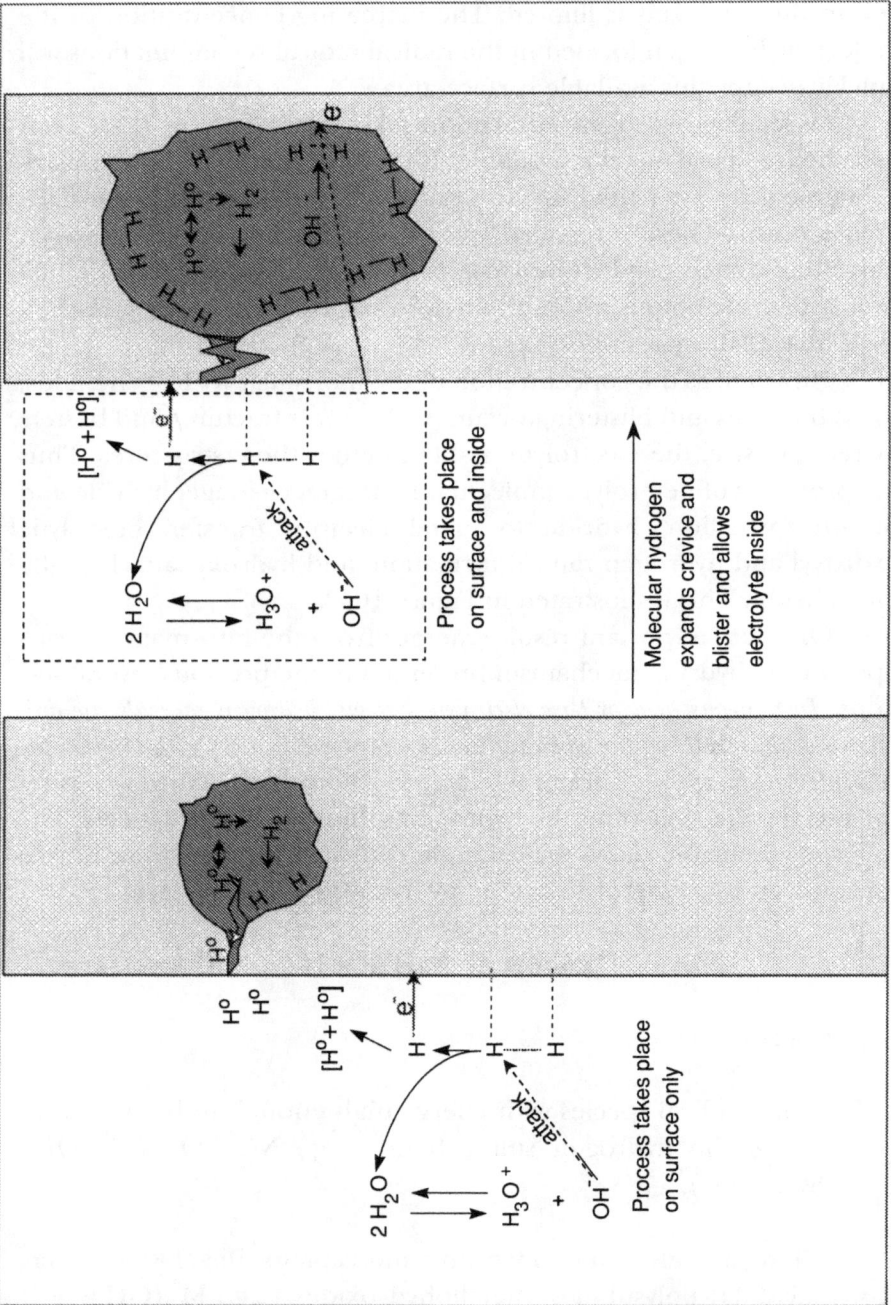

Fig. 10-2 Hydride/proton formation on the interior of a hydrogen blister as the external electrolyte enters the enlarged cracks in the metal

because they are so reactive, but as long as there are available free radical forming species, the presence of free radicals will be maintained. This sequence is illustrated in the following equations.

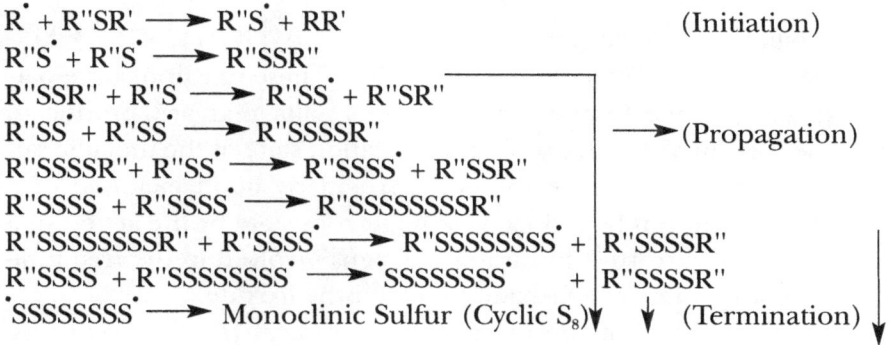

$$R^{\bullet} + R''SR' \longrightarrow R''S^{\bullet} + RR' \qquad \text{(Initiation)}$$

$$R''S^{\bullet} + R''S^{\bullet} \longrightarrow R''SSR''$$

$$R''SSR'' + R''S^{\bullet} \longrightarrow R''SS^{\bullet} + R''SR''$$

$$R''SS^{\bullet} + R''SS^{\bullet} \longrightarrow R''SSSSR'' \qquad \longrightarrow \text{(Propagation)}$$

$$R''SSSSR'' + R''SS^{\bullet} \longrightarrow R''SSSS^{\bullet} + R''SSR''$$

$$R''SSSS^{\bullet} + R''SSSS^{\bullet} \longrightarrow R''SSSSSSSSR''$$

$$R''SSSSSSSSR'' + R''SSSS^{\bullet} \longrightarrow R''SSSSSSSS^{\bullet} + R''SSSSR''$$

$$R''SSSS^{\bullet} + R''SSSSSSSS^{\bullet} \longrightarrow {}^{\bullet}SSSSSSSS^{\bullet} + R''SSSSR''$$

$${}^{\bullet}SSSSSSSS^{\bullet} \longrightarrow \text{Monoclinic Sulfur (Cyclic } S_8) \qquad \text{(Termination)}$$

Another series of reactions that are of particular significance concerns the radical reaction of molecular hydrogen with free radical species.

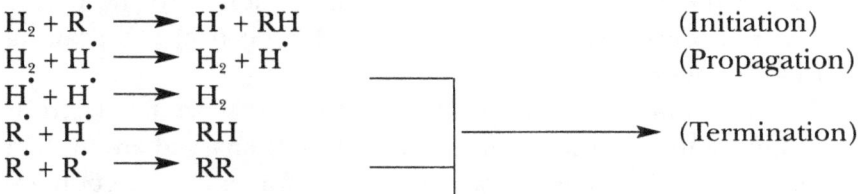

$$H_2 + R^{\bullet} \longrightarrow H^{\bullet} + RH \qquad \text{(Initiation)}$$

$$H_2 + H^{\bullet} \longrightarrow H_2 + H^{\bullet} \qquad \text{(Propagation)}$$

$$H^{\bullet} + H^{\bullet} \longrightarrow H_2$$

$$R^{\bullet} + H^{\bullet} \longrightarrow RH \qquad \text{(Termination)}$$

$$R^{\bullet} + R^{\bullet} \longrightarrow RR$$

These are of particular interest because they provide a mechanism by which the molecular hydrogen generated from metal corrosion process can be depleted. (This is illustrated by the last two termination reactions above.) Thus, the incorporation of a radical-forming species into the electrolyte could have a significant effect on the corrosion rate of the system.

Metallohydroxides and Radicals

Metallohydroxide radical reactions (e.g., $M^{n+}(OH^-)_n + R^{\bullet} \rightarrow M^{(+n-1)}[(OH)_{(n-1)}O^{\bullet}] + RH$) take place in solution. These reactions have particular significance to the form of metal salts that are produced, because they involve a change in the oxidation state of the metallic salt complex. Ferric hydroxide offers a particularly good example of a metallic salt that can be reduced to the ferrous state by the action of a free radical. Figure 10–3 illustrates the steps involved in the free radical reduction of ferric hydroxide to ferrous hydroxide.

The possibility of the occurrence of radical reactions with the atomic metal arises as an interesting consequence of the oxide radical formation from the metallic salt. If a free radical is capable of displacing the hydrogen from the metallic salt, and reducing the metal, then additional questions arise. Is it possible for a radical to blend with the *d* orbitals of the metal, thus becoming a partial anion to the partial cation of the metal? The answer is yes, if the electron withdrawing power of the radical is sufficient.

The withdrawing power must be able to cause *d* orbital electrons to spend more time in the vicinity of the partially charged anion than in the normal *d* orbital configuration. London forces of induction are responsible for the formation of the induced partial charges in the adsorbed molecular hydrogen and the coincident opposite partial charges in the metal. As a matter of fact, this mechanism explains the appearance of the hydride ion on the metal surface quite nicely. Figure 10–4 illustrates how the hydride anion might arise at the surface of the metal.

Thus, the hydride anion can be seen as means of partial metallic oxidation. However, the hydride anion is very unstable and gives up an electron to the metal to reestablish charge neutrality. The electron donated from the hydride anion excites electron transport through the metal to a sink site on the metal's surface. There it can combine with water and produce hydrogen radicals and hydroxyl anions. As the electron emerges, a properly oriented water molecule must be at a distance within the van der Waals radius to accept the electron to form a

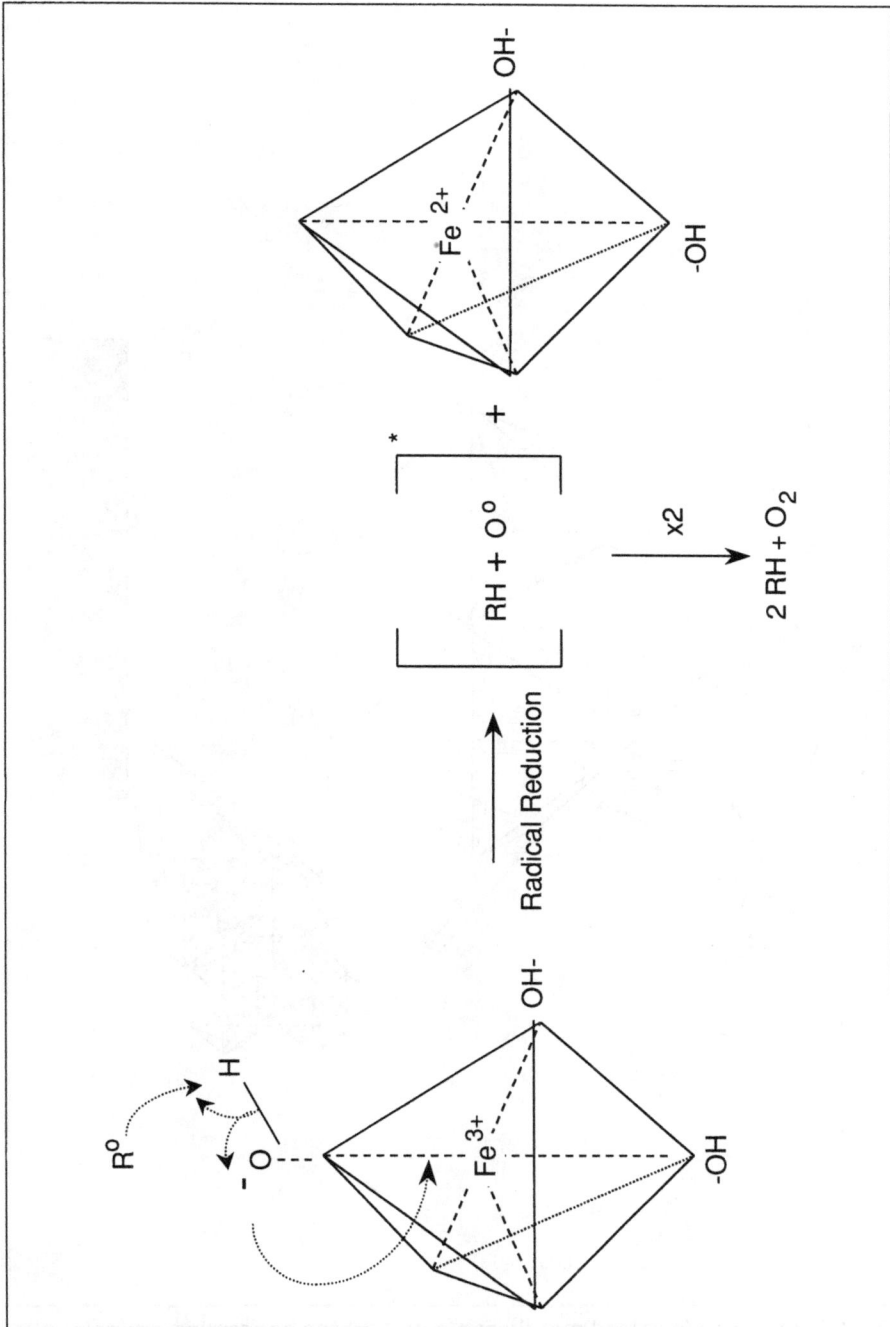

Fig. 10-3 Reduction of ferric hydroxide by reaction with a free radical

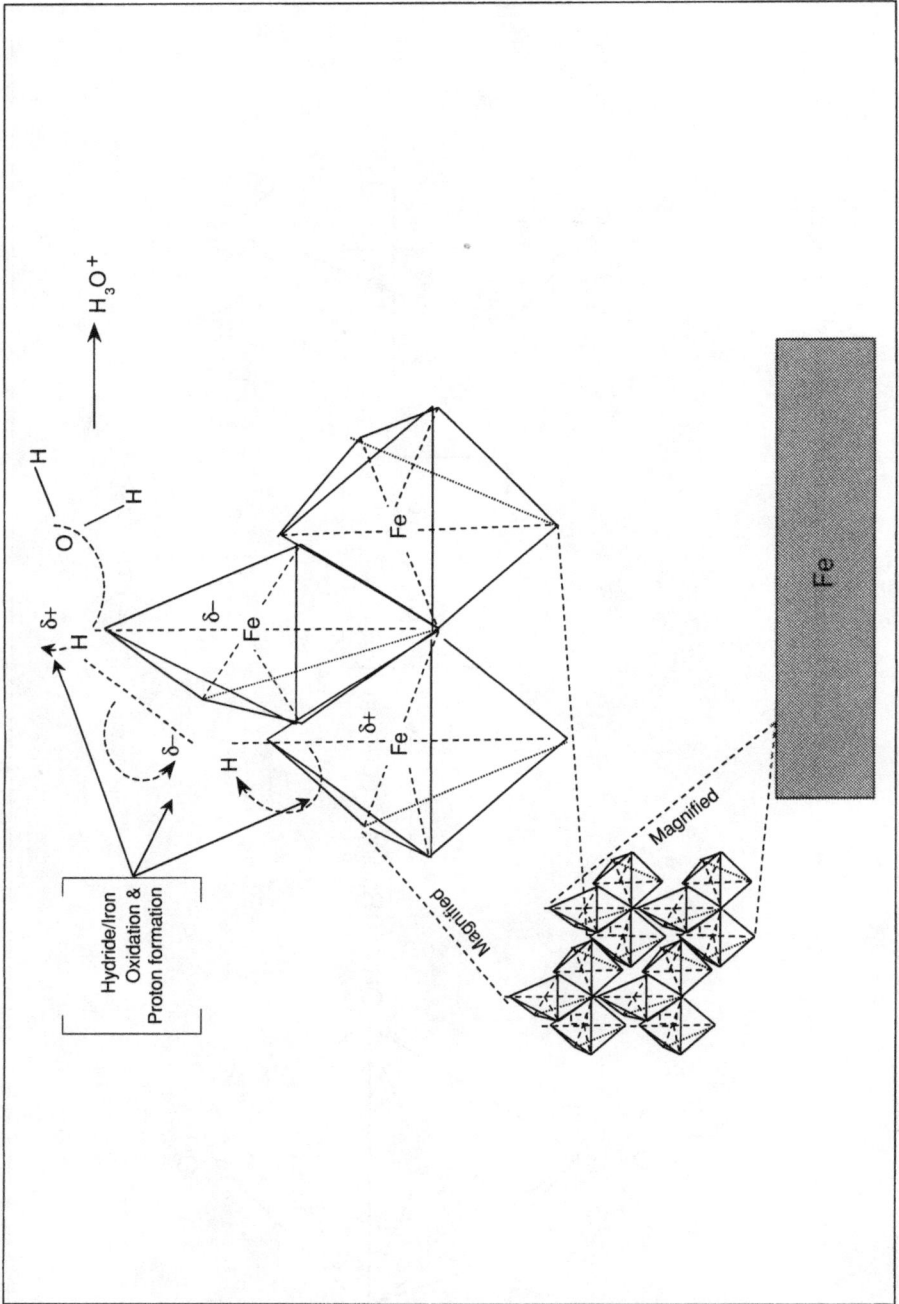

Fig. 10-4 Hydride/oxidized metal couple and proton generation on metal surface

hydrogen radical and hydroxyl. The electron-deficient metal complex (oxidized iron complex at the sink location) can then be attacked by the hydroxyl anion to yield the metallic hydroxyl salt. Figure 10–5 gives a schematic view of these sequences.

Hydrogen and Scale

Thus far the discussion of the special case of hydrogen has been confined to its effects on corrosion processes. However, scale formation is also affected by its various forms. The metallo-hydroxide radical and hydride reactions indicate that radical processes can lead to either oxidation or reduction products of the atomic metal and metal salts, respectively. These oxidations and reductions can change the coordination number of the metal and metallic salt complexes.

If the anionic ligand leaving the complex is high in electronegativity, it can depart the metal by a heterolytic electron cleavage, taking some electron density from the metallic complex. Conversely, if the ligand leaving the complex is low in electron density, it departs, leaving behind a complex with increased electron density. Thus, a rational explanation for the variability of complex coordination changes can be illustrated as in Figure 10–6. The results of these coordination number changes is then reflected in the nature of the scale formed by the continued aggregation of the coordinate salts to form crystals.

Group 2A Elements

The Group 2A elements usually occur as metallic salts in petroleum fluid systems. Consequently the mechanism for iron metal corrosion, as proposed above, would not be operative. However, the radical reduction process illustrated in Figure 10–3 is still possible, and the coordination changes illustrated in Figure 10–6 can still occur. Figure 10–7 illustrates some of the consequences for the Group 2A elements.

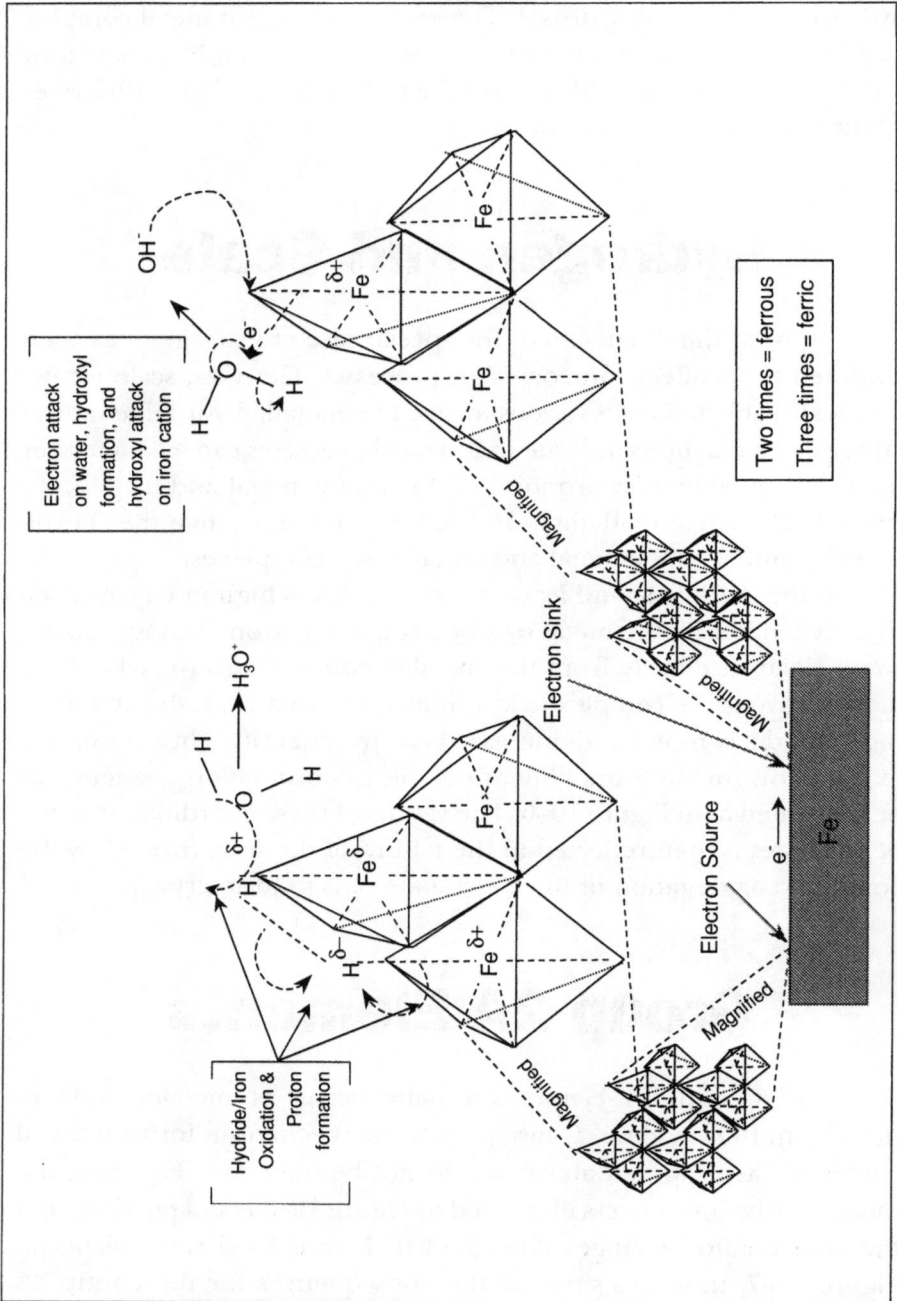

Fig. 10-5 Electronation (electron source) and de-electronation (electron sink) reactions leading to corrosion of iron in an aqueous system

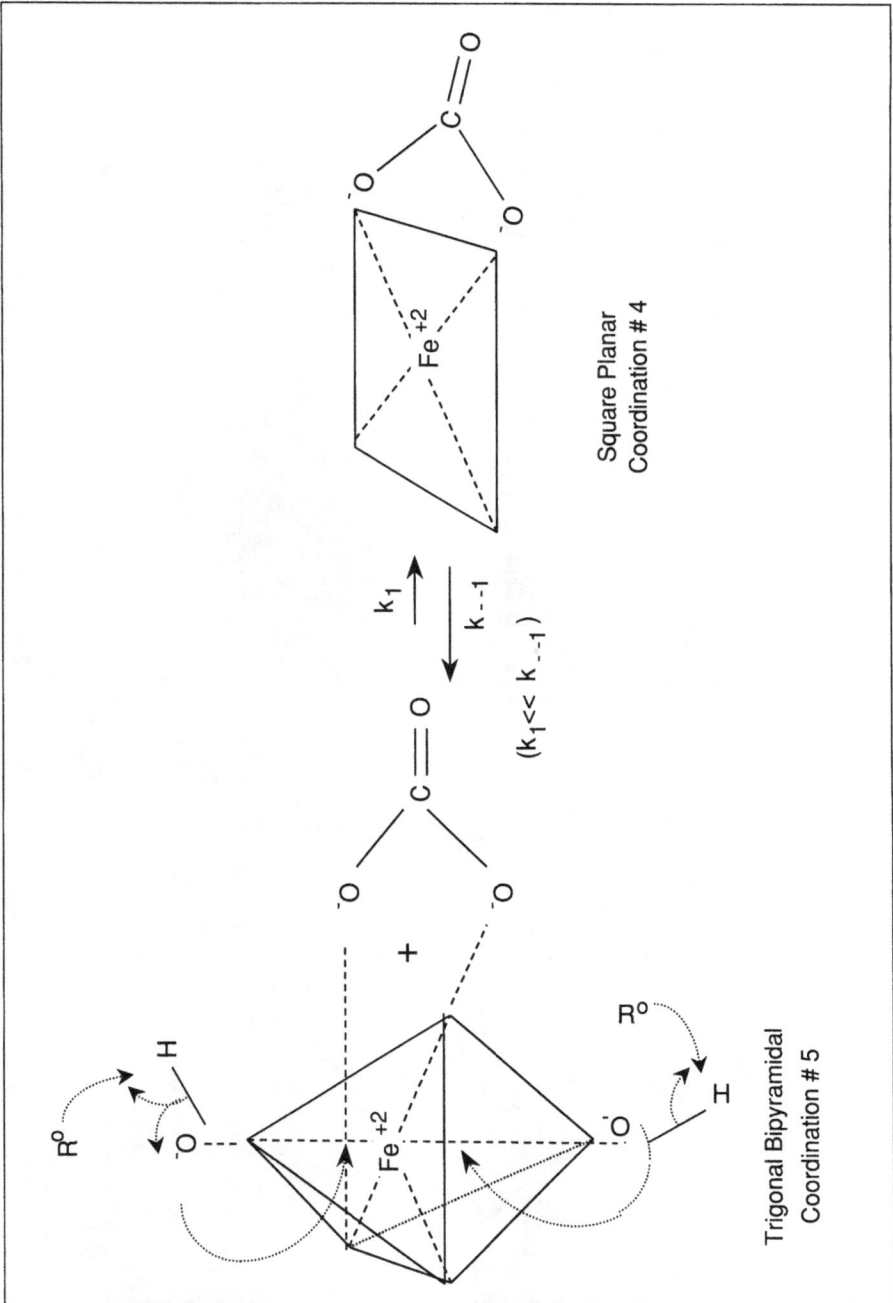

Fig. 10-6 The ferrous salt coordination number changes caused by high (hydroxyl) and low (carbonate) electronegativity leaving groups.

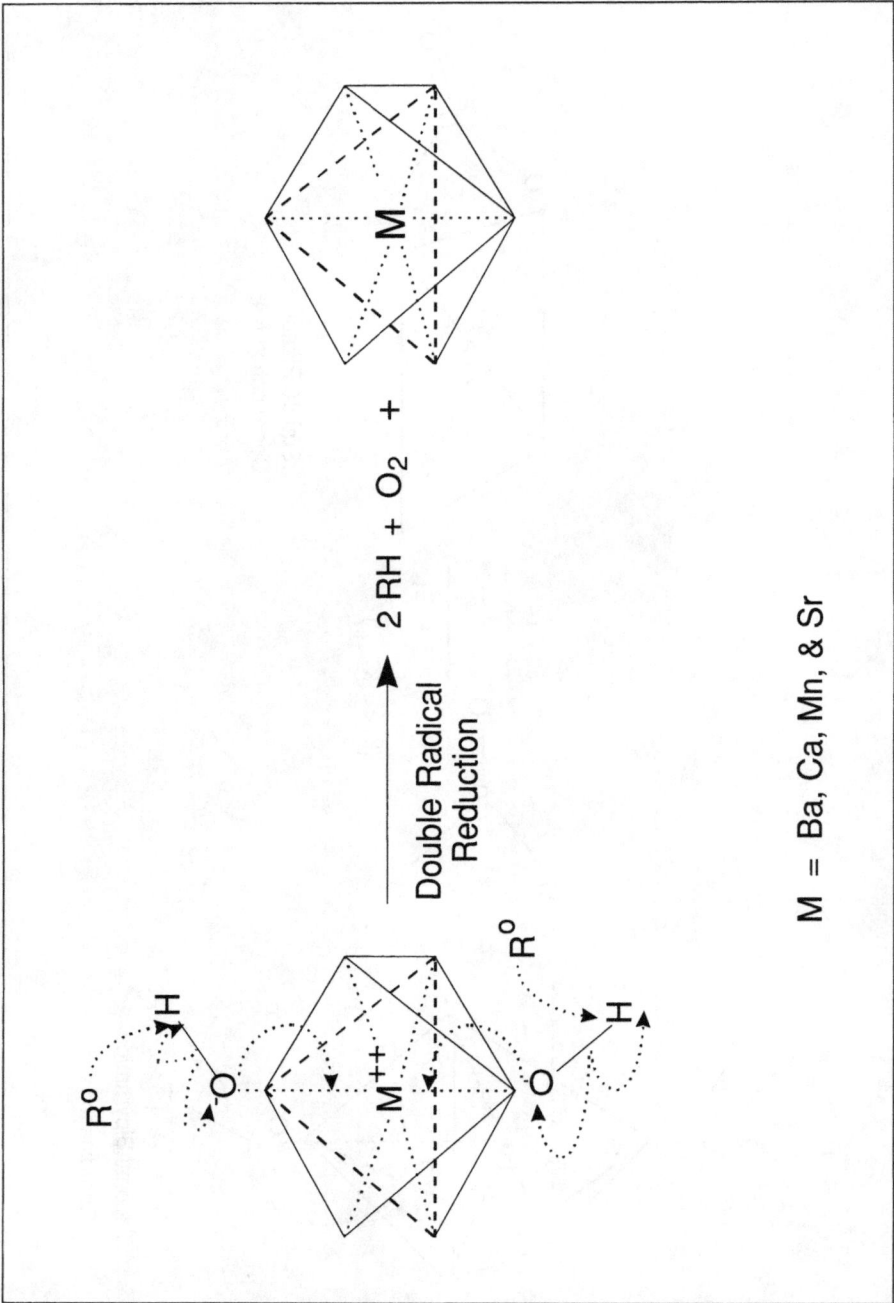

Fig. 10-7 The double radical reduction of the Group 2A elements as a possible route to the metallic element

The possibility of reduced metals of the 2A elements occurring in the petroleum liquid systems exists. However, the presence of substitute ligand groups makes the appearance of the metal unlikely. Furthermore, the solubility of the resulting salt forces the equilibrium reaction in the direction of the new coordinate complex. Figure 10–8 shows how this might take place.

The K_{sp} values of all but the $Mn(OH)_2$ favor the k_1 equilibrium direction, since all of the carbonate salts, except $MnCO_2$, are precipitated more readily than the hydroxyl salts. Both coordination complexes in Figure 10–8 are six coordinate. However, the geometry of the complex may be expected to change from octahedral to trigonal prism. Thus the crystal forms can become mixed, resulting in crystal imperfections.

Although it is unlikely that the Group 2A elements will get reduced to the metal, it is possible via the double radical reduction mechanism. The ability of the metal to adsorb molecular hydrogen then determines whether or not it will be oxidized. Thus, the molecular hydrogen adsorption properties of the Group 2A metals are important in the secondary formation of scales (complexed salts). It is difficult to find reported values for the adsorption of molecular hydrogen on the Group 2A elements, but the hydrides of these metals do exist.

It is interesting that these hydrides exist as cationic white powders that release molecular hydrogen upon reaction with water. Further, the Group 2A metals are so reactive that they typically do not exist in nature and generally occur only in the salt forms. Thus, it is very likely that these metals will adsorb molecular hydrogen, and that, because of their high reactivity, they will go on to form complexed salts.

Summary

The preceding chapter deals with the special properties of hydrogen and the effects of its presence in corrosion and scale processes. Many of the special properties of hydrogen are the result of its ability to form radicals. The formation of these radicals is intimately involved

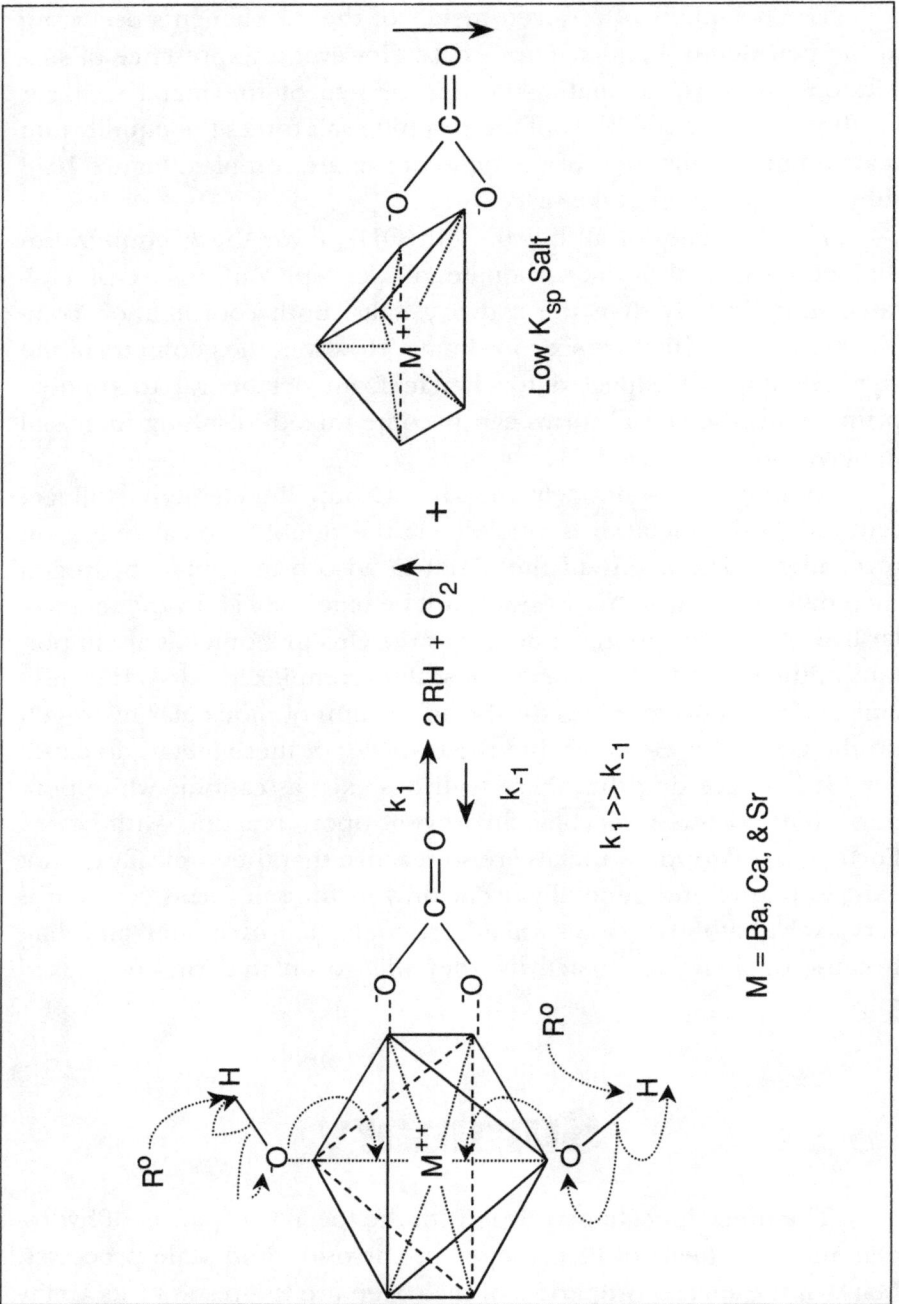

Fig. 10-8 The double radical reaction displacing the more electronegative hydroxyl ligands to allow displacement by the carbonate anion

with the metal surface and the mechanism of oxidation and reduction in aqueous systems. The importance of the inductive influences of molecular hydrogen at the metal surface was discussed. Also introduced was the requirement of appropriately orientated molecules for the electronation and de-electronation reactions.

The London dispersion mechanism and the resultant charge dispersal effects were discussed in context with the electron transfer reaction leading to metallic oxidation. The occurrence of free radicals in the electrolyte phase, and the effects they produce, were related to metal complex salt reductions. The pathway created by radical reactions with metal salts was also correlated with changes in coordination number and geometric changes.

References

Barrow, Gordon M. 1966. *Physical Chemistry.* 2d ed. New York, St. Louis, San Francisco, Toronto, London, and Sydney: McGraw Hill Book Company.

Bockris, J. O'M., and A. K. N. Reddy. 1973. *Modern Electrochemistry.* New York: Plenum Publishing Company.

Dickerson, R. F., H. B. Gray, and G. P. Haight, Jr. 1970. *Chemical Principles.* 1st ed. New York: W. A. Benjamin, Inc.

Huheey, James E. 1978. *Inorganic Chemistry Principles of Structure and Reactivity.* 2d ed. New York, Hagerstown, San Francisco, and London: Harper & Row.

Jones, Loyd W. 1988. *Corrosion and Water Technology for Petroleum Producers.* 1st ed. Tulsa: OGCI Publications, Oil & Gas Consultants International, Inc.

Wagner, C., and W. Trand. 1938. Z. Elektrochem. 44:391.

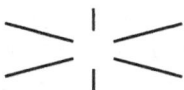

11
Corrosion and Scale Forms

Scale Formation

Much of the preceding discussion has pointed to an intimate connection between the processes of corrosion and the type of scale formed. As we have seen, scales are the products of anionic combinations with oxidized metals (cations). The oxidation state of these metal cations and their combination with particular anions determine their solubility. They also determine their subsequent tendencies to scale. It is important to realize that the oxidation states of the metal salts are the result of the process of oxidation and reduction reactions (corrosion).

In a broad sense, all inorganic scale can be viewed as the product of corrosion, since scales form from low solubility anionic combinations of oxidized metals. Iron salts of hydroxides, sulfides, oxides, and carbonate can be considered scale products of the combination of these specific anions and the corrosion products ferrous and ferric iron. Likewise, the hydroxide, sulfide, oxide, and carbonate salts of the Group 2A elements should be considered the scale forms of their respective corrosion products. It is important, therefore, to understand the process of corrosion leading to the metallic oxidation products that form scales.

The primary focus of the present discussion is aimed at the scale products commonly found in petroleum fluid systems. However, it is important to recognize that the Group 2A elements are not the only scales that result from oxidation and anionic combinations. In addition to iron, aluminum, cobalt, copper, cadmium, lead, manganese, mercury, nickel, silver, and zinc all form low solubility products from various anionic combinations.

Each of these combinations involves d orbital configurations of the central metal cation, while the Group 2A elements are confined to s orbital arrangements. The Group 2A elements are also restricted to the 2+ oxidation state, and consequently only form divalent salts. Thus, the oxidation of the Group 2A elements would be expected to proceed either by the attack of a divalent anion or the combined attack by two monovalent anions.

Since the most common metal salt reactions take place in the presence of water, it is most likely that the hydroxyl salts are the first products formed. It is only after the hydroxyl salts are formed that divalent anions enter into equilibrium reactions that lead to the lower solubility salts or scales. As we have seen, the Group 2A metals do not appear in nature in the reduced metallic state, since they are so reactive. Additionally, the hydroxyl forms of these metals exhibit low solubility characteristics, and in one case the hydroxyl form is lower in solubility than the divalent anionic salt. (The magnesium hydroxide is much less soluble than the carbonate salt.)

Which Came First— Corrosion or Scale?

To a first approximation, it would appear that corrosion must occur before scale formation can take place. However, what corrosion mechanism can bypass the requirement for counter ion involvement in achieving charge balance with the emerging cation? It would appear that the two processes must be interconnected, but what is the connection? Thus far the discussion of corrosion has centered on the following:

- The adsorption of molecular hydrogen
- Hydride/proton formation
- Hydride electron donation
- Hydroxyl formation
- Hydroxyl attack of the metal

One part of the picture is missing, and that part is the loss of electrons by the metal to form the metallic salts. This is illustrated in Figure 11–1.

Figure 11–1 illustrates the sequence of events proposed as the means of metallic salt formation. This sequence is broken down into the following order:

1. Molecular hydrogen adsorption on metal surface causing inductive partial charges in metal (London forces at van der Waals radii)
2. Heterolytic cleavage of adsorbed hydrogen to proton and hydride anion
3. Hydride electron donation to metal (electron source) converting hydride to hydrogen radical
4. Electron conduction through metal to a point where water is within van der Waals radius (electron sink)
5. Electron attack on water to form hydroxyl and hydrogen radical
6. Hydroxyl attack on electron sink metal to form departing metallic salt
7. Electron conduction through metal to adsorbed proton, converting it to hydrogen radical
8. Hydrogen radical recombination to form molecular hydrogen
9. Radical reduction of metallic salt to metal (return to step 1)

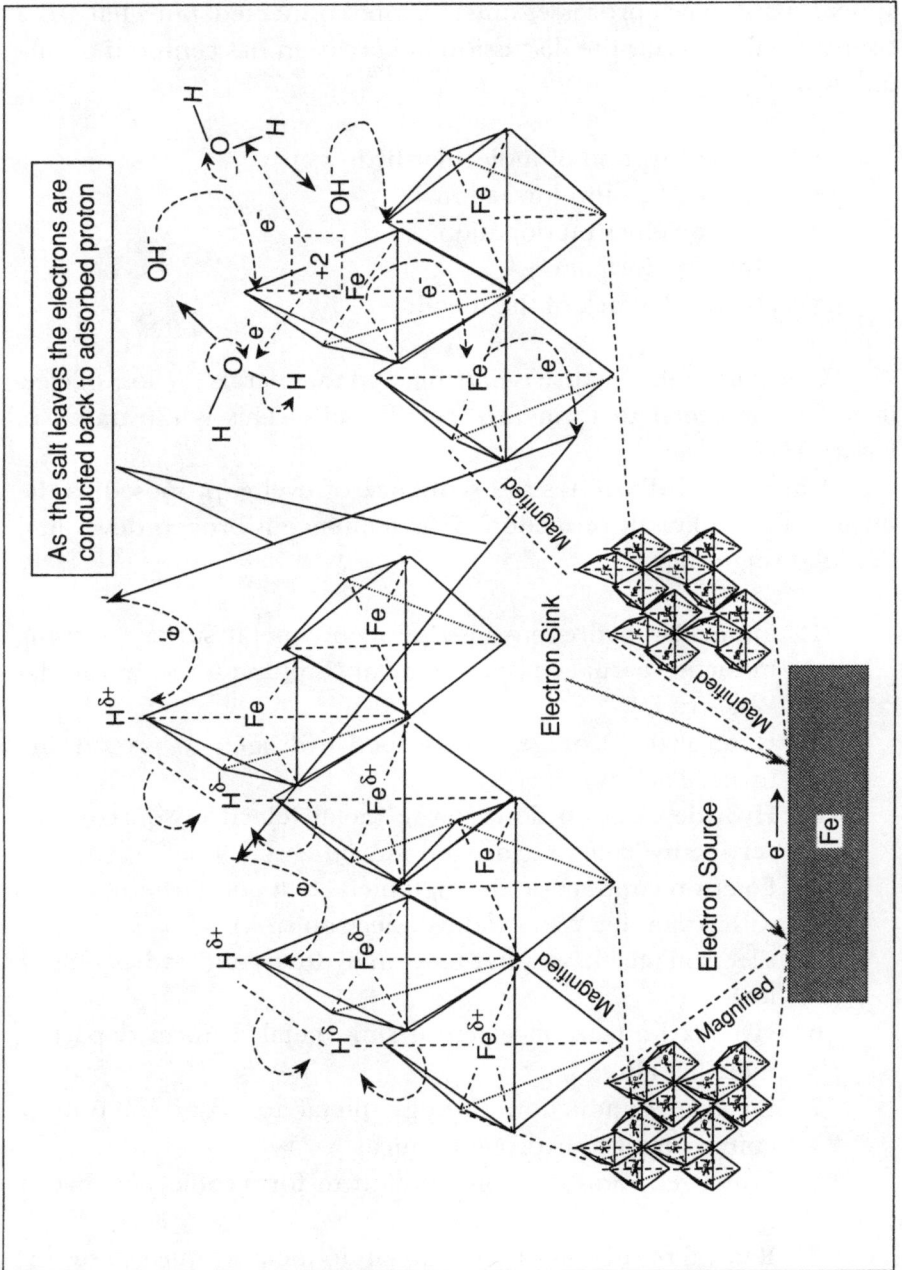

Fig. 11-1 Two-step adsorbed molecular hydrogen A. going to hydride and proton electron donation B. electron attack on water, hydroxyl attack on metal, and return of electron to proton of adsorbed hydrogen

From the first eight steps, it would appear that our initial inclination to suggest corrosion takes place before salt formation is correct. However, since radical reduction of metallic salts has been shown to take place, the question now becomes moot. This is because the process can be seen as cyclical, and the point in the cycle chosen as starting point determines the order. Thus, the cycle requires one final step, the radical reduction of the metallic salt to the metal (step 9).

The ninth step finishes the cycle, and the process begins again. Thus, the processes of corrosion and scale are cyclic, and they take place within and without the metal and metallic salt surfaces.

Radicals

The reaction of pure metal immersed in pure water can be represented by the following equation:

$$M + [\ 4\ H_2O \leftrightarrow 2\ H_3O^+ + 2\ OH^-] \longrightarrow M^{2+}(OH)_2 + 2\ H_2O + H_2$$

However, if the reaction in the brackets is ignored, and only the water and metal are considered, there would appear to be no reason for the reaction to proceed. Since the water is in equilibrium with the hydronium and hydroxyl (H_3O^+ and OH^-), there would appear no reason for the hydroxide to react with the metal. However, oxygen would be present under most environmental conditions. If oxygen is added to the system, it is possible to see that the equilibrium of the hydronium hydroxyl salt can shift due to the following reaction:

$$O_2 + H_3O^+ + OH^- \longrightarrow H_2O + HOOH + {}^1\!/_2\ O_2$$

The HOOH continues to react as a peroxide and homolytically cleaves to yield radical hydroxide molecules as follows:

$$HOOH \longrightarrow 2\ OH^{\bullet}$$

The reaction then continues by abstracting a proton from the hydronium ion, which shifts the aqueous equilibrium in favor of the hydroxide anion.

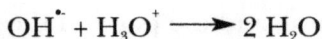

$$OH^{\bullet} + H_3O^+ \longrightarrow 2\,H_2O$$

Although the sequence of reactions depicted above may only take place to a very small extent, it is the major catalyst for the initiation of corrosion in aqueous systems. Given the formation of the hydroxide radical, the presence of other reactive species can continue the process of radical propagation. Hydrogen sulfide is a good example of this effect.

$$HSH + OH^{\bullet} \longrightarrow H_2O + HS^{\bullet}$$

Thus, the radical effects instigated by the presence of oxygen in an aqueous system are instrumental to the process of corrosion.

Oxidation Films and Corrosion

Passivation of metallic surfaces is a process that can be considered as a natural mechanism for the preservation of the metallic surface. It involves the in-situ deposition of an insoluble oxidation product of the metal. The ferrous and ferric oxide corrosion products of iron can be considered as iron metal passivators, but due to their poor adhesion properties, they are very ineffective. What factors determine the adhesion properties of the scale (oxidation films) or corrosion products? If the intermediate oxidation products are the hydroxide salts of the metal, then the metal salts are soluble under acid conditions. The soluble salts proceed away from the metal surface and into the electrolyte, where they are converted to iron oxide (insoluble) scales.

This procession of ferrous and ferric hydroxides away from the metallic surface accounts for the ineffectiveness of the metal hydroxides as passivation agents. Under base conditions of aqueous iron corrosion, the hydroxide salts of the metal are nearly insoluble in the electrolyte. They would, therefore, be expected to remain as an oxidized film at the metal surface. It is not clear why the base corrosion products of iron are unstable as passivators, but the different solubilities of the ferrous hydroxide ($K_{sp} \approx 10^{-14}$) and ferric hydroxide ($K_{sp} \approx 10^{-36}$) may have something to do with their instability. A possible clue may be available in the base corrosion product of iron and sulfide ions.

Under aqueous base conditions, the reaction of hydrogen sulfide salts with iron involves attack of the metal by anionic sulfhydrides (HS_n^-, where n = 1, 2, 3,...8). It also involves production of ferrous sulfide (FeS_2, $K_{sp} \approx 10^{-19}$) and ferric sulfide (Fe_2S_3). Ferric sulfide decomposes to ferrous sulfide and sulfur. Oxygen and sulfur both possess one unfilled p orbital. However, oxygen's p orbital vacancy is in the $2p$ orbital and closer to the nucleus and more energetic (held tighter) than the $3p$ orbital of sulfur.

The p orbital interaction of sulfur with the metal d orbitals changes the geometry of the iron complex. In contrast, the hydroxide complex leaves the metal surface in the same geometric configuration as it possessed in the metal. Thus, it is likely that the stability of the passivation product is determined by its ability to alter the geometry of the metal complex.

Radical-Induced Complex Geometry

Previously the effects of radicals on metal complex salts were discussed in connection with geometric changes to the complex. The mechanism proposed was that of proton abstraction by the radical on hydroxyl salts of a metal. This mechanism can help explain the changes in stability of the passivation species occurring at the metal surface.

As explained earlier, the reactant (metallic salt complex) must be activated by radical abstraction of a proton from the existing complex before the substitution of a less reactive ligand can proceed. Thus, the hydroxide salts of the metal have to be converted to a reduced metallic salt (e.g., ferric to ferrous cation) in order to accommodate the carbonate anionic ligand group. This reduction and subsequent reaction with the carbonate ligand changed the geometry of the complex salt.

The same mechanism can be invoked to explain the reactions of sulfhydrides with iron, but the resulting geometric changes in the metallic complex can take several forms. Figure 11–2 illustrates one of these effects and proposes a mechanism for the change. The geometric changes and *in-situ* deposition of the geometrically altered iron sulfide provide a passive ferrous sulfide film surface that resists further oxidation. Strong aqueous acid conditions, however, attack the sulfhydride complexes to yield hydrogen sulfide and the conjugate anionic salt.

An example of this is the reaction between the sulfhydride metallic salts and aqueous hydrochloric acid. The resulting chloride salt of the metal is soluble in aqueous acid solution, while the hydrogen sulfide possesses very little solubility and escapes as a gas. Figure 11–3 illustrates an example of this reaction and the products formed.

The aqueous base reaction would be similar to the acid. However, the attacking species would be the hydroxide anion rather than the chloride anion. Further, the stronger base character of the hydroxide favors attack on the more electropositive metal cation. Thus, all of the alkali Group 1A and alkali earth metals (Group 2A) form more electropositive cationic salts than the ferrous iron. They would be expected to predominate as the hydroxide derivatives. This predominance then suggests that the ferrous sulfide salt is more effective as a passivation agent in an aqueous base environment.

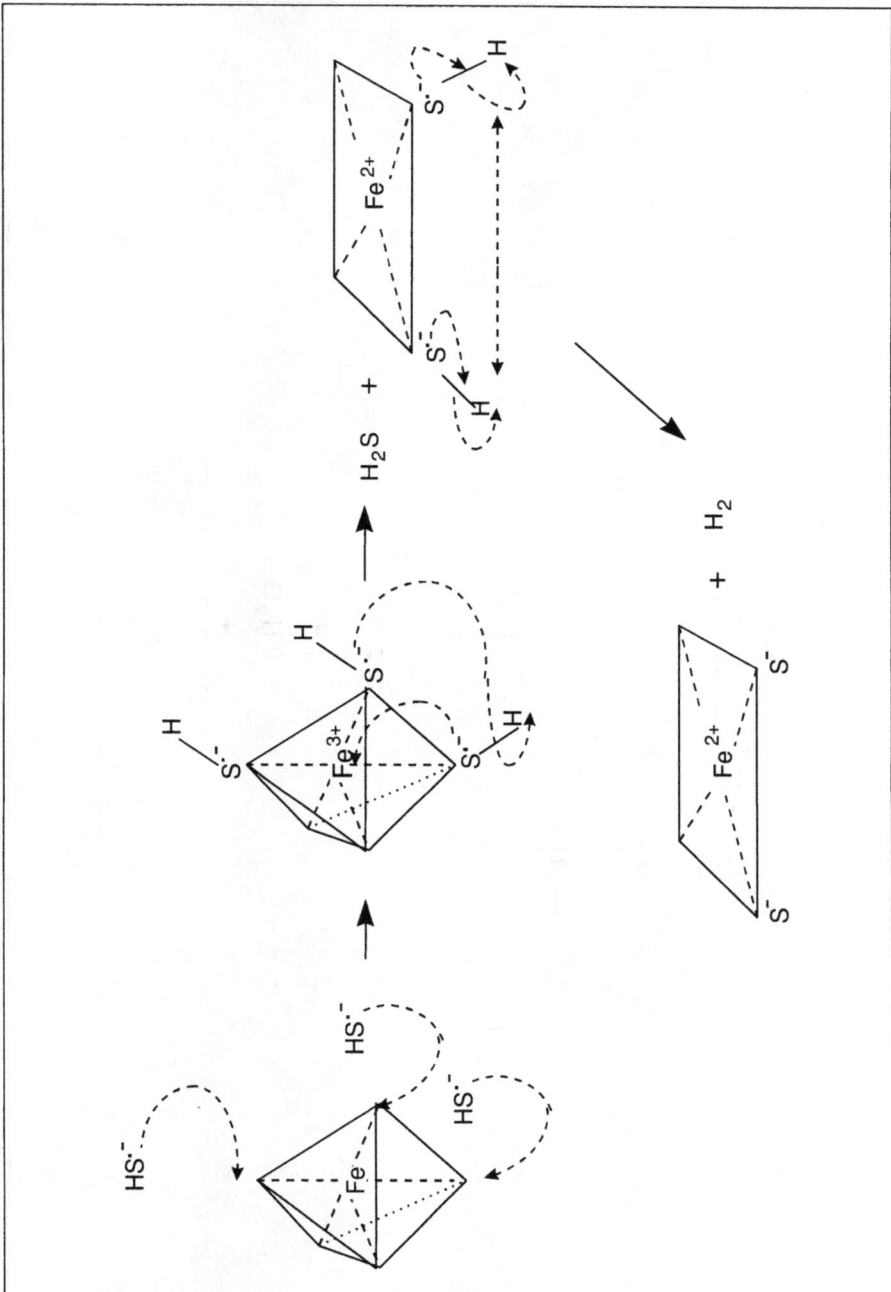

Fig. 11-2 The effect of the sulfhydride radical on the geometry of the iron salt

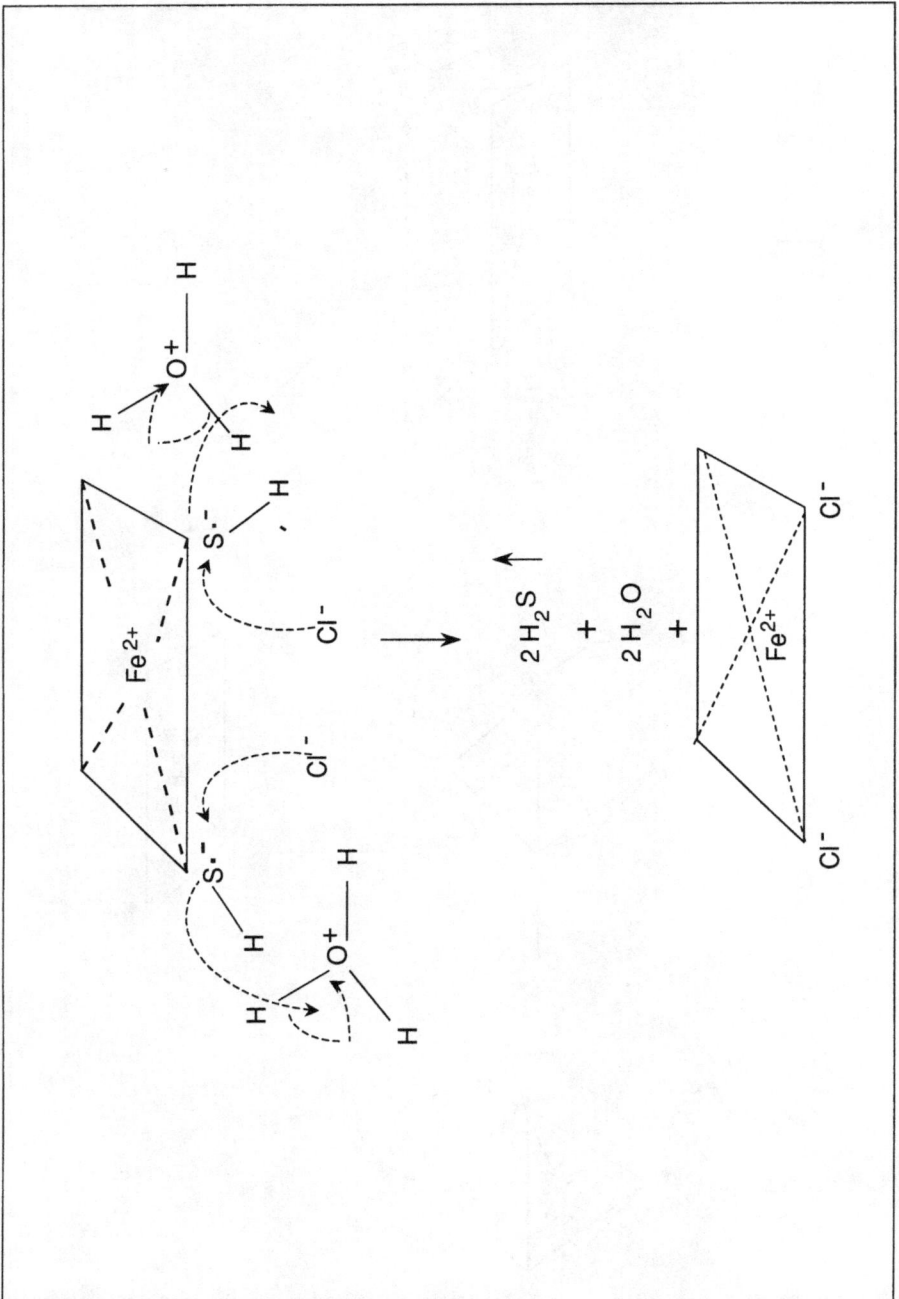

Fig. 11-3 Ferrous hydrosulfide reaction product with aqueous hydrochloric acid

Mixed Metallic Salts

Our discussion has centered on the single anionic component salts, and no mention has been made of the possible combinations of salts within a single complex. Of course there are conditions required for such salt combinations. The conditions would need to be such that the greater reactivity of one anion versus the other could be overcome by an increase in the concentration of the lower reactivity anion. Thus, combination reactions like sulfides and hydroxides would require that the sulfide anion be statistically favored over the hydroxide. Figure 11–4 shows the mixed anion salt of ferric iron.

Filling the Non-ionic Ligand Sites

Although water coordination of the remaining ligand sites is favored in aqueous systems, it is possible for other non-ionic ligand groups to form in the presence of water. The main criterion for their substitution for water is that they be more reactive than water (e.g., more polar, or more charge diffuse). Hydrogen sulfide is larger than water, because the radius of the sulfur atom is greater than that of oxygen, and shielding effects of the $3p$ orbital in the sulfur is less than that of the $2p$ orbitals of the oxygen. It therefore holds its electrons less strongly than water and can form non-ionic ligands even in the presence of water. Figure 11–5 shows some possible structures for multiple ligand salts.

The combinations of mixed anionic and mixed non-ionic ligand salts produce mixtures of salts that exhibit anomalous crystal formation. These anomalous crystal forms aggregate either as single component or mixed component crystals. Thus, mixed complexes can co-crystallize or crystallize separately. The results depend upon the solubility of the complex in the solvent system. If the complexes cocrystallize, the resultant crystal forms will exhibit crystal imperfections. These imperfections will either result in crystal dislocations and grain bound-

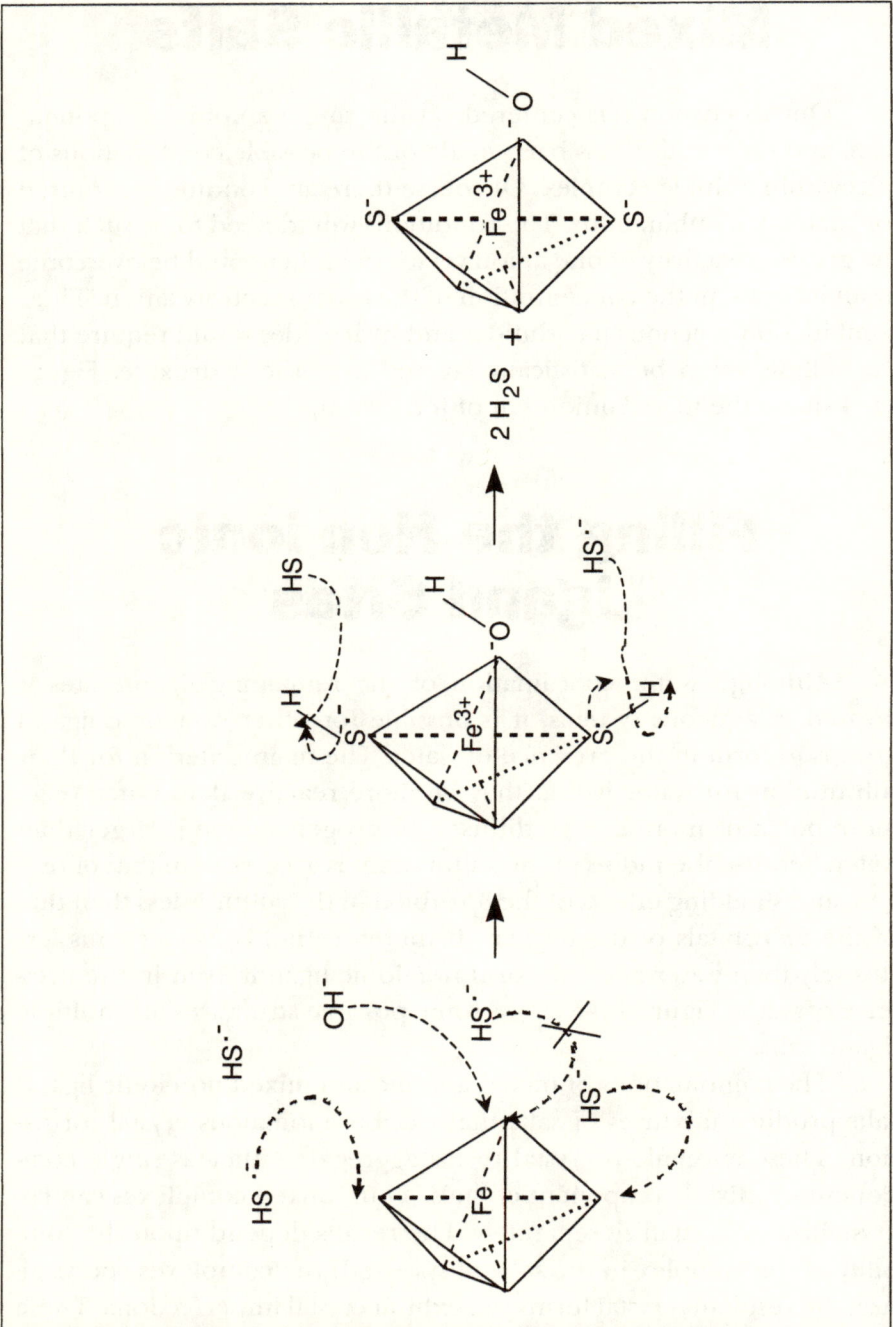

Fig. 11-4 Mixed anion metallic salt formation in mixed systems

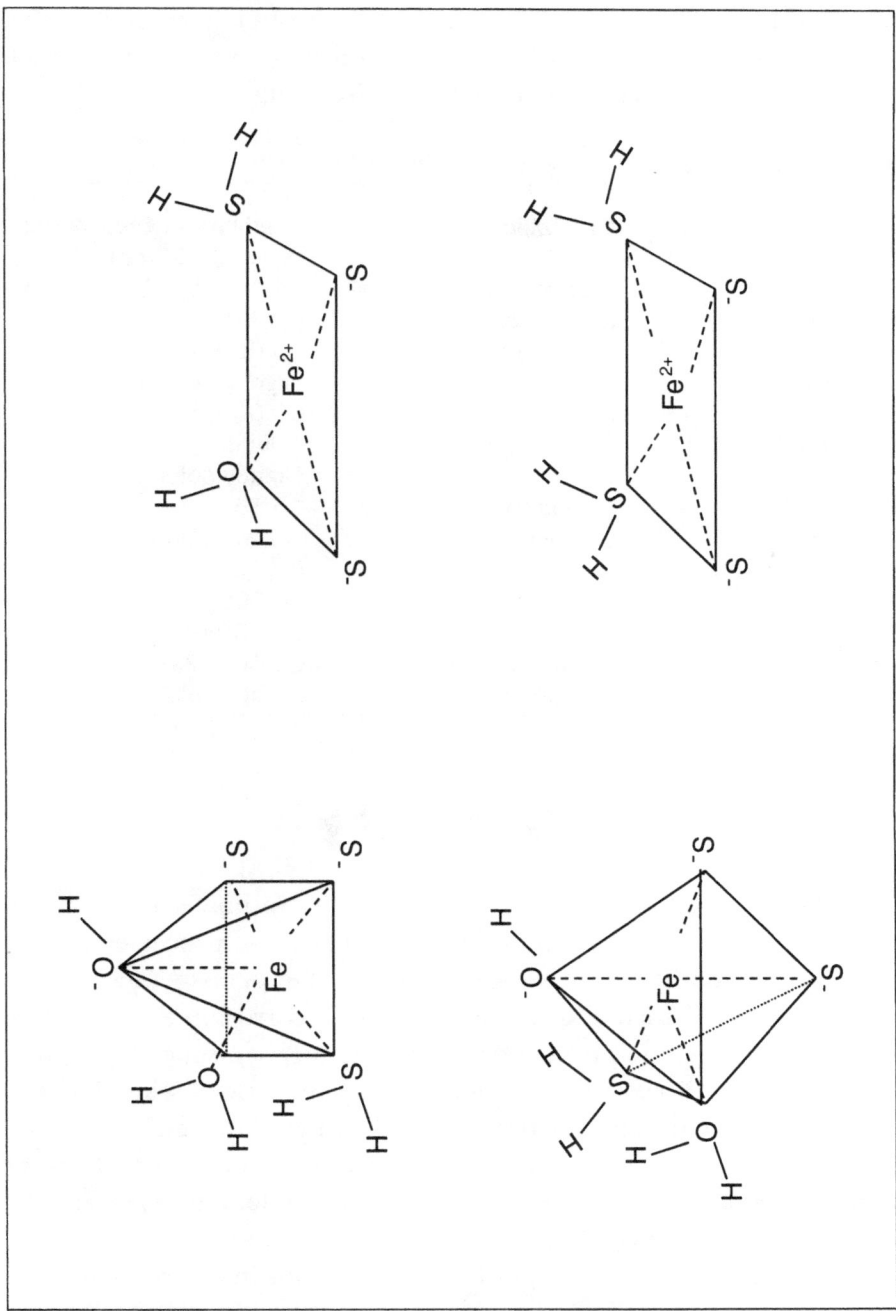

Fig. 11-5 Multiple salts and multiple nonionic ligand complexes of ferric and ferrous salts

aries, or they will take on a different crystal morphology than the pure complex crystals. Table 11–1 is a compilation of some of the crystal types and solubilities exhibited by certain iron salts.

Table 11–1 Crystalline Form and Solubility of Iron Salts

Chemical Formula	Crystalline Form	Solubility in Cold Water (grams/100 cc)
$Fe(Cl)_3$	black-brown hexagonal	74.4
$Fe(Cl)_2 \bullet 2H_2O$	green monoclinic	not listed
$FeCO_3$	gray trigonal	00.00067
$FeCO_3 \bullet H_2O$	amorphous	slightly soluble
$Fe(OH)_2$	pale green hexagonal	00.00015
$Fe_2O_2 \bullet 4SO_3 \bullet 9H_2O$	white to pink powder	soluble
$Fe(SO_4) \bullet H_2O$	white monoclinic	slightly soluble
$Fe(SO_4) \bullet 7H_2O$	blue green monoclinic	15.65
$Fe_2(SO_4)_3$	yellow rhombic	slightly soluble
$Fe_2(SO_4)_3 \bullet 9H_2O$	rhombic	not listed
FeS_2 (pyrite)	yellow cubic	00.00049
FeS_2 (marcasite)	yellow rhombic	00.00049
FeS	black-brown hexagonal	00.00062
Fe_2S_4	yellow-green	slightly soluble

Summary

The preceding chapter discussed the intimate connection between scale and corrosion and suggested that the processes of one do not proceed without the processes of the other. A tie-in between the types of scales formed and the mechanism of corrosion suggests that solubilities of the end-product scale feed the mechanism of corrosion. Passivation was discussed in connection with the type of scale deposited in-situ within the metal lattice. The solubility of the deposited scale in the external electrolyte was implicated as a prime factor in determining whether the surface continued to corrode, and at what rate corrosion would proceed.

The importance of free radical mechanisms in the reduction of (and ligand group occupation of) metallic salts was discussed. The geometric conversion of complex salts, and the role of anions and free radicals in these geometric conversions was surveyed. Finally, multiple anion and ligand group complexes were discussed, and a brief look at some of the iron salt crystal forms was given in Table 11–1.

References

Barrow, Gordon M. 1966. *Physical Chemistry.* 2d ed. New York, St. Louis, San Francisco, Toronto, London, and Sydney: McGraw Hill Book Company.

Dickerson, R. F., H. B. Gray, and G. P. Haight, Jr. 1970. *Chemical Principles.* 1st ed. New York: W. A. Benjamin, Inc.

Huheey, James E. 1978. *Inorganic Chemistry Principles of Structure and Reactivity.* 2d ed. New York, Hagerstown, San Francisco, and London: Harper & Row.

Jones, Loyd W. 1988. *Corrosion and Water Technology for Petroleum Producers.* 1st ed. Tulsa: OGCI Publications, Oil & Gas Consultants International, Inc.

Handbook of Chemistry and Physics. 1976. 56th ed. Cleveland: CRC Press.

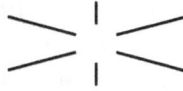

12
Electrochemical Behavior of Petroleum Fluids

Applied Electromotive Forces

All the discussions to this point have dealt with a system that provides all the oxidation and reduction potential required for corrosion and scale processes within the systems. No attempt has been made to discuss the effects of an externally applied electric potential. This chapter will examine those effects and attempt to analyze them in terms of the models previously discussed. It is our intention to rationalize the behavior of the systems in terms that relate to present-day techniques used to analyze corrosion and scale processes.

We will begin by designing an electrolytic cell that uses the mild steel metal (iron) as the anode and an inert platinum wire as the cathode. The platinum wire is considered inert because it serves only to transfer electrons to or from solution without contributing to the electrolyte. It does not accept any metal ions from the electrolyte. A voltage source (battery, or seat of EMF), switch, and a galvanometer are placed between the anode and the cathode. Prior to immersing the electrodes in pure water (neutral pH and 20 mega-ohm resistance), a vacuum is pulled on the entire system. The meter is set to zero.

After the electrodes are immersed in the water, periodic readings from the galvanometer are recorded. The system is maintained at room temperature (25°C). If corrosion takes place at the anode (mild steel), the galvanometer readings should show an increase in potential as electrons leave the anode and proceed to the platinum cathode. Figure 12–1 shows the arrangement of this device.

A potential change should be evident as soon as the electrodes are immersed in the water, since the water exists in stages of equilibrium between the hydronium and hydroxyl ions. Although platinum has a greater ability to adsorb hydrogen from the water than the mild steel, the increased surface area of the mild steel electrode compensates. Thus, a potential difference should still develop across the electrodes if the mild steel undergoes corrosion.

Given that hydroxide of the hydronium/hydroxide pair attacks the iron to produce the ferrous hydroxide salt, a potential should develop between the anion (mild steel) and the cathode. When this occurs, the anode becomes negatively charged and transfers its excess charge through the galvanometer to the cathode (platinum wire). The excess charge on the cathode then is neutralized by the abstraction of a proton from the hydronium ion. Figure 12–2 illustrates this sequence.

Figure 12–2 indicates that as the hydroxide attacks the anode, it forms the iron hydroxide salt, and the electrons flow from the anode to the cathode. Once at the cathode, the electrons abstract a proton from the hydronium ion, producing water. The cathode is neutral until another hydroxide pair attacks the anode and starts the cycle anew. The cathode also converts the protons at its surface to monatomic hydrogen by supplying an electron. The monatomic hydrogen is free to migrate through the electrolyte, since its valence shell is filled, but remains on the cathode's surface.

The continued supply of electrons repels the monatomic hydrogen by exciting the valence electrons to a higher energy level. These conditions favor bonding. The excited monatomic hydrogens then form a bond between one another and are converted to molecular

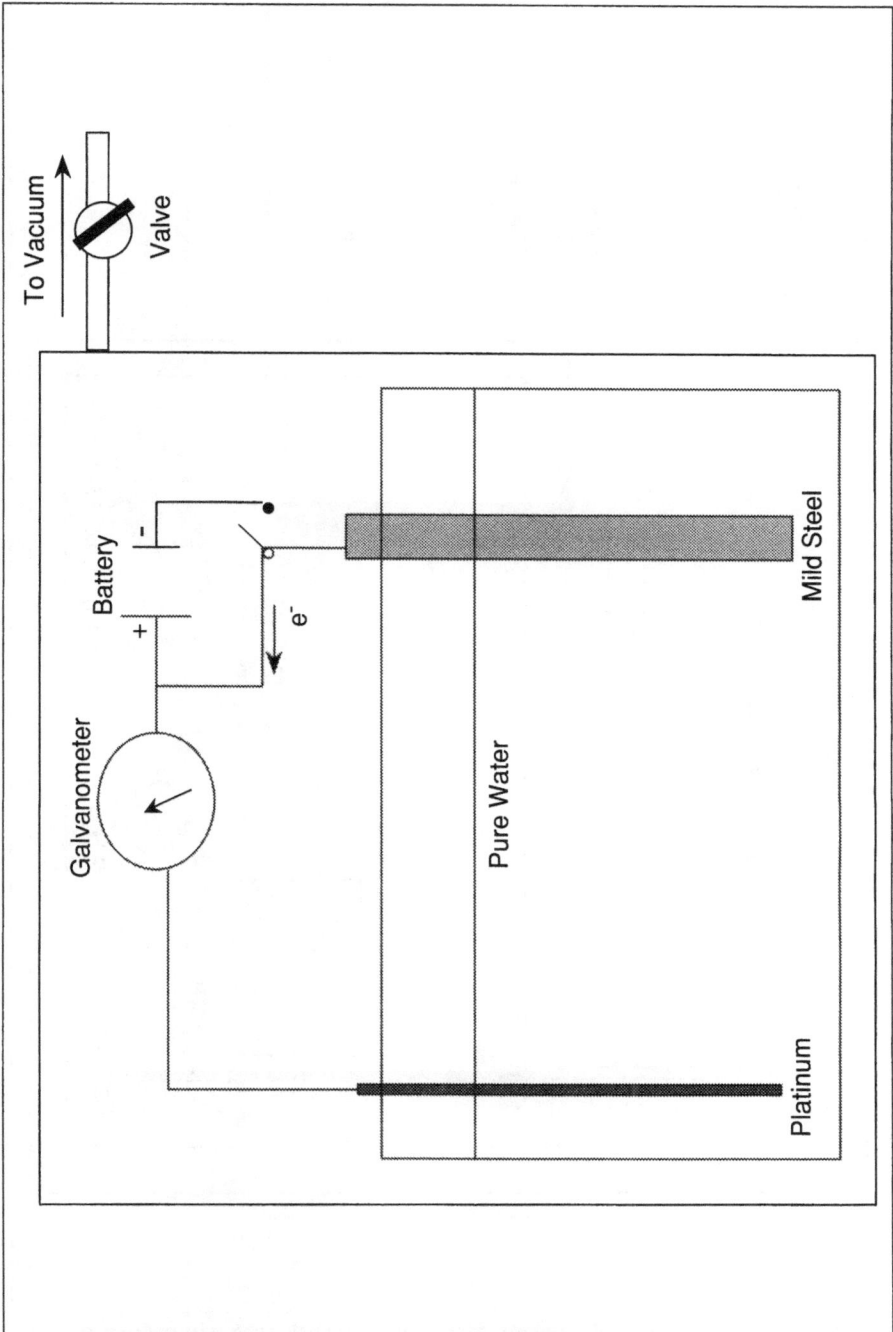

Fig. 12-1 Device for measuring corrosion potential

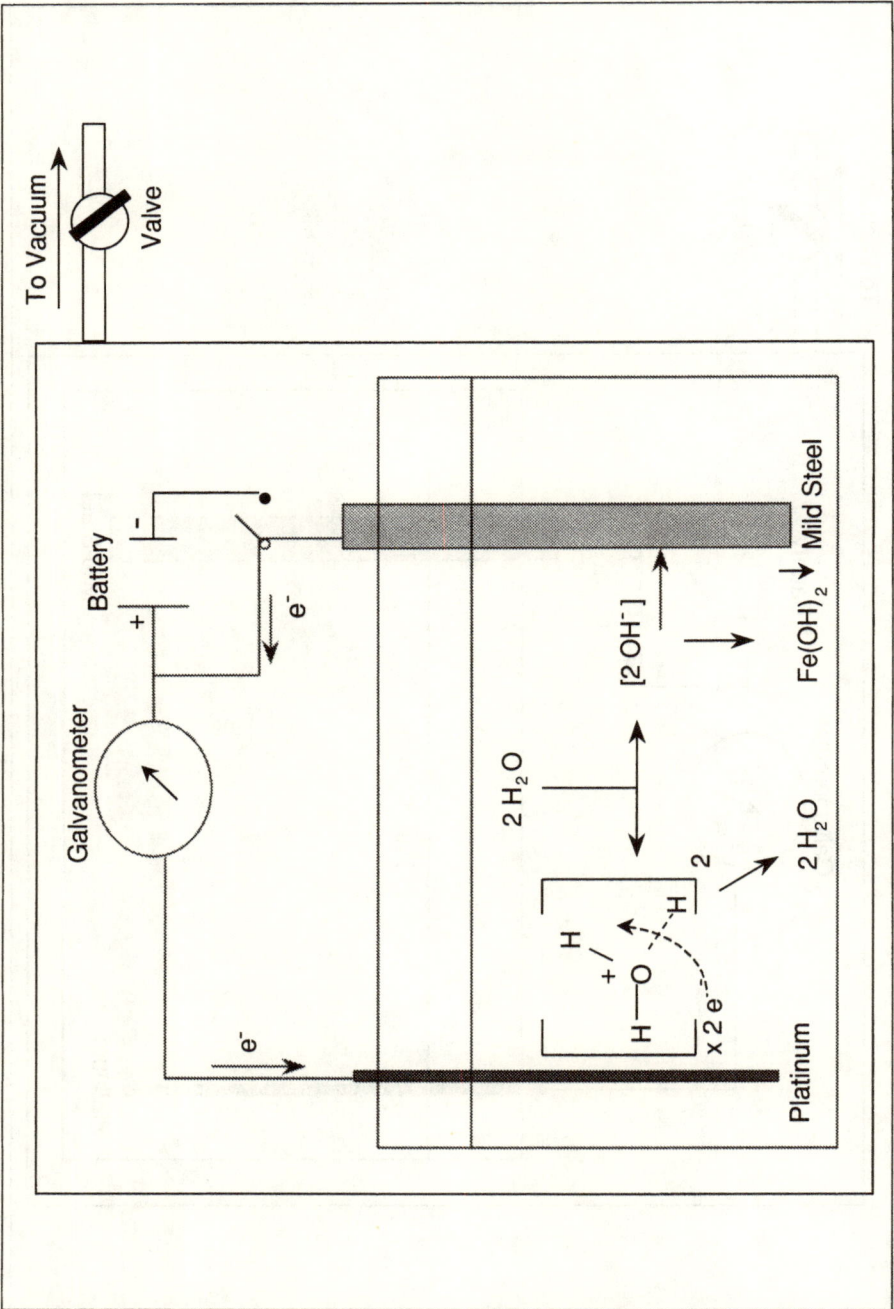

Fig. 12-2 The operation of the corrosion potential device while corrosion of the mild steel proceeds

hydrogen. They then leave the charged platinum wire (cathode). Figure 12–3 shows the migration of H_2 to the anode.

As the molecular hydrogen migrates to the vicinity of the anode, the continued electron removal from the mild steel (de-electronation) results in a diffuse partial charge couple to the molecular hydrogen. When the molecular hydrogen is within the van der Waals radii for the mild steel anode, London dispersion forces result between it and the anode, and surface adsorption occurs. This process continues, and one molecular hydrogen is produced for every molecule of metallic hydroxide or every pair of electrons provided to the cathode.

However, this process cannot continue indefinitely, because the molecular hydrogen at the anode's surface has nowhere to go. It continues to build until the entire surface area of the anode is covered. Unless the molecular hydrogen at the anode can be caused to disproportionate, no further corrosion current will arise (e.g., no electron release from the anode occurs).

Hydroxyl Ions and Hydrogen Disproportionation

As the molecular hydrogen builds at the anode, the corrosion current drops. This drop in current is due to the inaccessibility of the metal surface for continued hydroxyl attack. However, if the hydroxyl can attack the molecular hydrogen and cause its disproportionation by producing water and monatomic hydrogen, it can again access the metal surface. However, in the process of proton abstraction from the adsorbed molecular hydrogen, an intermediary hydride anion must be produced. The hydride ion then must donate its excess electron to the metal, which is conducted to the platinum cathode, producing more molecular hydrogen.

Another path is open to the electron donated by the hydride anion, and this path is the radicalization of hydrogen from water. The hydride-donated electron is conducted through the metal to a point where water is oriented within a distance of the van der Waals radii. The electron tunnels to the hydrogen oxygen bond and combines with

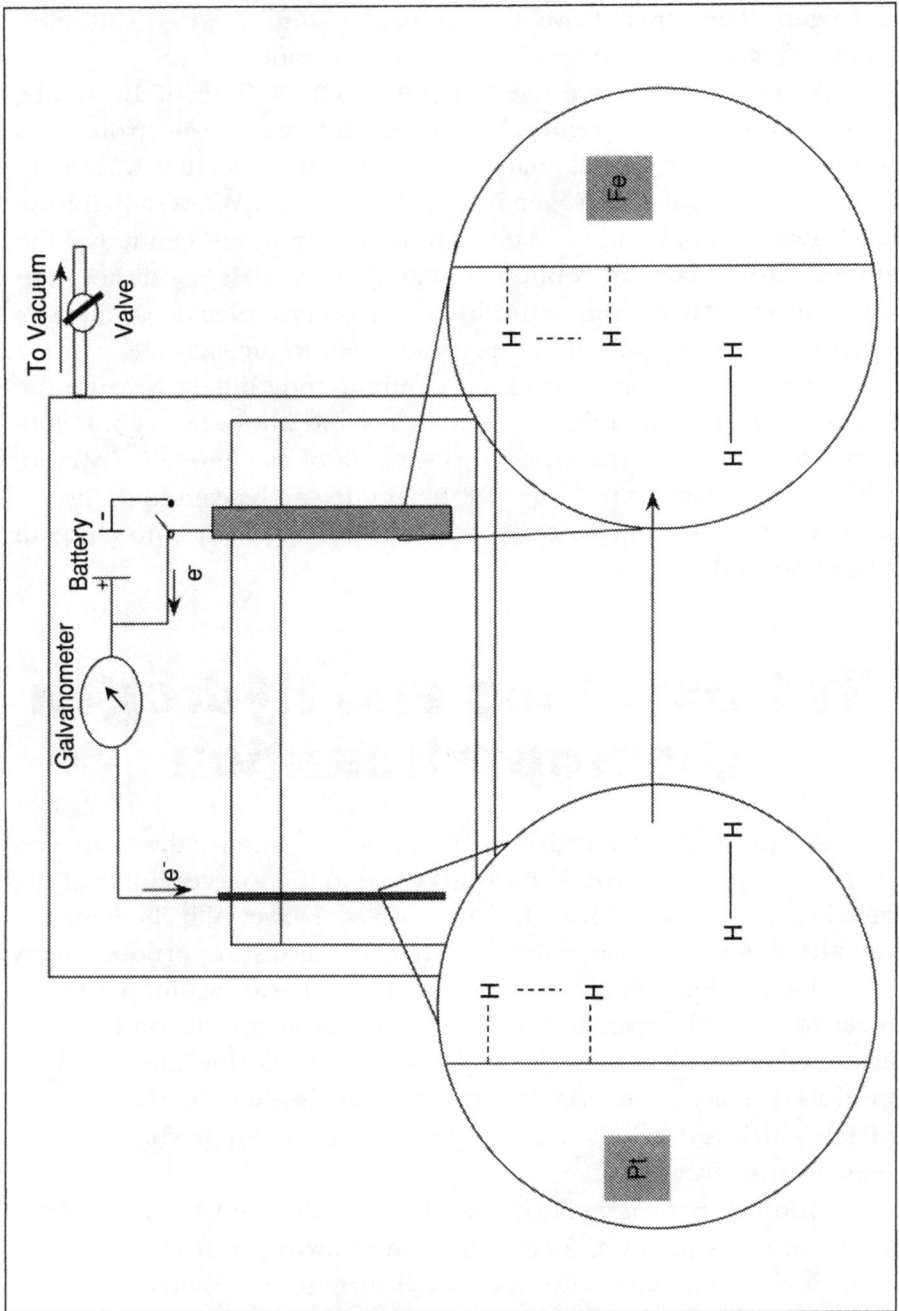

Fig. 12-3 The de-adsorption of molecular hydrogen from the platinum electrode and the adsorption of molecular hydrogen at the mild steel anion

the oxygen's electrons. This produces a monatomic hydrogen and a hydroxyl anion, and the hydroxyl attacks the metal. Figure 12–4 is a graphical depiction of how current effects change given the conditions described thus far. As the hydroxyl is again able to attack the metal, the current begins to flow, but at a slower rate. This current is then maintained as long as corrosion continues and can be thought of as a steady state.

Adding an External Current

Applying an external current to the system that is in opposition to the current flow of the corroding metal (mild steel) impedes the flow of electrons to the cathode. The extent to which this impedance resists the applied current can be measured. If the applied current is set such that the galvanometer reading is zero, it is reasonable to assume that the flow of electrons from the anode to the cathode (corrosion cell) has ceased. Thus, the measured applied current should be equal to the corrosion current. However, it is important to apply the current at the appropriate point in the corrosion current versus time curve.

If the applied current is started during the initial stage of corrosion, the measured value will be high compared to one measured at the steady state condition. Furthermore, the excess applied current will reverse the function of the mild steel to a cathode and that of the platinum wire to a noncorroding anode. Thus, the adsorbed molecular hydrogen will be released from the mild steel electrode, and corrosion versus applied current will achieve balance. This procedure will give abnormally high values for the corrosion rate of mild steel.

The New Anode

If an external current is applied too early in the corrosion process, the molecular hydrogen adsorbed to the mild steel is driven off before it can impede corrosion by electrode coverage. At the other electrode (platinum wire) the withdrawal of electrons from the aque-

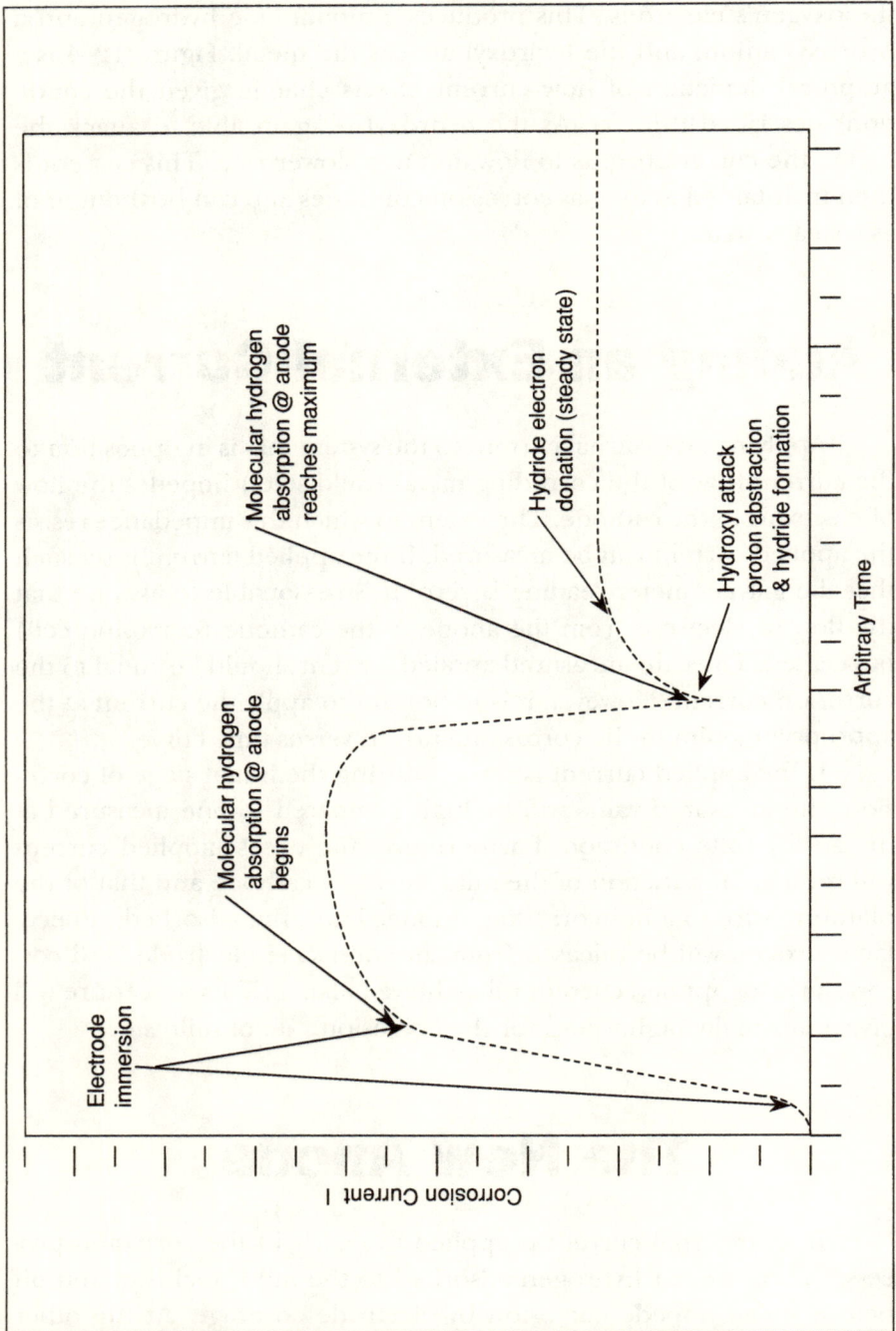

Molecular hydrogen absorption @ anode reaches maximum

Hydride electron donation (steady state)

Hydroxyl attack proton abstraction & hydride formation

Molecular hydrogen absorption @ anode begins

Electrode immersion

Arbitrary Time

Corrosion Current I

Fig. 12-4 The corrosion current derived from the mild steel anode

ous electrolyte occurs. Figure 12–5 illustrates a mechanism for electron withdrawal from the electrolyte by the application of an excess externally applied current.

The reversed roles of the electrodes cause the corrosion reaction to stop, but the continued supply of current causes de-protonation of hydronium ions. It also results in de-adsorption of molecular hydrogen from the mild steel electrode. Monatomic hydrogen is adsorbed by the mild steel electrode and subsequently released as molecular hydrogen as current application is continued. Thus, the mild steel electrode serves as a source for molecular hydrogen to the platinum electrode. The supplied molecular hydrogen is then attacked by the hydroxyl anion, removing a proton to form water and releasing monatomic hydrogen (radical hydrogen). Figure 12–6 shows what happens to the current versus time curve as constant external current is applied to the system.

It should be noted that the current applied is maintained at constant amperage and that the voltage (potential) must be varied as the electrolyte resistance changes in accordance with Ohm's law. Thus, current changes in direct proportion to voltage and inversely with electrolyte resistance. If we hold voltage constant instead, we can examine the change in the ratio of current to electrolyte resistance, or resistance as the ratio of current to voltage. By doing this, a feel for how the resistance changes with electrolyte, anode, and cathode behavior can be obtained. The result of this procedure can be examined in Figure 12–7.

It can be seen that the resistance increases under a constant external voltage application as the mild steel becomes coated with molecular hydrogen. When the molecular coating of the mild steel reaches a maximum, the disassociation of molecular hydrogen begins. The resistance drops to a steady state value reflecting the corrosion potential. The behavior of the resistance curve and the behavior of the electrolyte systems suggest another testing methodology involving the inverse relationship of resistance to conductivity.

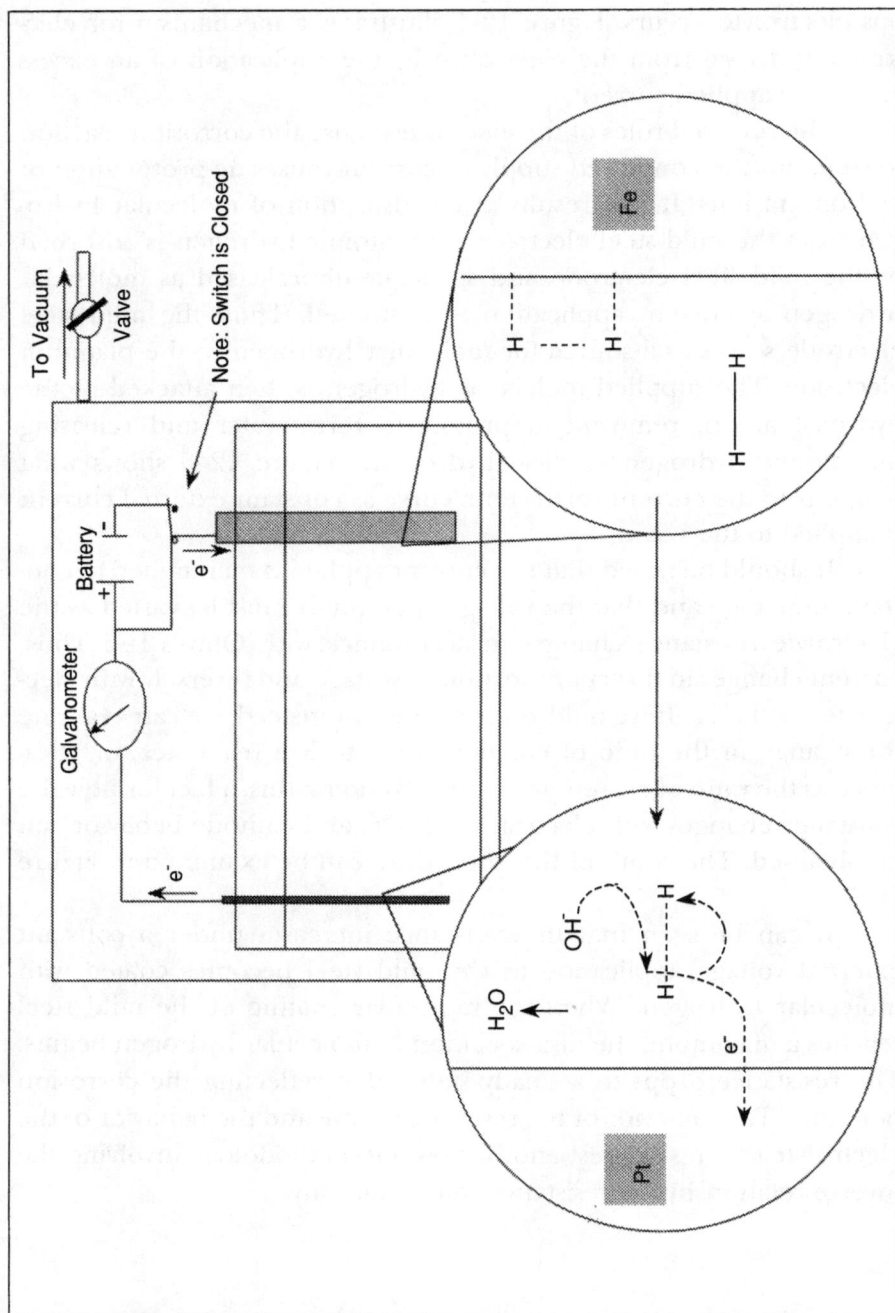

Fig. 12-5 The application of an externally supplied current and the reversed roles of the electrodes

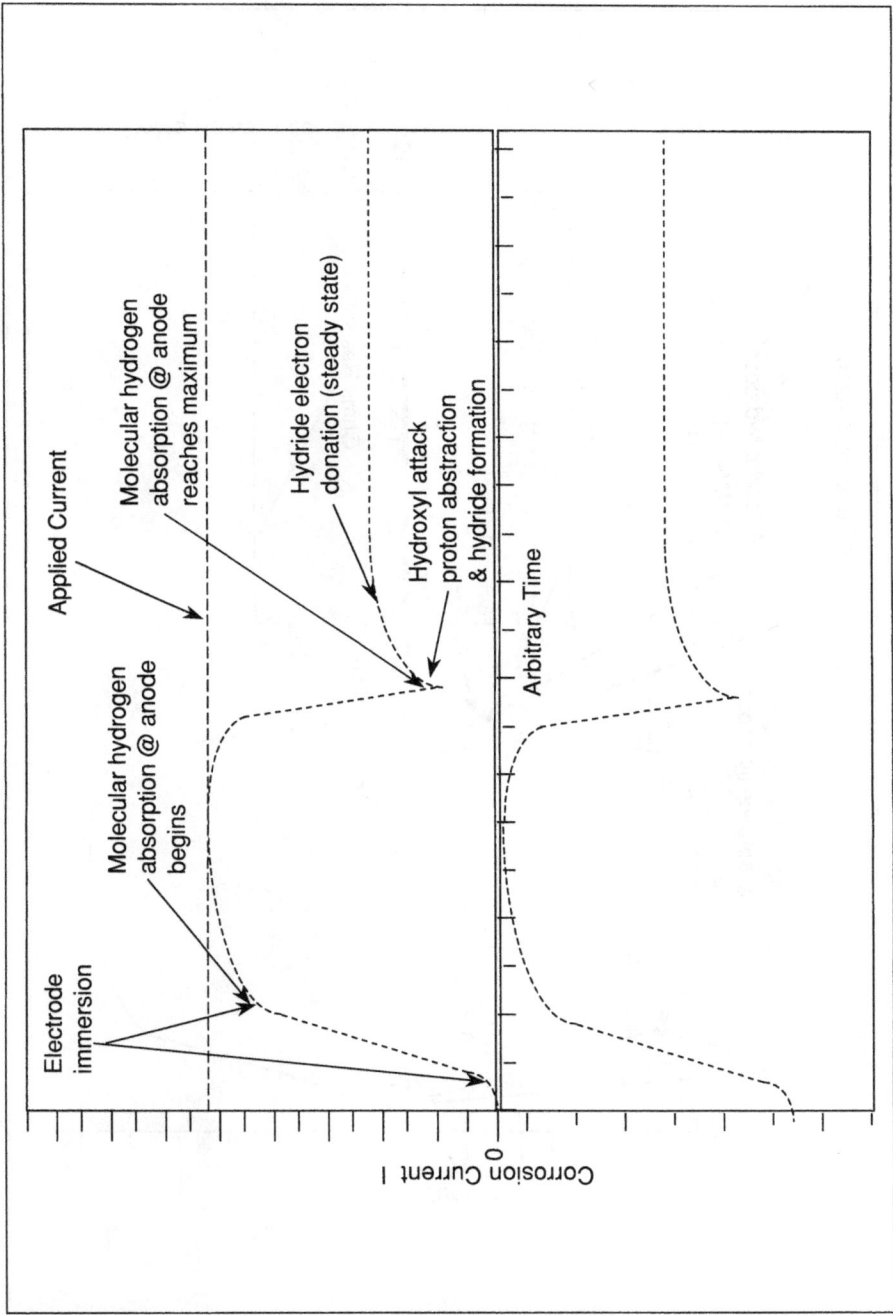

Fig. 12-6 The application of an external constant current to a corrosion cell

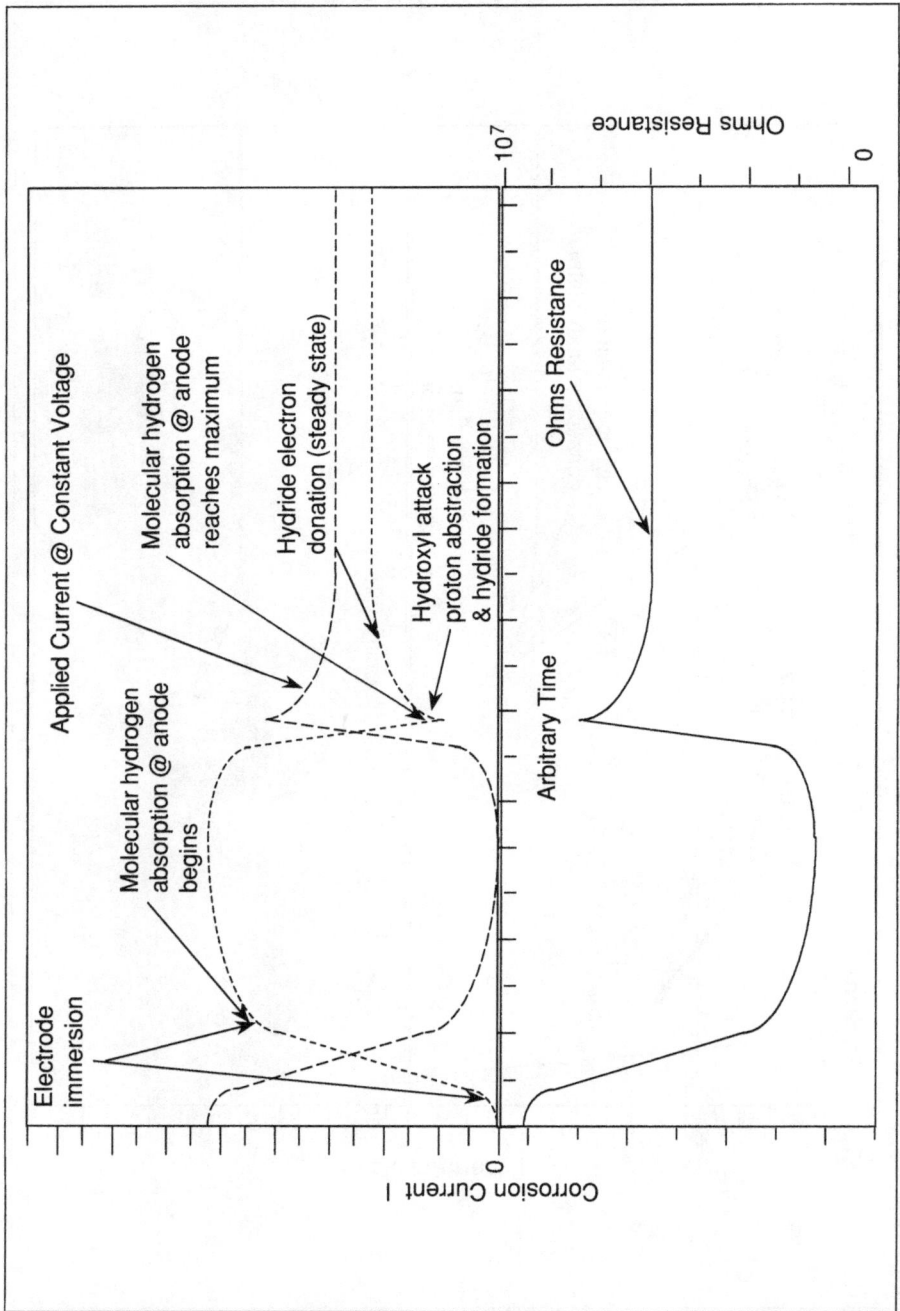

Fig. 12-7 Resistance change as a function of a constant applied voltage to a corroding mild steel electrolytic cell

Combined Conductivity and Current Profiles

In the preceding arrangement, the potential drop between the anode and the cathode of the electrolytic cell is measured as it relates to the behavior of the electrolyte. Figure 12–8 shows a diagram of a conductivity cell. The measurement is indirect with respect to the electrolyte. If electrolytic conductivity and resistance are measured simultaneously, a direct measurement of ionic effects can be obtained. This measurement is accomplished using a conductivity cell and a Wheatstone bridge. The cell is constructed of two parallel platinum electrodes (inert) spaced 1 centimeter (cm) apart, with a surface area of 1 cm^2 each.

A voltage is applied to the bridge resistors, and the resistance is measured, without the degradation of either of the electrodes. Thus, it is just a measure of the electrolyte's ability to conduct an electrical current across the gap separating the two bridge plates. Furthermore, the ability of an electrolyte to conduct current is a measure of the ionic strength and concentration of ions in the electrolyte. Figure 12–9 shows a plot of conductivity versus time of the mild corrosion cell.

As the mild steel corrodes, the ferrous and ferric ion concentrations of the electrolyte increase, and with these ionic increases comes an increase in conductivity. Because of the low solubility of the ferric and ferrous hydroxide salts, the conductivity peaks for this electrolyte at a value of ionic concentration equal to their combined solubility products. The bridge-measured conductivity profile of the electrolyte shows a cumulative effect up to the solubility limit of the metal hydroxide salts. This behavior suggests a methodology for adjusting the applied current to the electrolytic cell. Figure 12–10 illustrates an apparatus for adjusting the applied current using conductivity as a feedback mechanism.

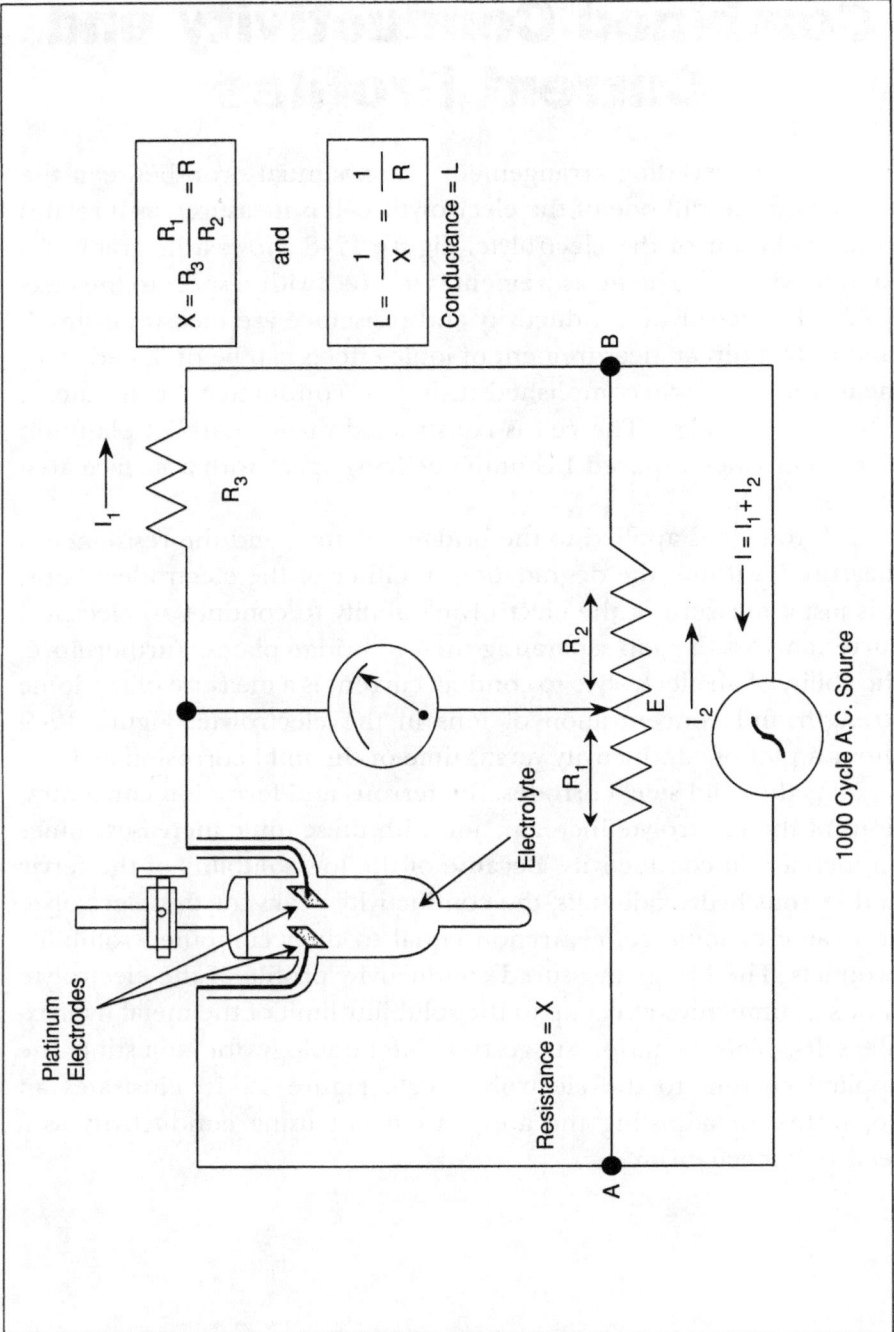

Fig. 12-8 Diagram of a conductivity device

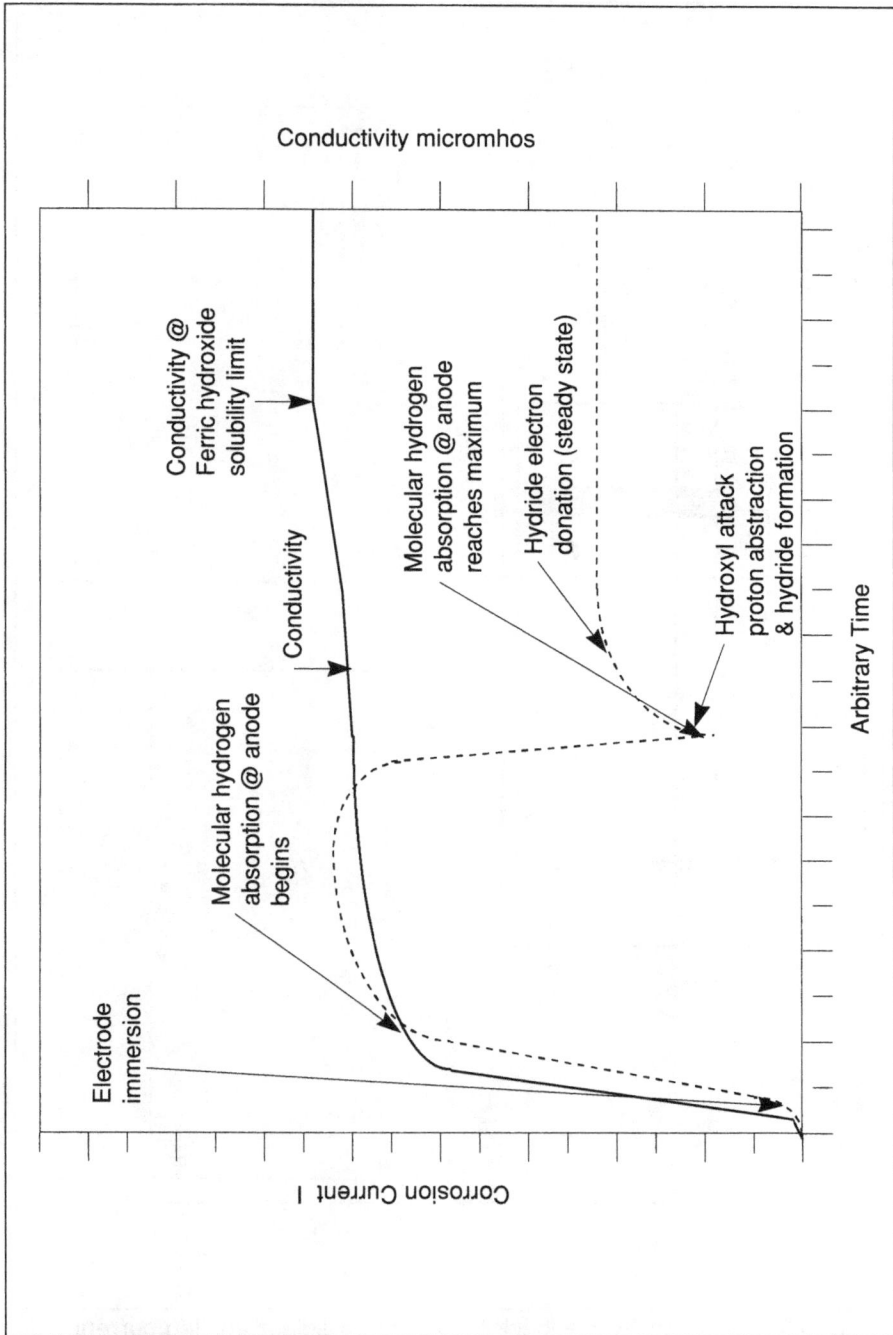

Fig. 12-9 Conductivity vs. time of the mild corrosion cell

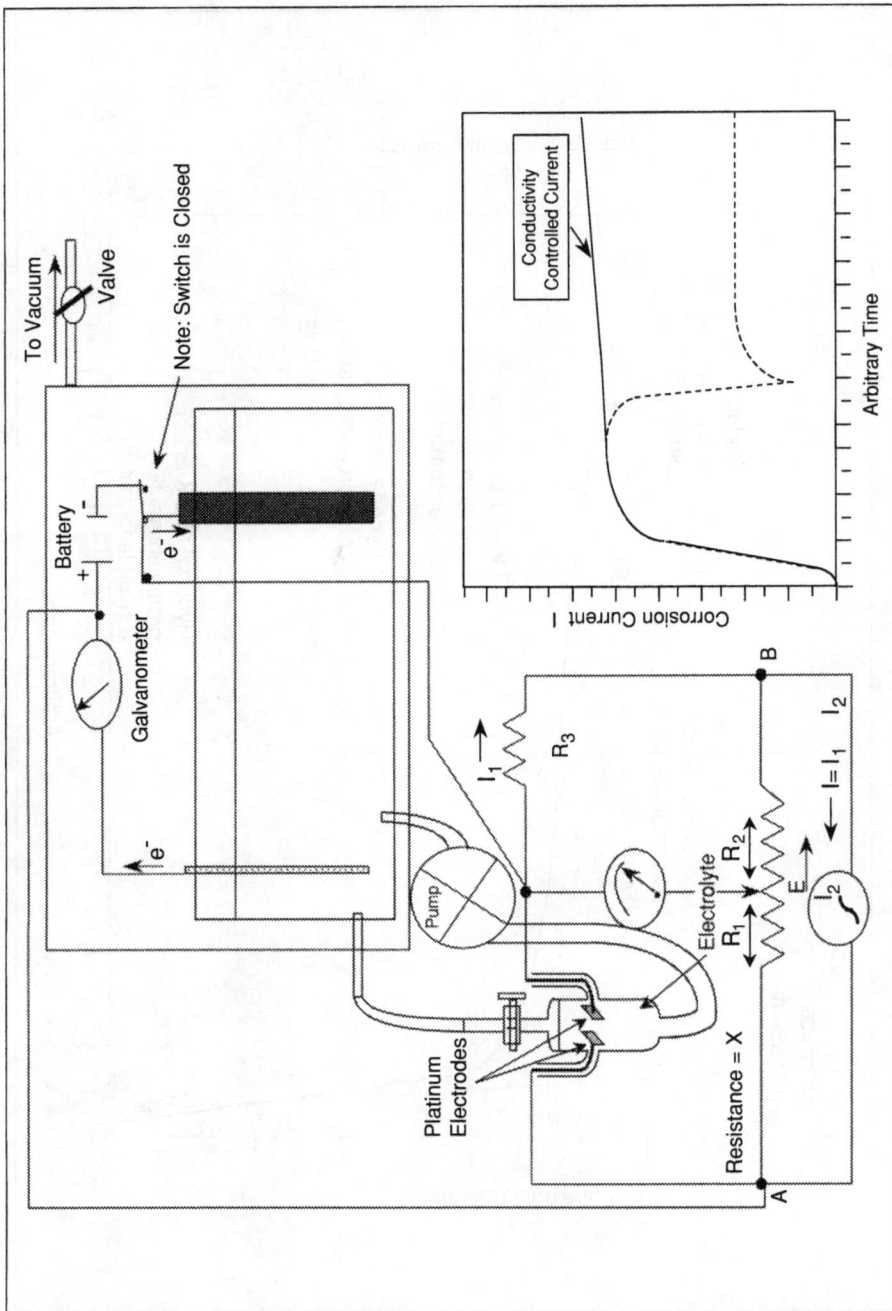

Fig. 12-10 A conductivity feedback loop used to adjust applied current to a corrosion cell

Potential-pH (Pourbaix) Diagrams

Since the ultimate product of metal-dissolution reaction is not always an ionic species (e.g., $M \longrightarrow M^{n+} + ne$), sometimes it is the solid oxide or hydroxide. If we assume that the intermediate product of dissolution of iron is ferrous hydroxide, then the solution can only dissolve ferrous ions up to a limit. This limit is given by applying the law of mass action to the reaction that follows:

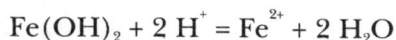

$$Fe(OH)_2 + 2\,H^+ = Fe^{2+} + 2\,H_2O$$

According to the law of mass action, for a constant concentration of $Fe(OH)_2$ in equilibrium with a solid phase:

$$[Fe^{2+}]/[H^+] = K = 10^{+13.29}$$

then

$$\log[Fe^{2+}] = 2\log[H^+] + 13.29 = 13.29 - 2\,pH$$

If the quantity of 10^{-6} is taken as the ferrous ion concentration from a corroding iron system, then:

$$pH = (13.29 + 6)/2 = 9.6$$

This means that ferrous hydroxide is stable at pH greater than 9.6. Consequently, the ferrous hydroxide to ferrous ion ratio depends on pH only—it is a proton transfer and does not depend on the potential due to electron transfer. Therefore, if potential is plotted versus pH, a vertical line parallel to the potential axis would be expected (see Fig. 12–11).

Potential-pH diagrams can be used to indicate whether a particular corrosion process is thermodynamically feasible (the pH that is referred to is that only in the immediate vicinity of the electrodes). Thus, bulk electrolyte pH values can lead to serious errors in deter-

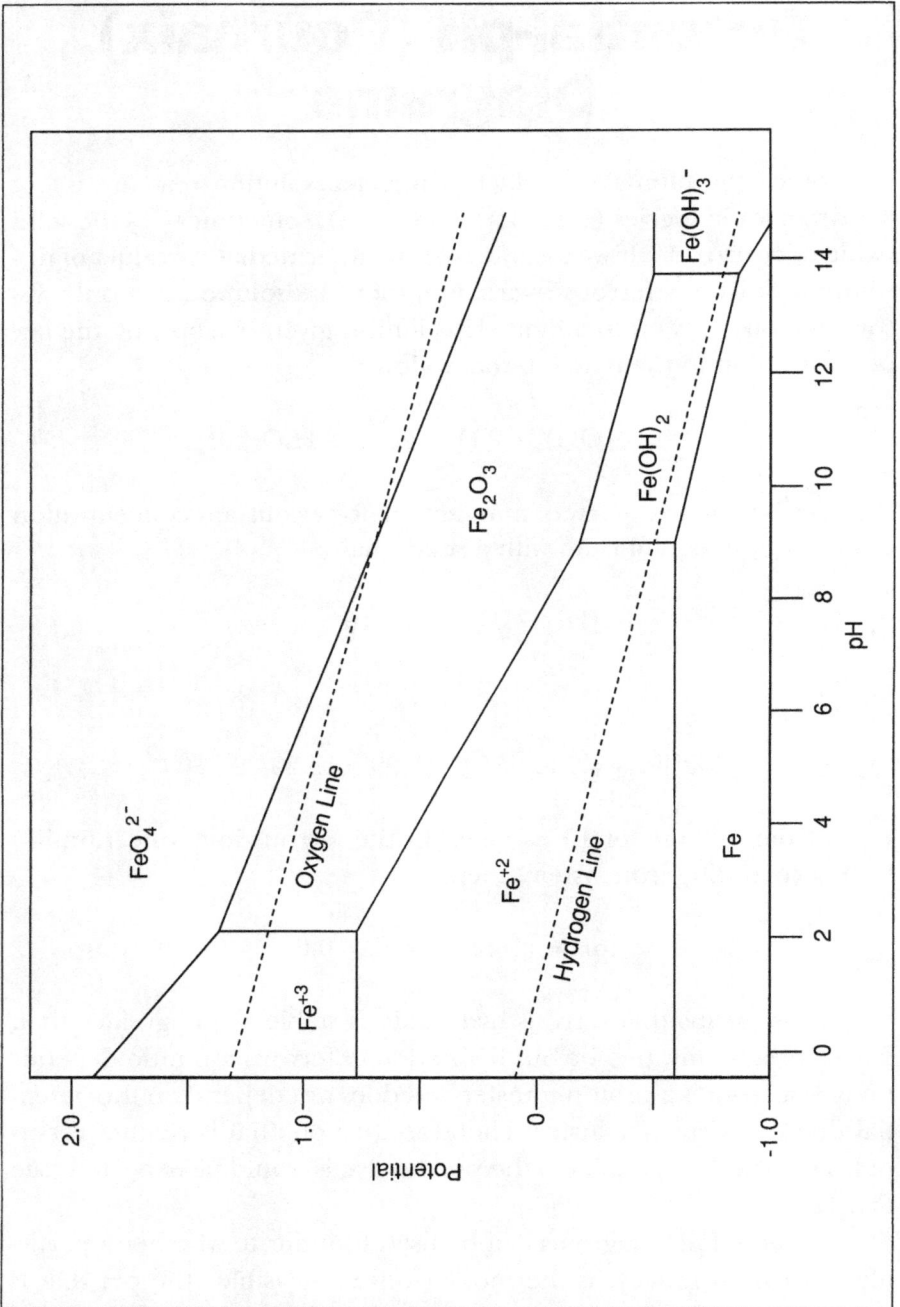

Fig. 12-11 A Pourbaix diagram for the corrosion of iron at different pH values

mining the thermodynamic feasibility of a reaction. Pourbaix diagrams provide a succinct visualization of proton, electron, and proton-electron transfer reactions that are thermodynamically favored when a metal is immersed in given solution. Figure 12–11 is a Pourbaix diagram for the potential versus pH of iron corrosion.

Thus, Pourbaix diagrams are helpful in determining the thermodynamic feasibility of a corrosion process, and determine the overall direction a corrosion reaction will tend. However, the rates of corrosion can only be found by a study of the electrodics of corrosion.

Bridge-Controlled Current: Cell Resistance and Bridge Conductivity

The conductivity bridge feedback loop devised previously for external current application can be used to compare bridge-measured conductivity to corrosion circuit resistance. Figure 12–12 shows the curves for these measurements.

It is interesting to note that the bridge conductivity [σ_b = 1/(Resistance x M), neglecting magnitude] could be viewed as the inverse integration function of the corrosion cell resistance over a period of time. Neglecting units, the bridge conductivity can also be viewed as an integration of the corrosion cell current over time. Thus, if the bridge conductivity is differentiated with respect to time, curves qualitatively resembling the corrosion current or the inverse of resistance result. Notice that the potential or voltage is a measure of energy/charge, and that resistance is a measure of (energy x time)/(charge^2).

If you divide voltage by resistance you get amperage = charge/time (Ohm's law). Likewise, if you multiply conductivity/meter by voltage, you get amperage. Thus, if you differentiate the inverse of conductivity/meter with respect to time, you should obtain the voltage curve as follows:

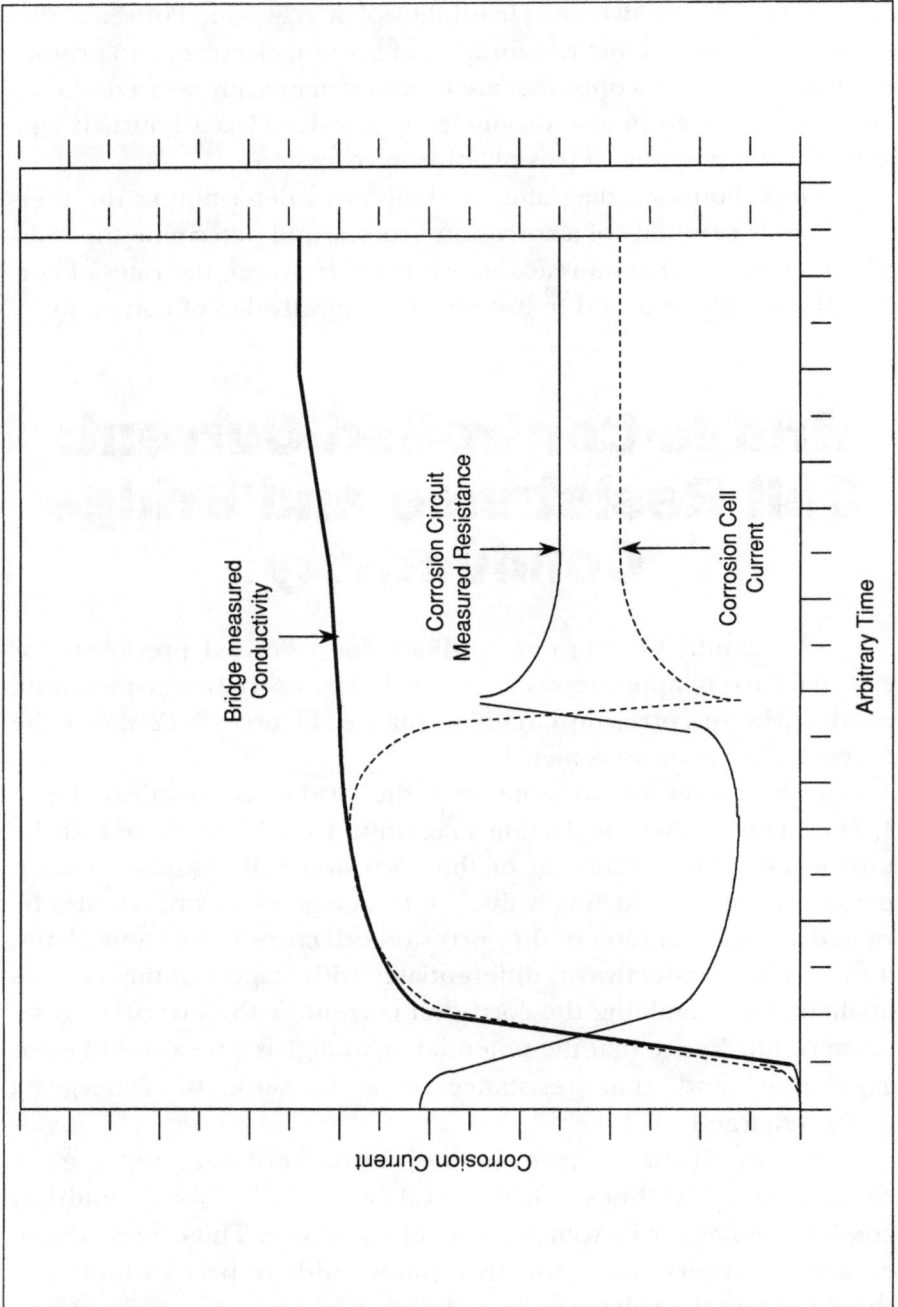

Fig. 12-12 Corrosion cell resistance versus conductivity bridge measurement

Voltage = $k$cd (meter/conductivity)/dt = $k$cd (Bridge Resistance)/dt

where

kc = constant of proportionality (1/coulombs).

Since we cannot measure corrosion resistance without the application of an external current, we must use the conductivity measurement for the corrosion potential or voltage determination. If the current is being applied in response to the conductivity measurement, then we can measure the resistance of the corrosion cell to the applied current. Knowing the corrosion cell and bridge resistances, the voltage of each can be determined. From these voltages, a plot of cell current versus corrosion potential or voltage can be constructed.

Although the applied voltage of the conductivity bridge is proportional to the applied cell potential, it may differ in magnitude. Consequently, the conductivity bridge potential and current is plotted on a separate axis from the applied corrosion cell potential and amperage. If we rearrange the expression above, and integrate the expression to obtain time, we can see how conductivity and voltage relate to corrosion rate as follows:

$$\text{time} = \int dt = k_c \text{ (meter/conductivity)/voltage} = k_c \text{ (Bridge Resistance)/voltage}$$

Since the bridge potential is proportional to the cell potential, the curve shape of cell voltage versus the log of cell amperage should be identical to bridge voltage versus the log of bridge amperage. Figure 12–13 shows a combined set of cell potential versus the log of cell amperage curves and bridge potential versus the log of bridge amperage curves. Thus, it is possible to determine the corrosion potential from the bridge potential versus the log plot of the bridge current, if the ratios of the currents are known.

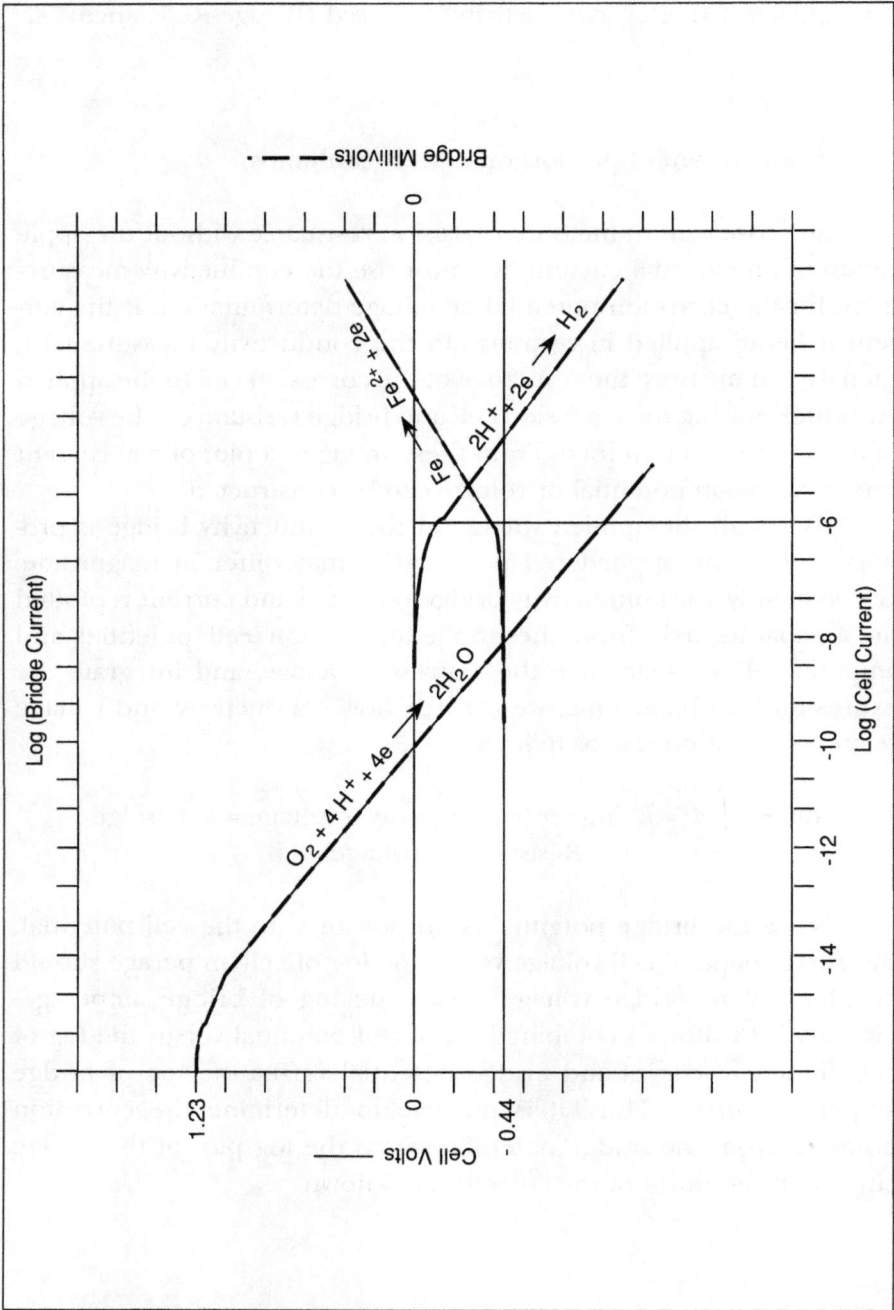

Fig. 12-13 Modified Evans diagram

Evans Diagrams

Many of the phenomena associated with corrosion rate can be understood by the superposition of current-potential curves for the metal-dissolution and electronation reactions. In the reaction $M^{+n} + ne = M$, one can construct a curve for the change in potential of the M electrode with the de-electronation current that crosses the electrode-electrolyte interface. Figure 12–14 shows how this curve is constructed.

Figure 12–15 shows the potential versus current relation for the reactions that occur at the corroding interface. The corrosion current and potential are defined at the point where the two currents are equal: $I_M = I_{sol}$. Figure 12–16 is the Evans diagram, or the plots of the potentials of the two reactions versus the log of their current. The intersection of the curves defines the corrosion current and the corrosion potential. Figure 12–17 shows how the electronation reaction curves change as the exchange-current density is changed.

Corroding metal is equivalent to a short-circuited energy-producing cell (battery) that has some specific characteristics. If the electron sink and electron source are chosen as equivalent to corresponding areas on corroding metal, the metal dissolution current I_M and the electronation current I_{sol} are equal in magnitude. However, they are opposite in sign, just as in energy-producing cells. Thus, the rate of corrosion of the metal is given by the metal dissolution rate:

$$I_M = I_{sol}$$

and

$$I_{corr} = I_M = -I_{sol}$$

The forms that the Evans diagrams take depend upon the reactions involving metal dissolution and electronation. Several situations arise that result in different Evans diagrams. Metal dissolution reactions are much greater than electronation reactions where $I_{o,M} \gg I_{o,sol}$, or vice versa: $I_{o,sol} \gg I_{o,M}$ (see Fig. 12–17). They can also bring out:

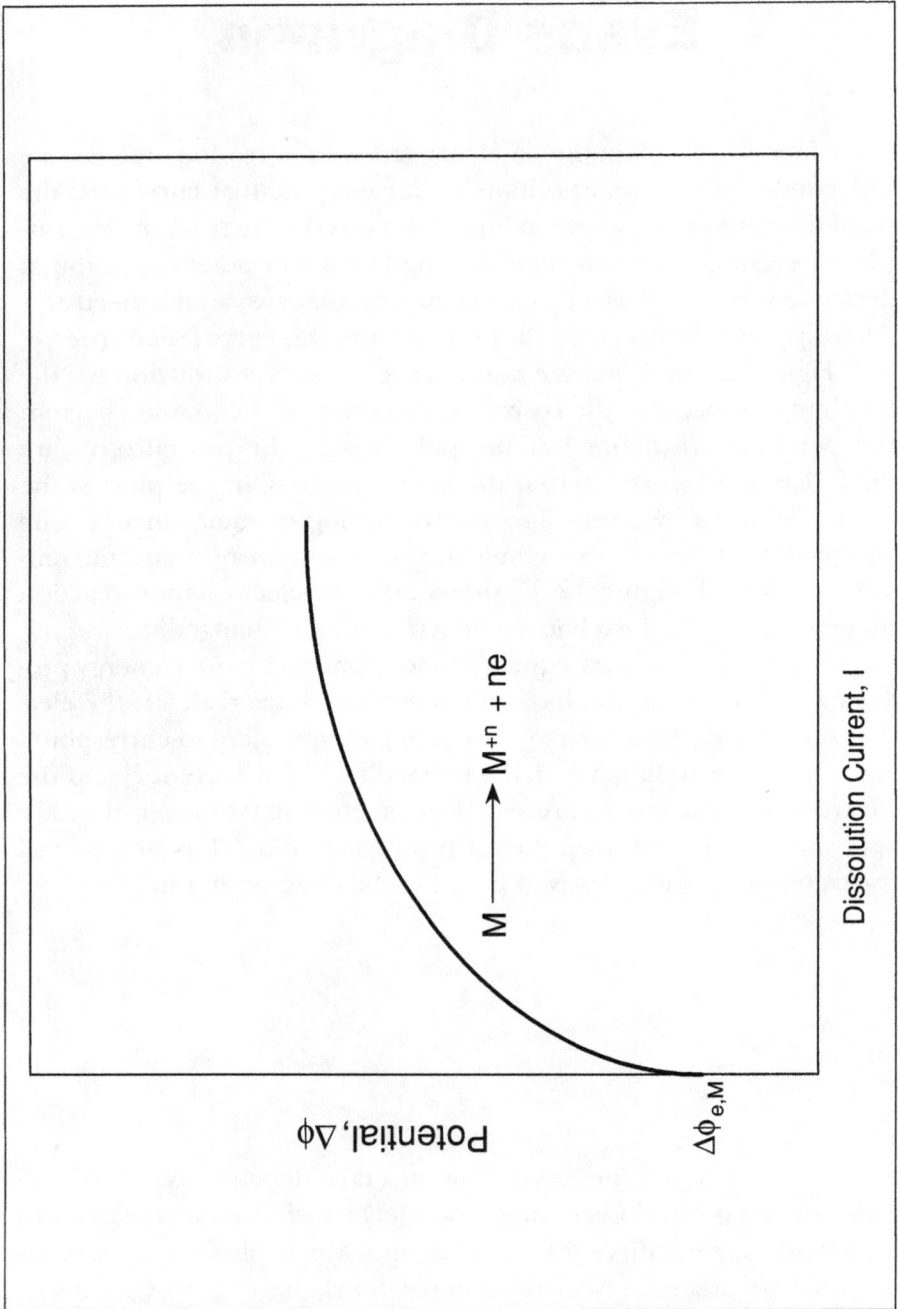

Fig. 12-14 Factors discernible from corrosion potential curves versus current or log of current plots involved in corrosion processes

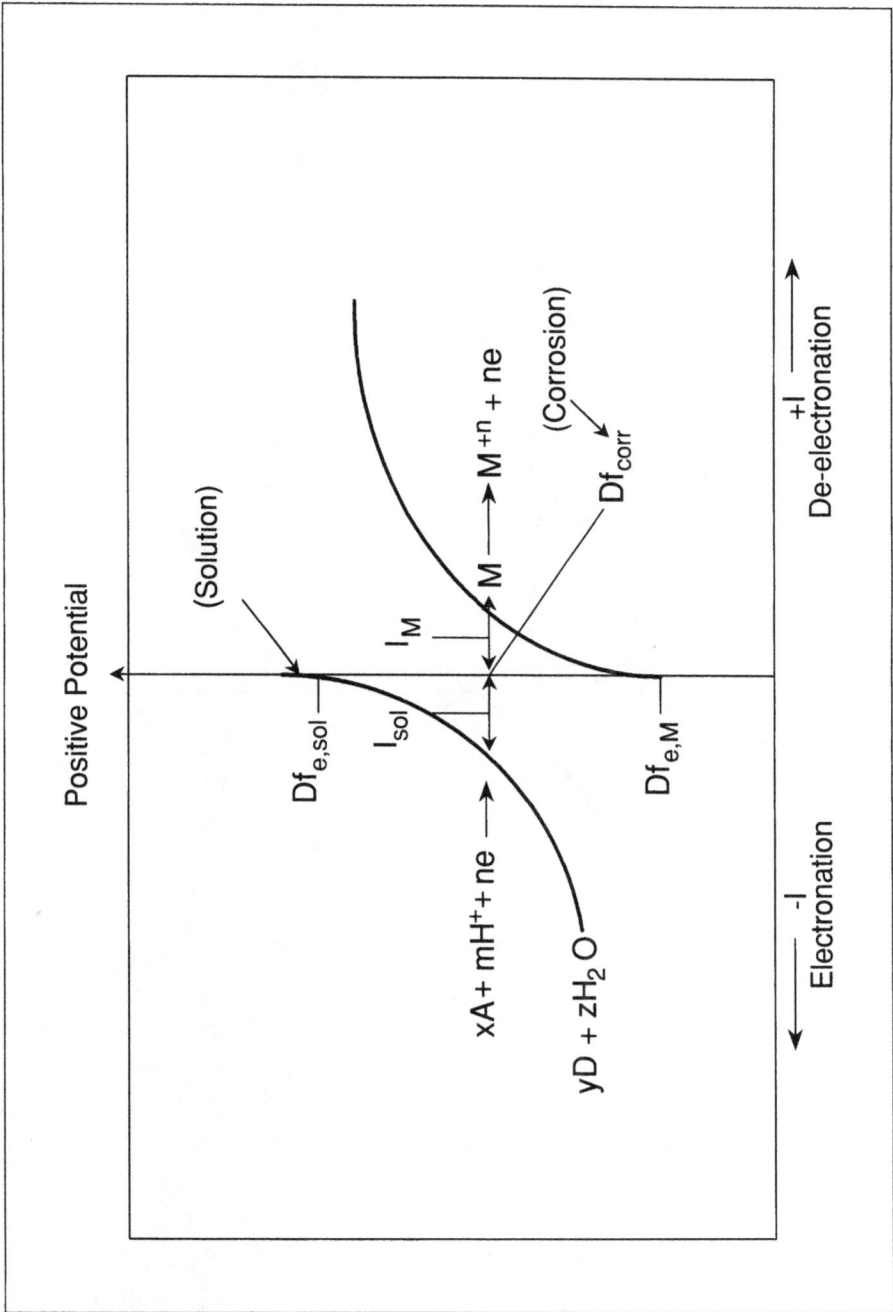

Fig. 12-15 Potential vs. current relation for reactions occurring at corroding interface

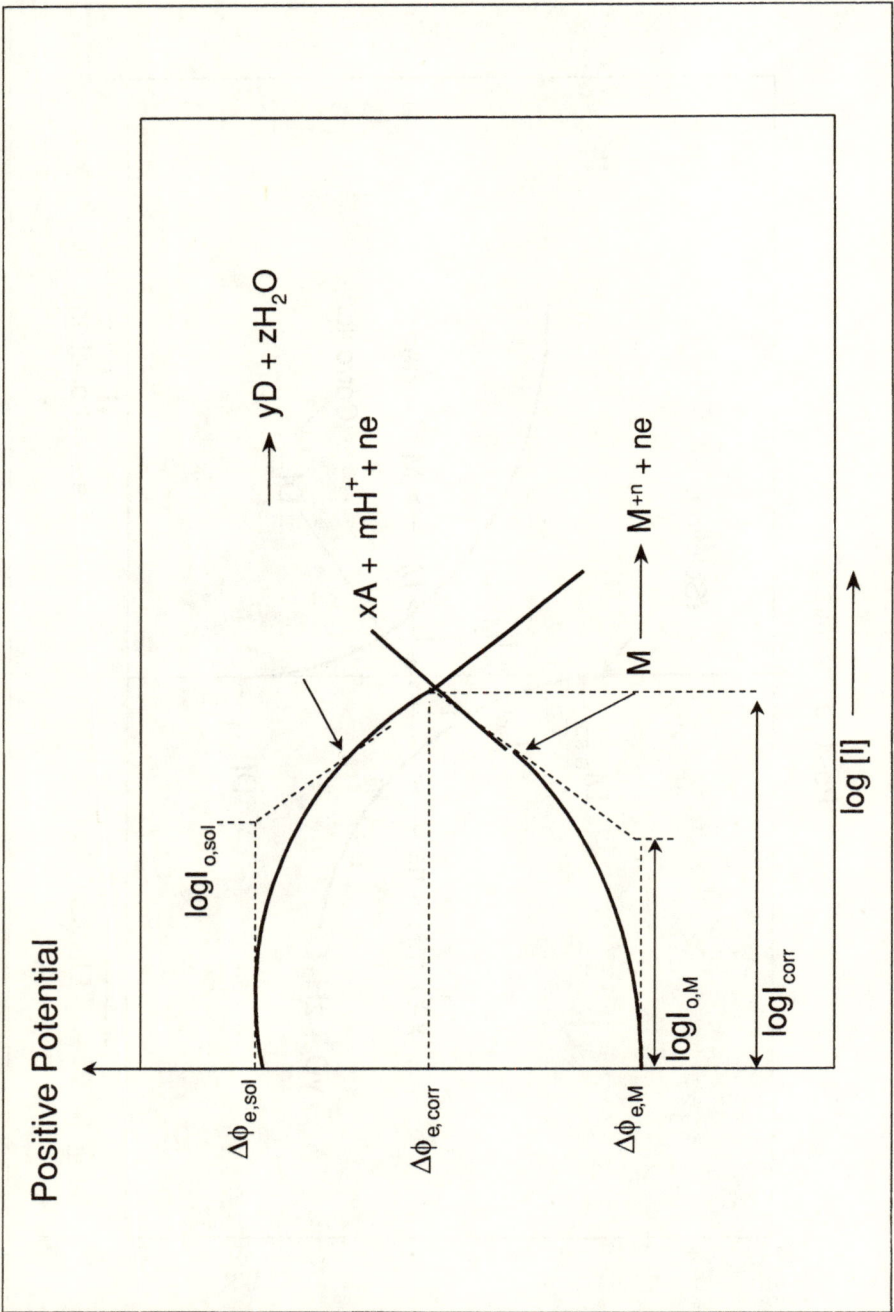

Fig. 12-16 Evans diagram

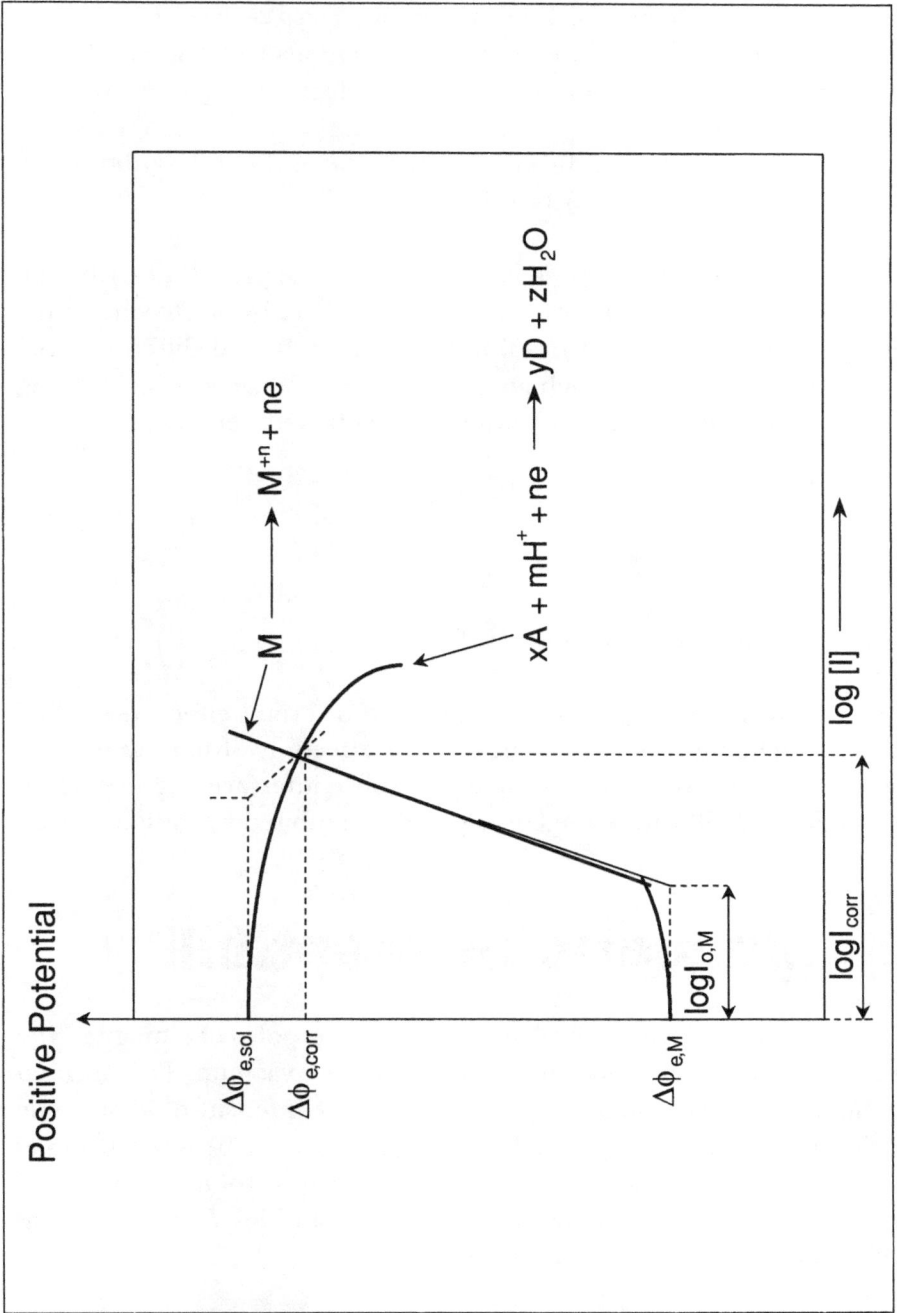

Fig. 12-17 Electron reaction curve changes in response to change in exchange-current density

- The influence of Tafel slopes (see Fig. 12–18)
- The influence of equilibrium potentials (see Fig. 12–19)
- The transport difficulties of the electron acceptor (see Fig. 12–20)
- The IR drop in the electrolyte between the sink (anode) and source (cathode) (see Fig. 12–21)

Before we look at these curves it will be helpful to describe the Tafel slope, perhaps one of the most used laws in electrochemistry. Bockris and Reddy (1970) define it as "the potential difference $\Delta\Phi$ across the interface at which an electrochemical reaction is occurring changes linearly with the logarithm of the current density i, or

$$\Delta\Phi = a - b \ln i$$

where

a and b are independent of $\Delta\Phi$ "

Thus, Evans diagrams give insight into several effects occurring in electrochemical cells, and one of these effects involves potential differences observed at the electrode-electrolyte interface. Figures 12–18 through 12-21 illustrate the curve profiles exhibited by these effects.

Magnetic Susceptibility

When any material is placed between two poles of a magnet, the magnetic field is changed from what it was in a vacuum. The magnetic field in the material is designated B, and the intensity of a magnetic field in a vacuum is designated by H. The difference between B and H is dependent on I, the magnetic moment per unit volume of material (or the intensity of magnetization). The quantities B, H, and I are related by the following expression:

$$B = H + 4\pi I$$

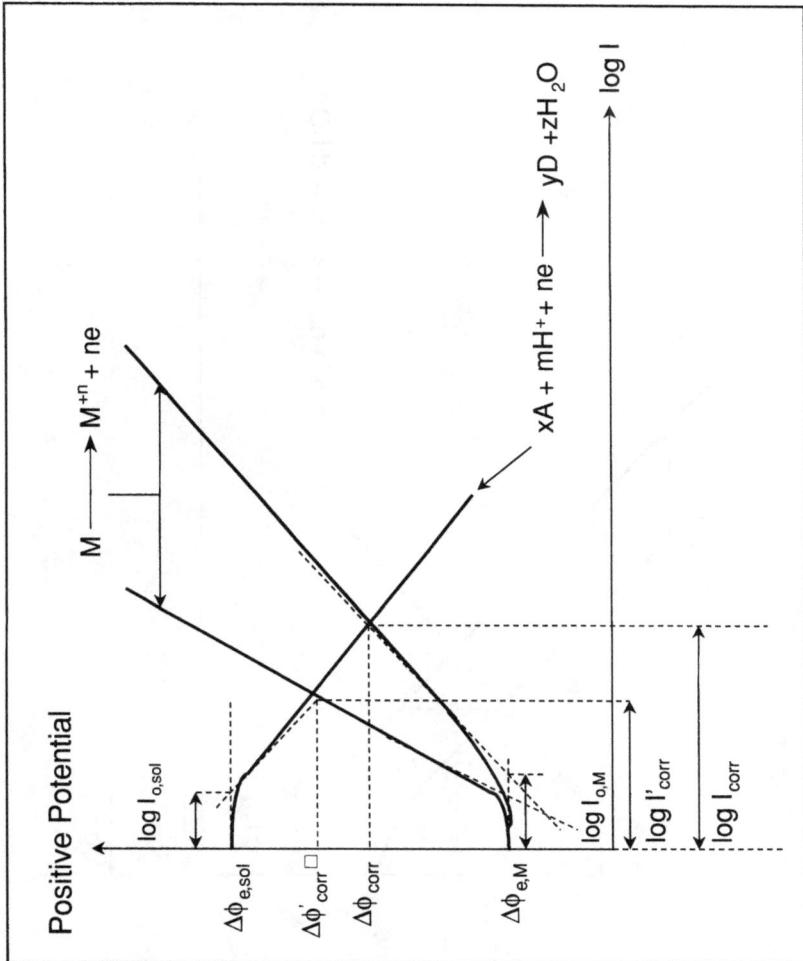

Fig. 12-18 Profile of metal-dissolution and electronation reaction (Tafel slopes influence)

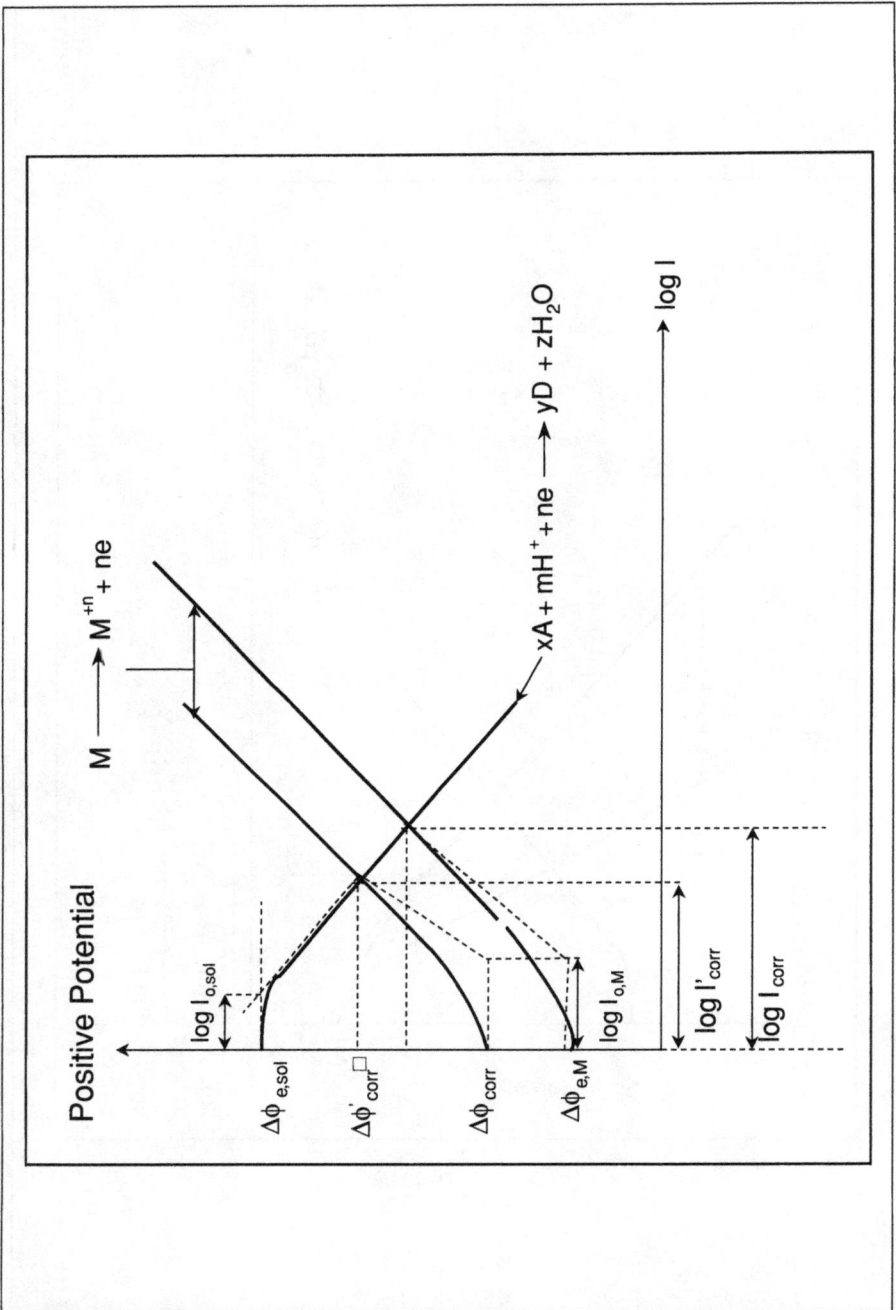

Fig. 12-19 Profile of metal-dissolution and electronation reaction (equilibrium potentials influence)

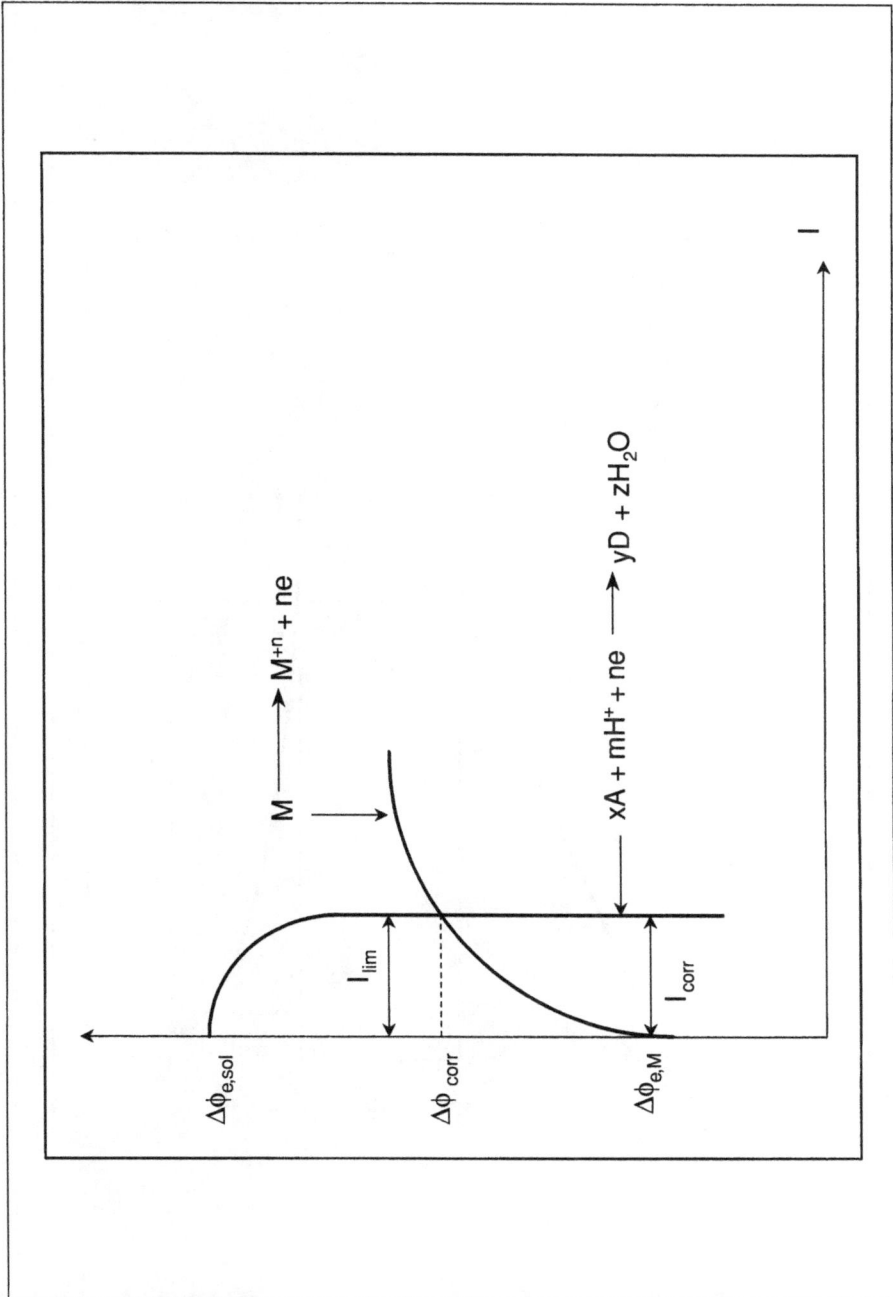

Fig. 12-20 Profile of metal-dissolution and electronation reaction (transport difficulties of electron acceptor)

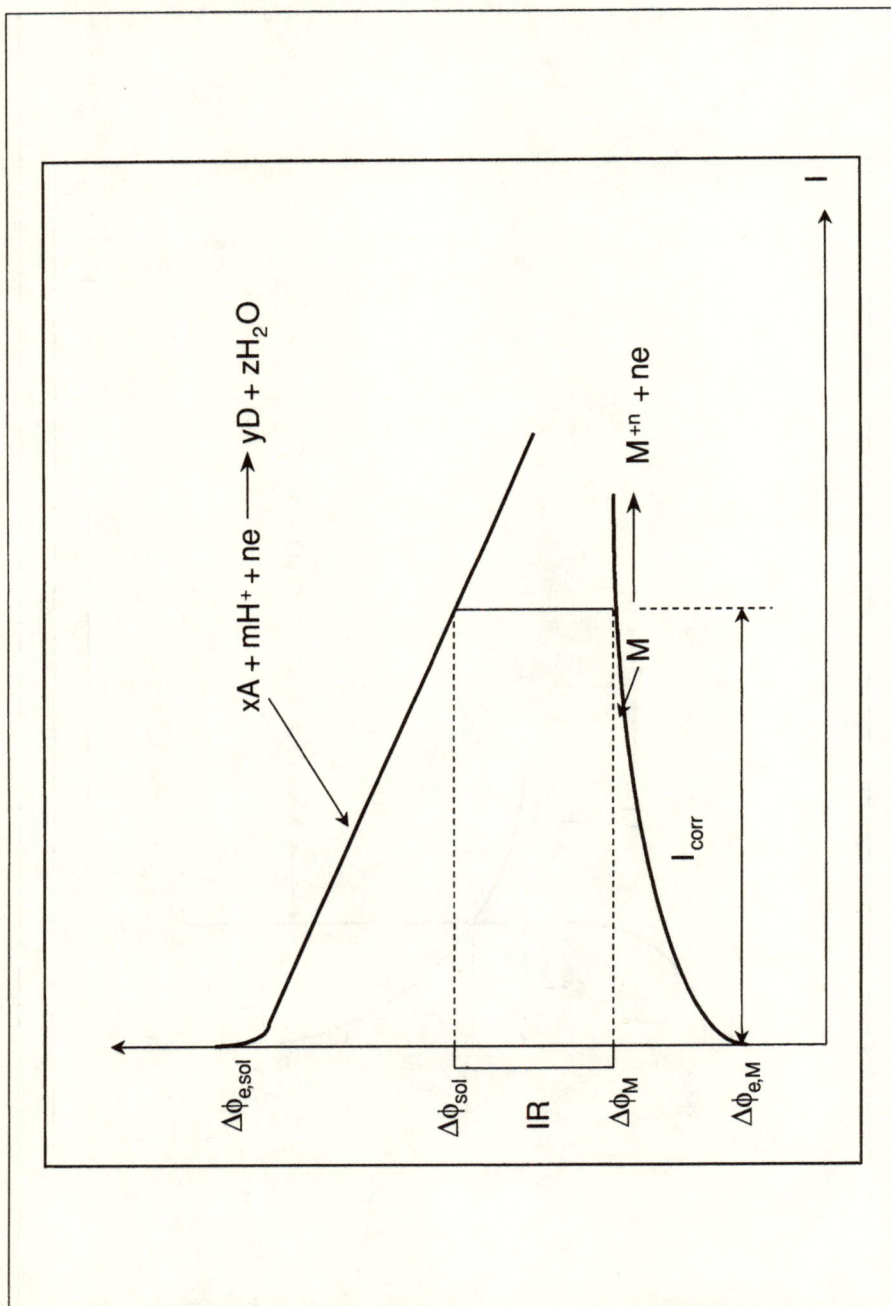

Fig. 12-21 Profile of metal-dissolution and electronation reaction (IR drop in electrolyte between anode and cathode)

Magnetic fields can be either increased or decreased by the presence of certain materials; therefore, *I* may be either positive or negative. The quantity that is obtained experimentally in magnetic measurements is the intensity of magnetization per unit field strength, or *I/H*. The magnetic susceptibility per mole is arrived at by multiplying the molecular volume M/ρ by the *I/H* value.

The molar magnetic susceptibility is the magnetization per mole induced by unit field strength. As such it can be recognized as the magnetic counterpart, on a molar basis, of the sum of the electrical polarizability a and the molecular-dipole term:

$$\mu^2/\,3kT$$

where

k = Boltzmann's constant, 1.38 x 10^{-23} Joules/molecule ° K).

It is convenient to deal with the corresponding terms α_M, the magnetic polarizability, and μ_M, the magnetic moment. This relation can be stated as in the following equation:

$$\chi_M = (M/\rho)\,(I/H) = X(\,\alpha_M + \mu_M^2/3kT)$$

where

X = 6.023 x 10^{23} molecules/mole (Avogadro's number).

Several different qualitative behavioral differences are recognized from sample to sample.

Most organic compounds have only the magnetic-polarizability contribution (μ_M is zero, thus χ_M and *I* are negative). Materials that behave in this way are called *diamagnetic.* Other materials characterized by possessing unpaired electrons line up in a magnetic field, which makes χ_M positive. Materials that behave in this way are called *paramagnetic.* Figure 12–22 illustrates the behavior of the magnetic lines of force in a vacuum, with diamagnetic and paramagnetic materials present. Table 12–1 (adapted from the *Handbook of Chemistry and*

Physics, 1970) lists the magnetic susceptibilities of several inorganic compounds found in petroleum fluids at 298° K.

Table 12–1 Magnetic Susceptibility of Some Inorganic Compounds*

Formula	Susceptibility (10-6 cgs)
Al_2O_3	-37.0
$Ba(OH)_2$	-53.2
$BaSO_4$	-71.3
$CaCO_3$	-38.2
$Ca(OH)_2$	-22.0
$Ca SO_4$	-49.7
$Fe CO_3$	+11,300.00
$FeCl_2$	+13,450.00
FeO	+7,200.00
$Fe SO_4$	+10,200.00
FeS	+1,074.00
$Mg CO_3$	-32.4
$Mg(OH)_2$	-22.1
MgO	-10.2
$MnCO_3$	+11,400.00
$Mn(OH)_2$	+13,500.00
MnO	+4,850.00
$Mn SO_4$	+13,660.00
MnS	+5,630.00
$Sr CO_3$	-47.0
$Sr(OH)_2$	-40.0
SrO	-35.0
$Zn(OH)_2$	-67.0
ZnS	+122.00

*adapted from *Handbook of Chemistry and Physics*. 1976. 56th ed. Cleveland: CRC Press.

From Table 12–1, it can be seen that of the compounds listed, only iron, manganese, and zinc exhibit paramagnetic properties. Although the ferric ion is not listed, it is also paramagnetic, because it is impossible to pair its odd number of electrons, and unpaired electrons impart a magnetic moment. Further, no matter what influence is exerted by the ligand groups attached to the ferric complex, it will remain paramagnetic. Thus, of the scale forms commonly found in

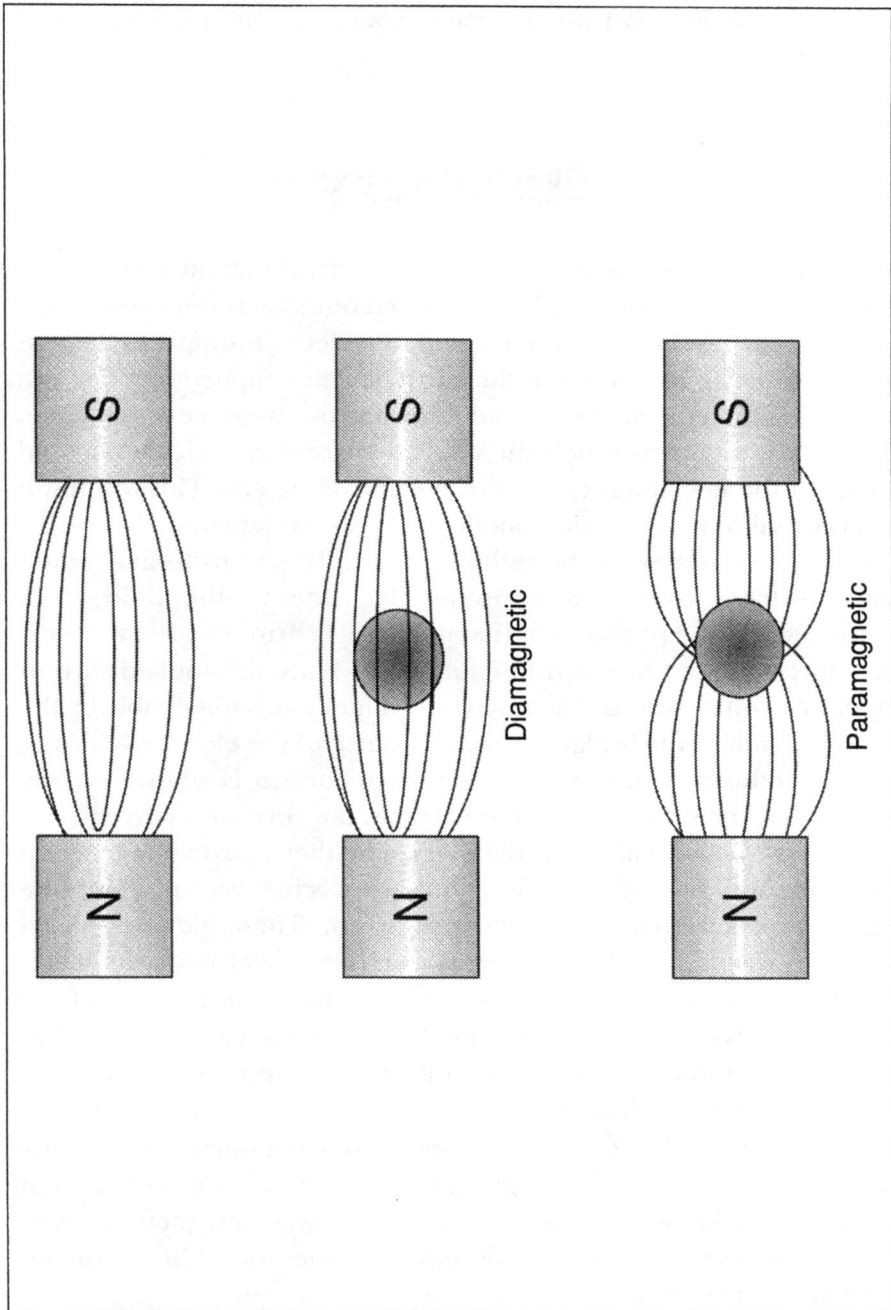

Fig. 12-22 Magnetic lines of force showing the effects of the presence of diamagnetic and paramagnetic materials in a magnetic field

petroleum fluids, only iron, manganese, and zinc show paramagnetic properties.

Summary

The preceding chapter discussed the effect that an applied current has on a corrosion cell. It was pointed out that as the applied current approaches the corrosion current, the electron supply to the corrosion anode(mild steel) is reduced. When the applied current and the corrosion current are equal, the corrosion process is halted. However, the time at which the applied current is instigated should not be at the maximum current of the corroding cell. This will result in a reversal of the corrosion anode to a cathode function.

Once converted to the cathode, the molecular hydrogen generation switches from the platinum electrode to the mild steel. Consequently, the potential of the mild steel is driven excessively negative (greater than the corrosion potential). Thus, the applied current should be controlled, and a good control method is obtainable by the use of a conductivity bridge. As the resistance of the electrolyte drops, or the conductivity increases, the supplied current is adjusted downward. Thus, the supply of current balances the corrosion current.

Plots of potential versus pH determine thermodynamic reaction feasibility, and plots of these functions are succinct ways of determining whether a reaction can take place or not. These plots are called Pourbaix diagrams. Although they can help determine the likely occurrence of a reaction, they can tell nothing about the rate of the reaction. Plots of potential versus the log of current can provide information about the rate of metal dissolution, and these plots are referred to as Evans diagrams.

Evans diagrams can be obtained from conductivity measurements, if the ratios of cell voltage and amperage to bridge voltage and amperage are known. Finally, paramagnetic and diamagnetic properties of some of the corrosion products and scales found in petroleum fluids were discussed.

References

Barrow, Gordon M. 1966. *Physical Chemistry.* 2d ed. New York, St. Louis, San Francisco, Toronto, London, and Sydney: McGraw Hill Book Company.

Dickerson, R. F., H. B. Gray, and G. P. Haight, Jr. 1970. *Chemical Principles.* 1st ed. New York: W. A. Benjamin, Inc.

Handbook of Chemistry and Physics. 1976. Cleveland: CRC Press.

Huheey, James E. 1978. *Inorganic Chemistry Principles of Structure and Reactivity.* 2d ed. New York, Hagerstown, San Francisco, and London: Harper & Row.

Jones, Loyd W. 1988. *Corrosion and Water Technology for Petroleum Producers.* 1st ed. Tulsa: OGCI Publications, Oil & Gas Consultants International, Inc.

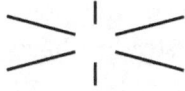

13
Corrosion Inhibition

Mechanisms: Cathodic Protection

In the previous chapter, cathodic protection was discussed in connection with the various Evans diagrams obtained from the application of current to the corroding metal, or by the placement of a sacrificial anode. Although these cathodic protection methods are practiced in some places, they have some serious drawbacks. In many cases when an external current source is used, the power consumption may be prohibitively large. The larger the exchange current density (the lower Tafel slope) of the electronation reaction, the greater the required protection current.

Shifting the potential difference between the entire corroding surface and electrolyte interface can be hampered by IR drops from one area to the next. Thus, inconsistent protection over the entire surface results. Because of this inability to apply uniform protection over the entire surface, some areas remain excessively positive (electronation), and others remain excessively negative with respect to the hydrogen equilibrium potential. If sufficient current is applied to successfully inhibit the dissolution of the metal, IR drops can cause local-ized hydrogen adsorption excesses.

These localized hydrogen evolution excesses lead to hydrogen entering the metal and causing hydrogen embrittlement. Thus, the use of cathodic protection can be problematic at best, and even dangerous at worst. (This is due to the possibility for ignition of excess hydrogen generated by excessive localized negative currents as well as hydrogen embrittlement of metals being protected.)

Mechanisms: Surface Passivation

Cathodic protection achieves corrosion protection by altering the double layer field (metal surface electrolyte interface) in such a way as to diminish the dissolution of metal. Suppose that it were possible to stabilize the metal by superimposing a double layer field that accelerated metal dissolution. In the process, it would passivate the surface toward continued dissolution. This turns out to be possible in some circumstances. In fact some corroding metals can be made passive by increasing the potential in the positive direction. This phenomenon is called *enforced passivation.*

Special equipment has been developed to measure some of the phenomena associated with metallic surface passivation. One such device is called a potentiostat. This device applies a source potential to a piece of corroding metal (the test electrode). Using a voltmeter, it measures the potential against a non-polarizable reference and auxiliary electrode. The potentiostat measures the potential (V) of the test electrode under study and compares this with the prechosen value (V^*) of the potential source. If there is a difference between the measured potential and the chosen potential ($\delta V = V^* - V$), the device sends a current I between the auxiliary and test electrode.

The direction of the current sent is determined such that $\delta V = V^* - V = 0$. Thus the potentiostat maintains the charge transfer rate at the metal solution interface by increasing or decreasing the electronation or de-electronation current to coincide with the preselected value V*. The design of the potentiostat includes a voltage comparator that is indicated in Figure 13–1. This comparator does much of the work of the potentiostat.

The potentiostat makes it possible to study the process of surface passivation. It allows one to hold the potential of the electrode at a selected value and make measurements (e.g., the steady state current at each value of the potential). If one chooses mild steel or iron as the corrodible electrode, the current potential plots show patterns reflecting the current imposed. These plots are informative, because they show the change in dissolution of the metal with the change in the imposed current. Figure 13–2 shows a plot of current density versus potential of a corroding steel electrode.

From Figure 13–2 it can be seen that as the potential is made more positive, the dissolution increases. However, as the potential continues to be even more positive, a point is reached where a current fall-off is observed (passivation potential). Once the metal is passivated, flowing currents are negligible (~ 100 to 1,000 times less than the dissolution current). Thus, the metal is for all appearances noncorroding and has been anodically protected. One major point should be made about anodic protection. It can provide corrosion protection by pumping electrons from the metal and requires only a fraction of the charge required for cathodic protection.

Corrosion Protection by Chemicals

Sometimes the addition of chemical substances including phosphorus, arsenic, or antimony compounds interferes with the hydrogen evolution electronation reactions. Figure 13–3 shows the effects of the addition of these compounds on the potential versus current plot. (Fig. 13-3B shows the effects of organic inhibitors on current versus potential plots of corroding systems.)

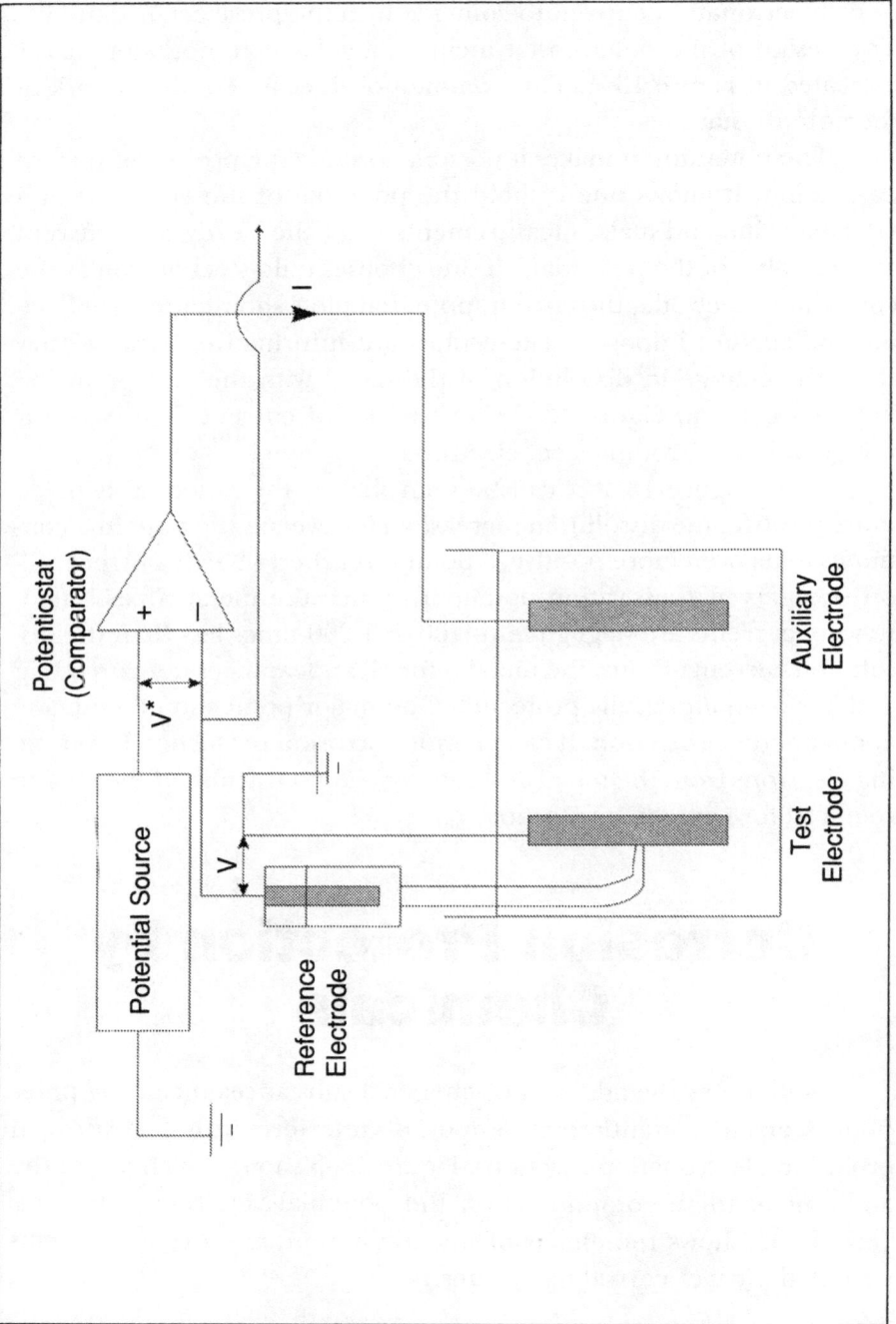

Fig. 13-1 Potentiostat diagram

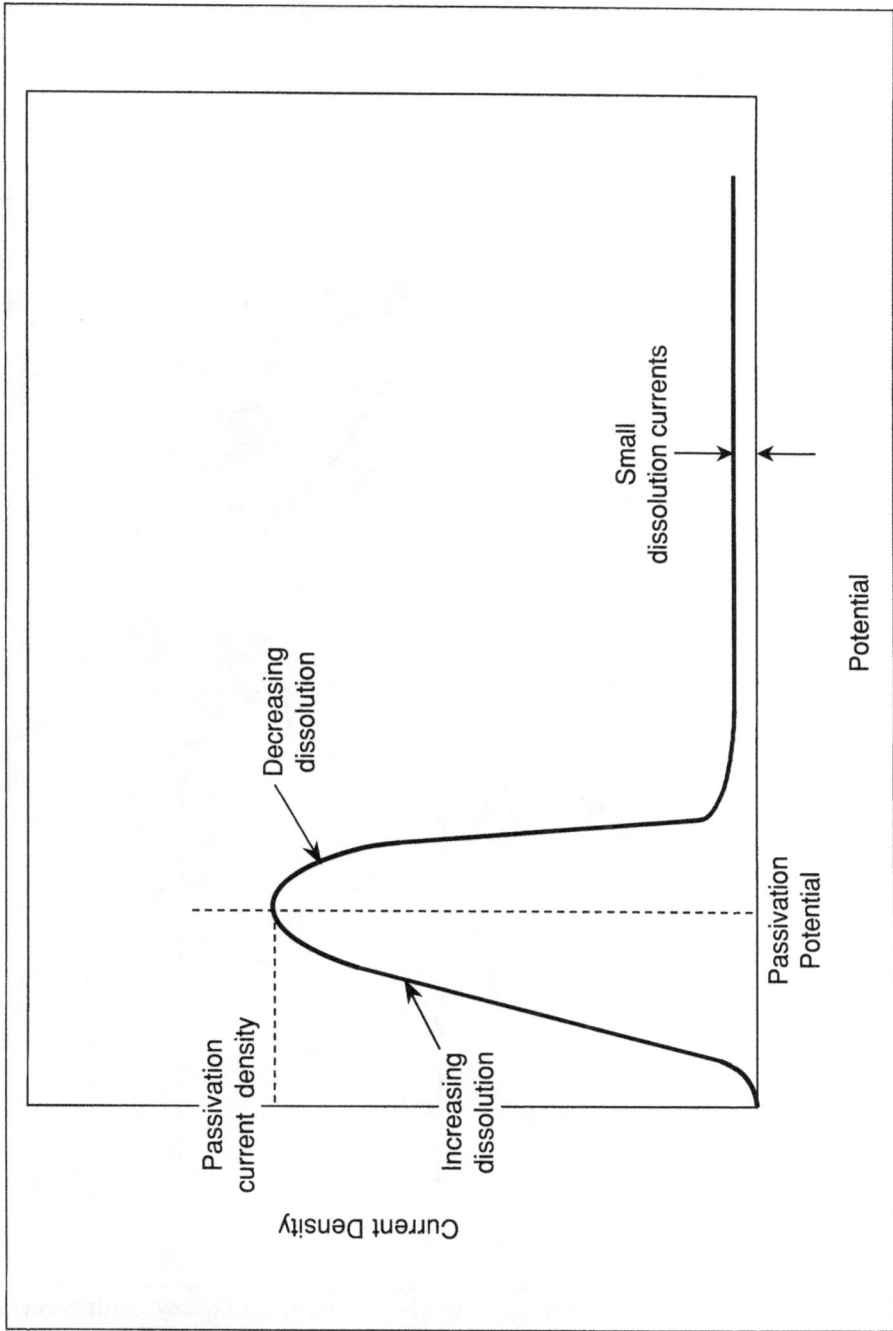

Fig. 13-2 The plot of steel potential

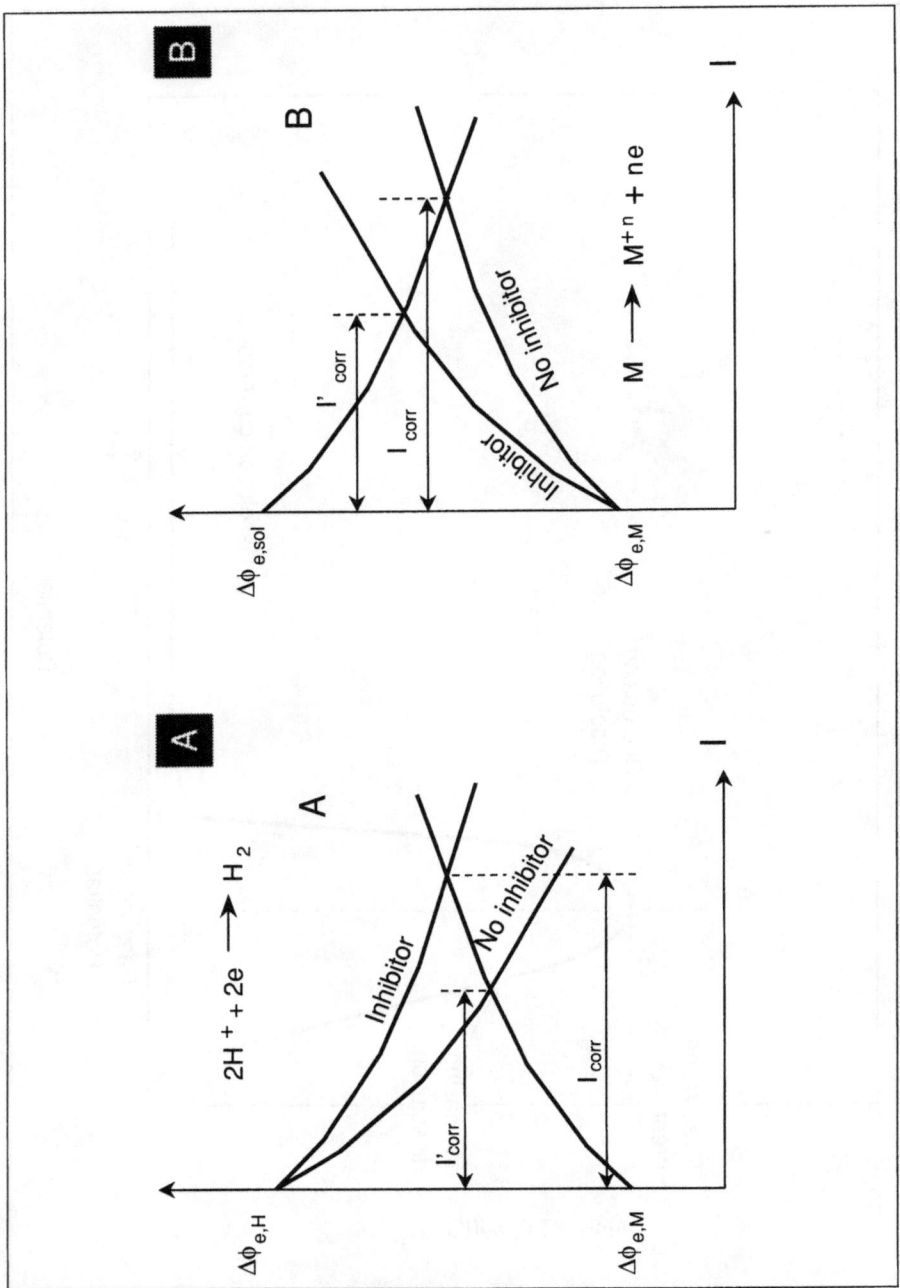

Fig. 13-3 A. The effect of phosphorous, arsenic, or antimony compounds **B.** The effects of organic inhibitors on current versus potential plots of corroding systems

Organic corrosion inhibitors act by adsorption to either the electron source or the electron sink of corroding systems. Potential versus log current plots show how the curves are shifted by the presence of these materials. Figure 13–4 shows the changes in the curves as a consequence of the addition of the inhibitors.

The effectiveness of these inhibitors is determined by the efficiency with which they react with either the electrolyte (hydrogen evolution or oxygen electronation reactions) or the metal surface through adsorption. Adsorption of organic compounds to the electron sink areas (de-electronation sites) usually involves compounds with excess electrons or unshared pairs of electrons. (These include aliphatic, aromatic, or aliphatic-aromatic amines, sulfur-containing compounds, or oxygen-containing compounds.)

Several proposed mechanisms exist for the adsorption efficiency of these derivatives, most of which involve the pi cloud bonding capability of the electron-rich heteroatoms of organic compounds. These heteroatoms are atoms other than the carbon contained in the organic derivative, and include nitrogen, oxygen, sulfur, and phosphorous. The notion that these heteroatomic compounds are adsorbed on the metal surface at the sink area of the corroding metal, and slow the rate of metal dissolution, is quite reasonable.

If the metal dissolution occurs by attack of the metal by anionic ligand groups, then it is reasonable to assume that materials that tie up ligand sites on the metal could provide protection. These sites would otherwise be susceptible to anionic attack. Thus, heteroatomic compounds containing unshared pairs of electrons will tend to be specific to the electron sink site adsorption.

The adsorption of chemicals at the electron source areas of the corroding metal is not as simple to reconcile with their modes of protection. The electron source area of a corrosion cell, it will be recalled from previous discussions, is the area of the cell that adsorbs molecular hydrogen from the electrolyte. This area is not formally charged, but takes on partial charge character as the molecular hydrogen approaches the metal to within a van der Waals radii. Thus, the source area adsorption involves a competition between molecular hydrogen and the would-be inhibitor substance for electron source site locations.

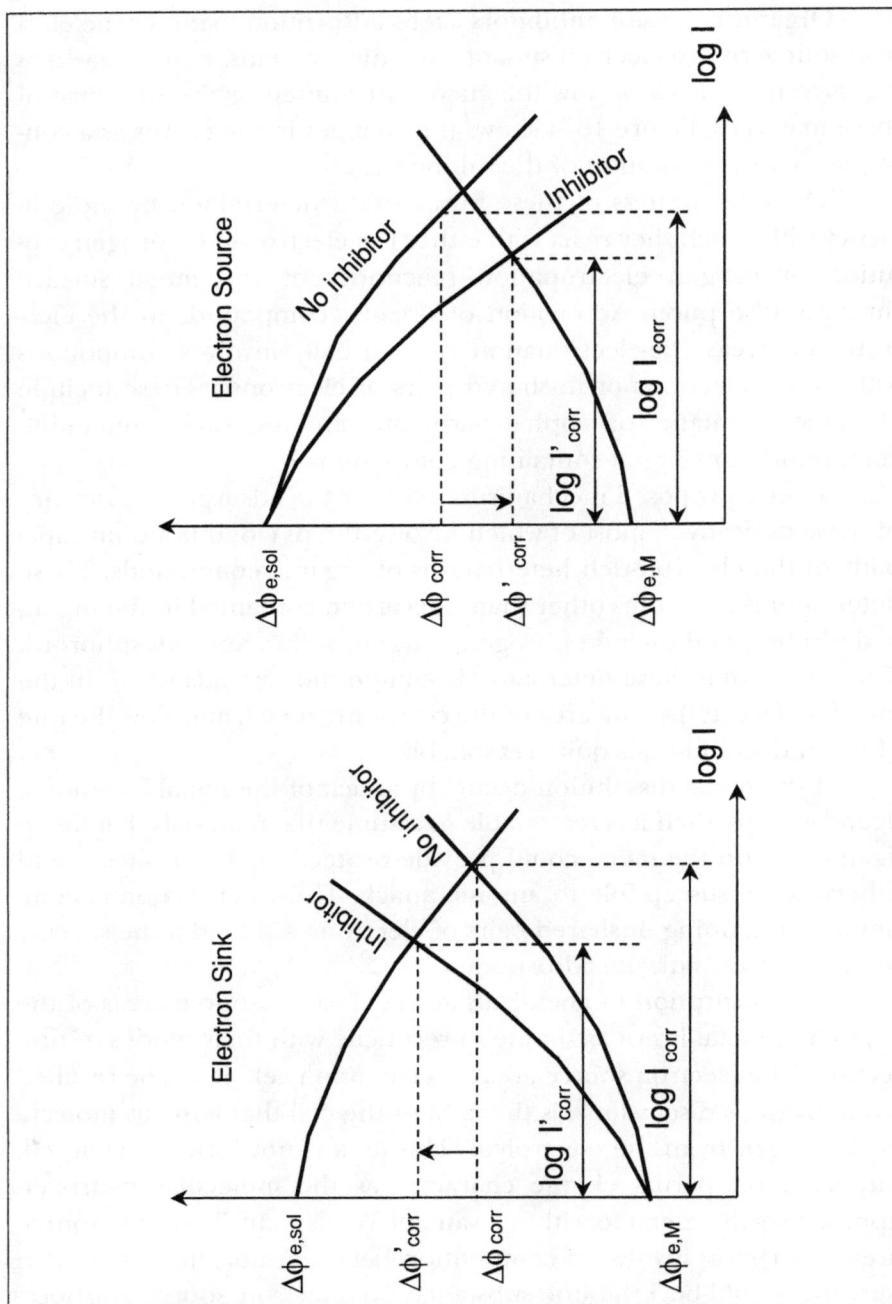

Fig. 13-4 The effect of organic inhibitors absorbing to the electron sink or source of corroding metal on the potential versus log current plots

This is not unlike the competition that occurs between the heteroatomic organic compounds and the anionic ligand groups at the sink area. However, there is one subtle difference, and that difference is the dual site occupation by the molecular hydrogen at the metal surface. This dual site occupation is the result of the partially induced charge and its conjugate nearby charge on both the metal and the molecular hydrogen. Figure 13–5 shows this arrangement.

As a result of this dual site nature, competition by chemical adsorption inhibitors will have one additional restriction placed on it: it also must be capable of dual site occupation. Some of these structures that are capable of dual site occupation include hydroxyl, sulfhydryl, carboxylic, and primary amines. It is also interesting to note that each of these types of molecules is capable of undergoing hydrogen bonding.

Both the electron sink and source adsorption phenomena involve orbital overlap between the metal and the chemical and metallic site competition between the corroding species and the protecting chemical. The source (electronation) site competition differs from the sink (de-electronation) site competition by requiring a dual point adsorption of competing chemical. Figure 13–6 schematically illustrates the sink and source adsorption differences.

Although the adsorption mechanisms differ between the source and sink areas of the corroding metal, both sites require approach to a minimum distance (van der Waals radii). The appropriate molecular orientation for successful site occupation is also necessary.

Electrolyte Chemical Interaction

Adsorption competition is not the only chemical method of interfering with corrosion processes. Sometimes, the addition of chemicals that react with dissolved oxygen interfere with the corrosion process, particularly if oxygen electronation is the reaction taking place at the source area. Thus, chemicals like hydrazine (N_2H_4) and

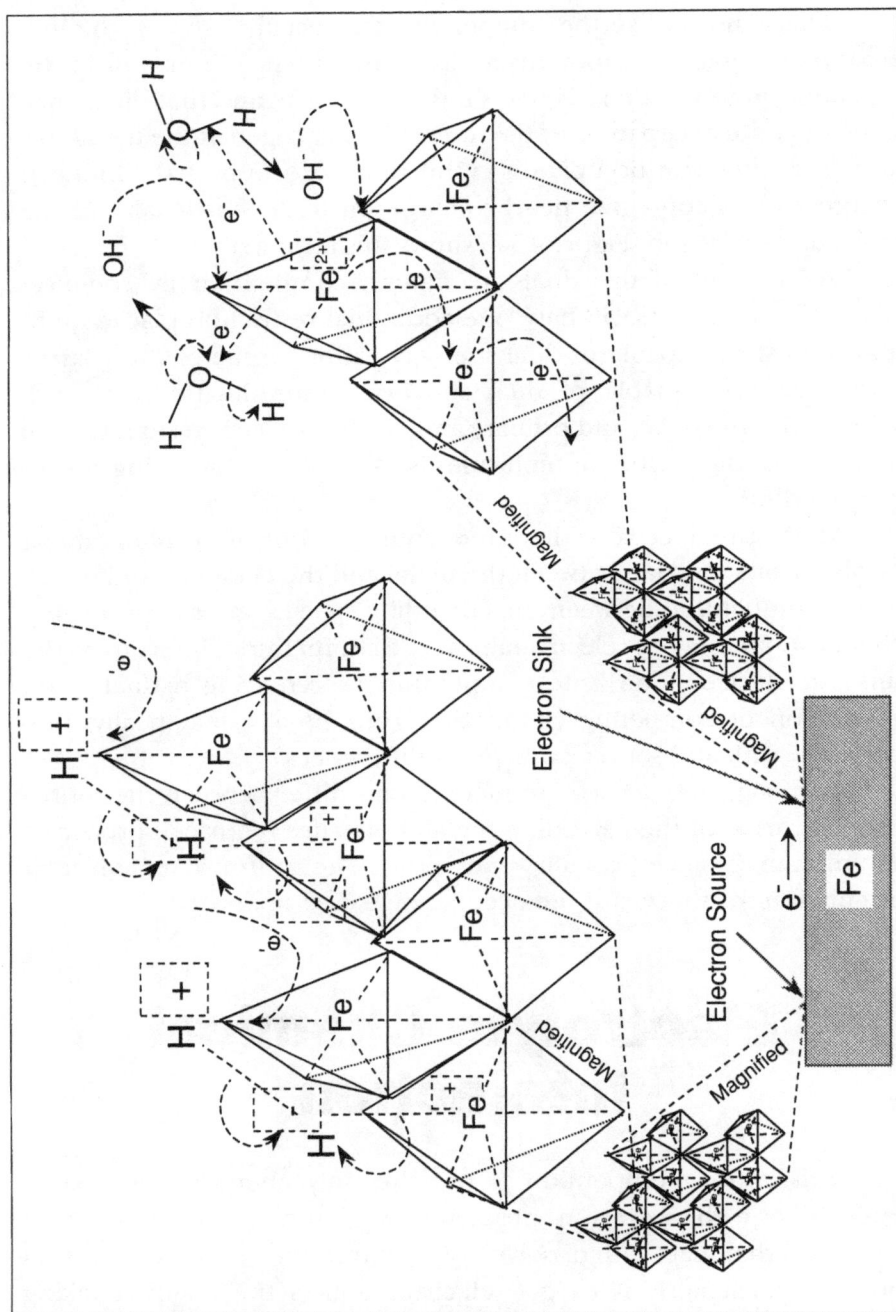

Fig. 13-5 The electron source and sink areas of iron metal, with the dual induced charge nature of the absorbed molecular hydrogen

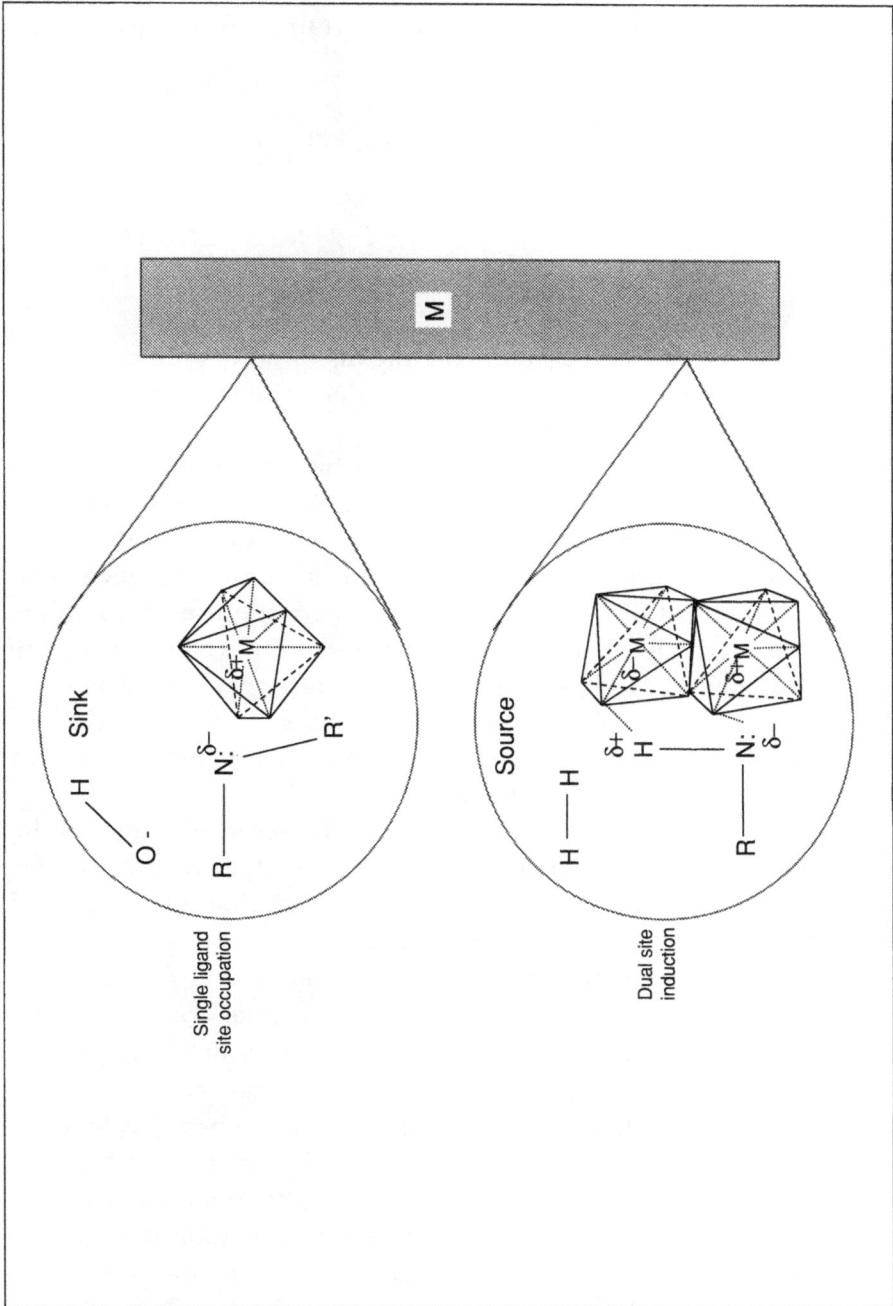

Fig. 13-6 Source and sink competition between chemical additives and the electronation and de-electronation species

sulfite (SO_3^{2-}) react with molecular oxygen (O_2) to reduce its concentration:

$$2\,N_2H_4 + 5\,O_2 \longrightarrow 4\,NO_2^{\cdot} + 4\,H^{+} + 2\,H_2O$$

and

$$2\,SO_3^{3-} + O_2 \longrightarrow 2\,SO_4^{2-}$$

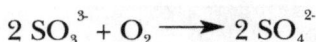

These reactions act by reducing the dissolved oxygen content of the electrolyte, and by so doing, they reduce the exchange current density for the oxygen reduction.

One area that has been discussed but has not been addressed in terms of chemical interference is that of radical reactions leading to corrosion. These mechanisms are much more important in the corrosion processes than is typically indicated in much of present-day corrosion literature. Considerable information on the use of quinoline and substituted quinoline structures for corrosion protection is available, and much of this information contends that the protection mechanism is due to adsorption. Further, substituted pyridines, imides, and imidazolines have also been reported to affect corrosion protection by adsorption mechanisms.

This is a very reasonable mechanism; however, it might not be the only mechanism these chemicals employ to interfere with the corrosion process. In the hydrogen adsorption mechanism developed throughout earlier sections of this book, molecular hydrogen was produced by the combination of radical or monatomic hydrogen. This monatomic hydrogen resulted from the electron attack on water at the sink site, the production of monatomic hydrogen (radicals), and formation of hydroxyl anions that attack the sink metal.

The radical hydrogen species then recombine by radical-radical reactions to produce molecular hydrogen that is adsorbed at the source site of the metal. If the substituted quinoline, alkyl pyridine, imides, and imidazolines act to inhibit the radical-radical hydrogen reactions, then an electrolyte mechanism is also operative. More precisely though, a hydrogen scavenger function is taken on by these chemicals. This mechanism is also suggested by the fact that the phosphorous, arsenic, and antimony derivatives interfere with the exchange current density for hydrogen evolution. Each of these atoms possesses half-filled p orbital configurations, as does nitrogen, and can provide five valence electrons for bonding. Thus, these orbital arrangements lend themselves to the formation of sp orbital hybrids.

In the discussion concerning hard and soft acids and bases, there was an increasing tendency for the elements nitrogen, phosphorous, arsenic, and antimony to complex with alkali metals (Group 1A). There was also an increasing tendency for them to complex with the alkali earth metals (Group 2A) and the lighter transition metals of higher oxidation states including Ti^{4+}, Cr^{3+}, Fe^{3+}, Co^{3+}, and H^+. Thus, the sink adsorption mechanism, proton abstraction, and electrolyte complex formation mechanisms can be invoked to explain interference with the corrosion process.

Aromatic Amines and Quinoline Structures

Molecular planarity is very interesting feature of the class of amine chemicals that prove to be effective corrosion inhibitors. Figure 13–7 shows some of the structural properties of the pyridine, isoquinoline, acridine, and some interesting reactions due to irradiation.

In the fused bi-cyclic compounds, the possibility of forming Dewar structures is determined by the resultant strain placed on the aromatic structure. (If a Dewar structure twists the aromatic ring out of a planer form, it is not allowed.) An additional consideration involving the nature of adsorption to the metal surface is the presence of a single heteroatom in these quinoline, isoquinoline, and acridine structures. Thus the adsorption mechanism must involve the pi orbitals of the planer complex, where the unshared pair on nitrogen is included in the delocalized pi orbital clouds above and below the aromatic rings. Figure 13–8 illustrates this delocalized pi bonding in a complex.

The possibility of Dewar phenanthridine is indicated in Figure 13–8 to show that the planarity of the fused heterocycle is still possible.

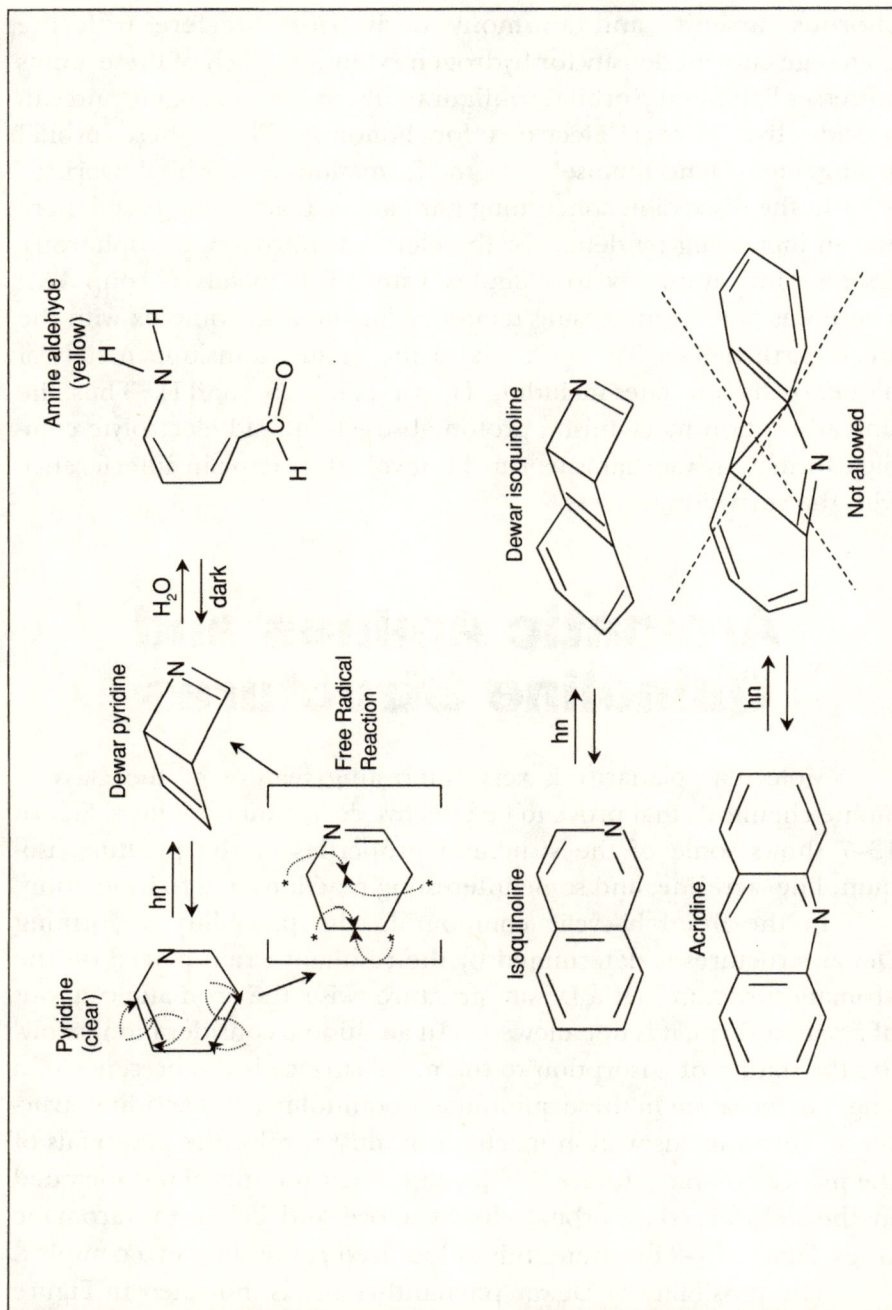

Fig. 13-7 The possibility of radical formation in six-member aromatic nitrogen compounds, and the need for planarity in the fused bi-cyclic compounds

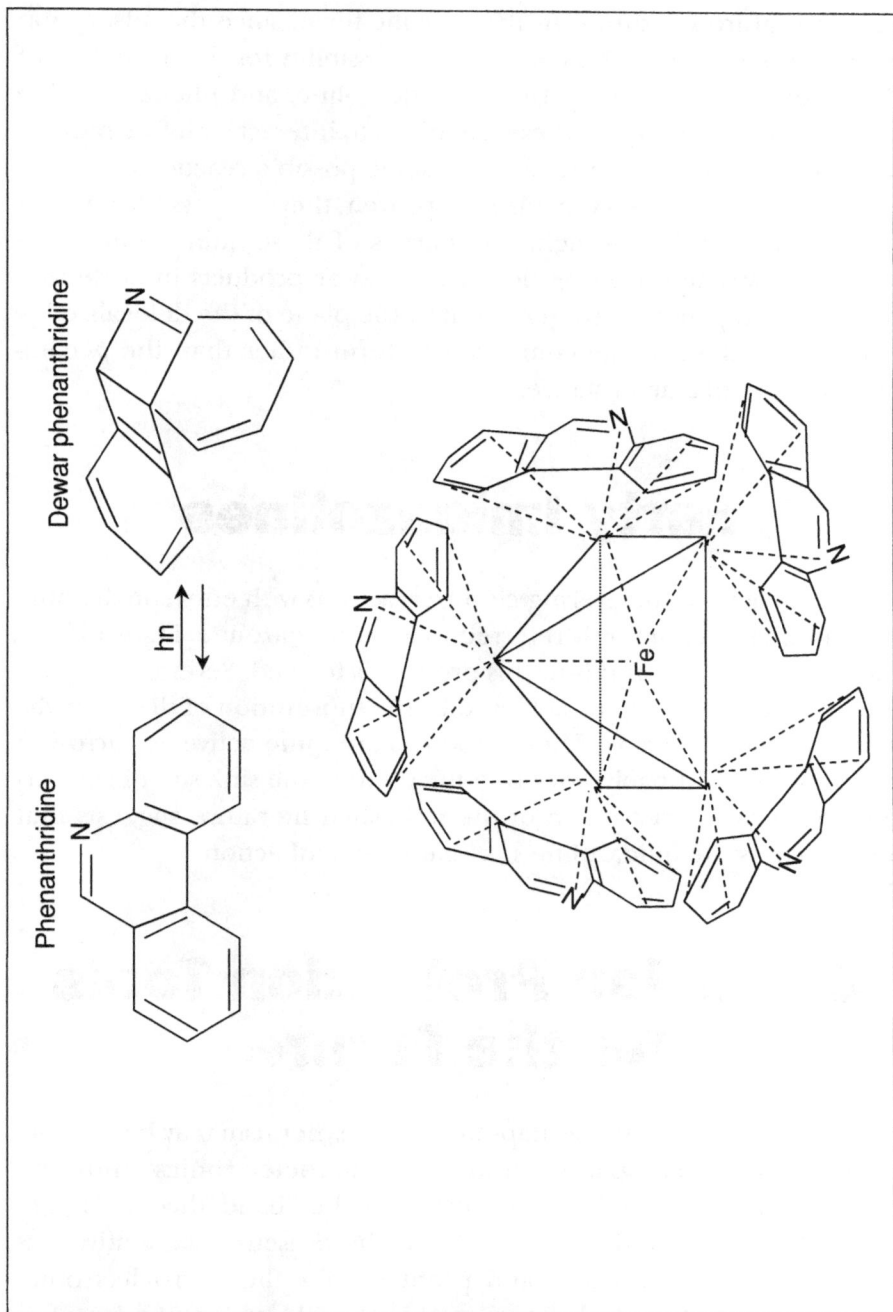

Fig. 13-8 The possibility of forming Dewar phenanthridine and the suspected pi-bonded metal complex structure

The central iron is shown in the metallic form, since the adsorption involves no oxidation of the metal. The possibility for the formation of Dewar structures for the pyridine, isoquinoline, and phenanthridine leaves the opportunity for these chemicals to interact with free radicals in the electrolyte. Figure 13–9 shows some possible reactions.

Though these reactions are not proven, they are possible and can account for radical-scavenging properties of these quinolidine structures. The structural properties of the Dewar products indicate that the unshared pair on nitrogen is out of the plane of the delocalized *pi* bond. This allows amine complexes to form rather than the *pi* complexes of the planar structure.

Fatty Imidazolines

Fatty acids undergo ring closure reactions with ethylene diamine to form what is commonly referred to as an imidazoline. Figure 13–10 shows this structure and how this product is formed. Several variations on the imidazoline can be achieved by the substitution of di-carboxylic and tri-carboxylic acids. These products are quite active as corrosion inhibitors and probably work by adsorption at the sink site of the corroding metal. In fact much of the published literature suggests that adsorption is the imidazoline's primary mode of action.

Corrosion Protection Tools for the Future

An interesting and perhaps unnoticed synergism may be possible between steel companies and the microelectronics industry. Semiconductor technology has advanced the "band theory" significantly during the period from 1960 to the present. These advances have proved both practical and profitable for the microelectronics industry. Some of the techniques used in the fabrication of microelectronic circuitry may have application in the steel industry. Silicone and

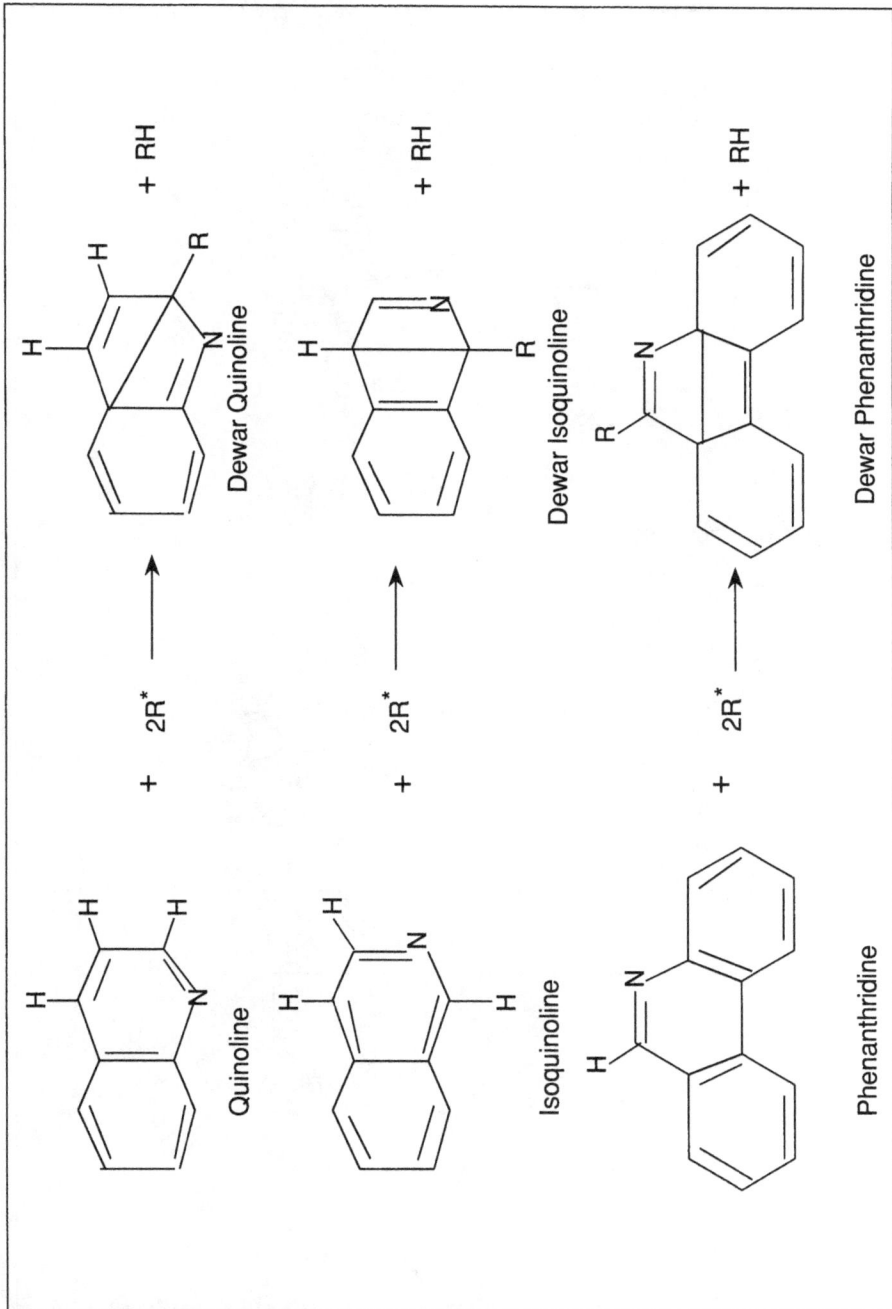

Fig. 13-9 Possible radical reactions with quinoline, isoquinoline, and phenanthridine

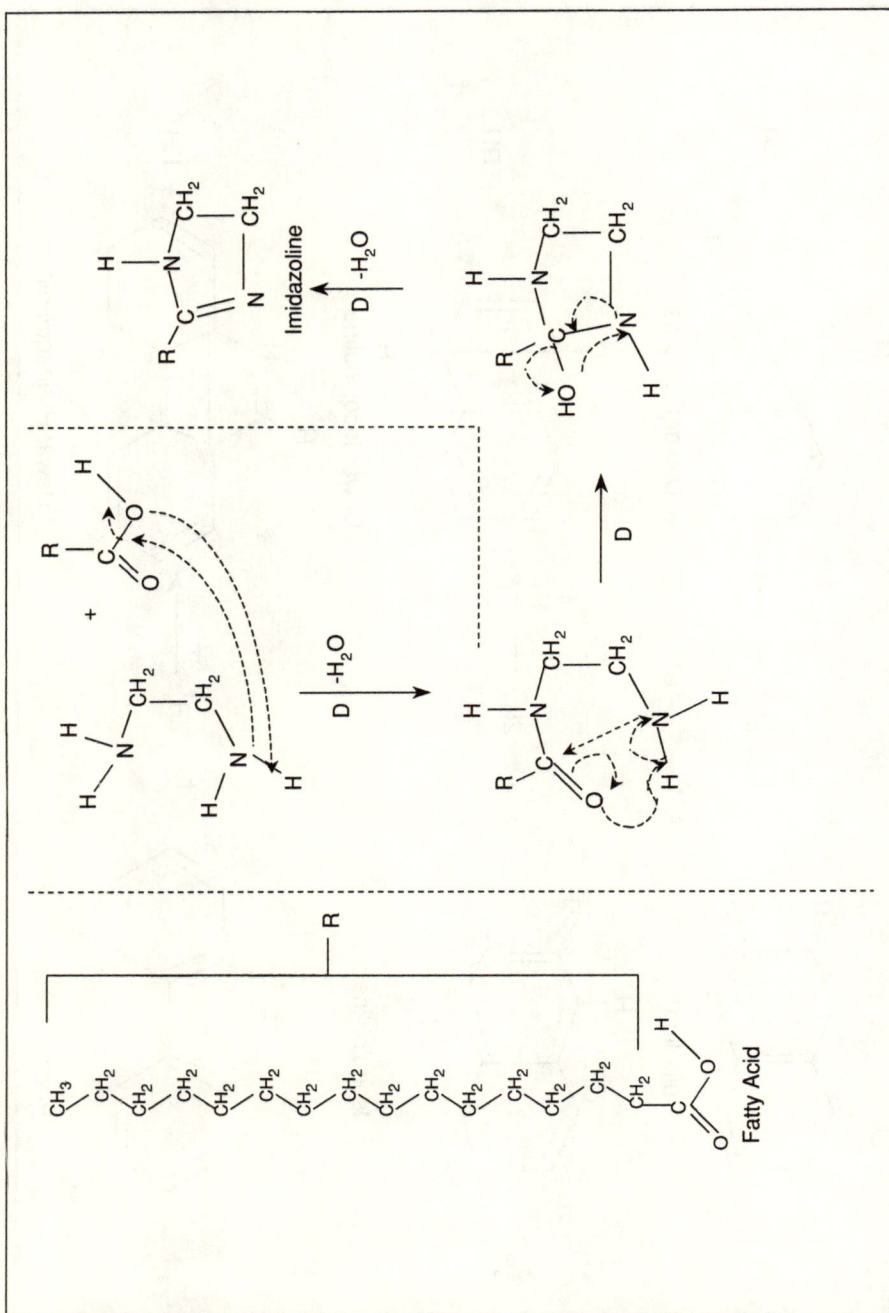

Fig. 13-10 The synthesis and structure for a fatty imidazoline made from ethylene diamine and fatty acid

germanium crystals are often diffused with gaseous arsenic, phosphorous, boron, and other elemental impurities to make them perform as semiconductors.

By selectively etching photographic patterns developed on a surface of vacuum-deposited aluminum, and connecting these microscopic metallic patterns, microelectronic circuits of magnificent complexity are possible. The potential of such technology for the purposes of protecting metals against corrosive environments is significant. Corrosion is a molecular process involving metallic crystals, ligand group interactions, and electron conduction. Consequently, the possibility of selective impurity addition by diffusion processes to sink and source areas may offer a significant improvement in corrosion protection methods. In fact it may be possible to electrify metals in diffusion ovens, diffuse gaseous impurities selectively to sink and source areas, and thereby passivate the metal surface.

Summary

The preceding chapter discussed both electrolytic and chemical modes of corrosion protection. Evans diagrams were presented to indicate change in the shape of the current versus potential curves due to anodic, cathodic, and chemical treatment. Potential versus current diagrams were also presented to show the different shapes of curves due to differing electronation reactions (oxygen versus hydrogen evolution). A discussion of the Dewar isomers of quinoline, isoquinoline, acridine, and phenanthridine developed the argument for their activity as radical scavengers. Finally, an example of the common imidazoline type product and its synthetic route was given.

References

Barrow, Gordon M. 1966. *Physical Chemistry.* 2d ed. New York, St. Louis, San Francisco, Toronto, London, and Sydney: McGraw Hill Book Company.

Becker, J. R. 1997. *Crude Oil Waxes, Emulsions, and Asphaltenes.* Tulsa: PennWell Publishing Company.

Bockris, J. O'M. and A. K. N. Reddy. 1973. *Modern Electrochemistry.* New York: Plenum Publishing Corporation.

Dickerson, R. F., H. B. Gray, and G. P. Haight, Jr. 1970. *Chemical Principles.* 1st ed. New York: W. A. Benjamin, Inc.

Hamill, William H. and Russell R. Williams, Jr. 1966. *Principles of Physical Chemistry.* 2d ed. Englewood Cliffs, New Jersey: Prentice Hall.

Huheey, James E. 1978. *Inorganic Chemistry Principles of Structure and Reactivity.* 2d ed. New York, Hagerstown, San Francisco, and London: Harper & Row.

Noller, Carl R. 1966. *Textbook of Organic Chemistry.* 3d ed. Philadelphia and London: W. B. Saunders Company.

14
Phase Behavior of Corrosion Chemicals

Petroleum Fluid Phases

Very early in the present discussion of corrosion and scale, an introduction to phase behavior was given. This introduction was not specific to the types of chemicals used for the purpose of corrosion or scale inhibition, but general to the phase behavior of bipolar molecular species. Over the last several chapters, a discussion of some of the chemistry and proposed mechanisms of action has been conducted. With this improved understanding of the properties of corrosion and scale chemicals, a discussion of the resulting phase behavior in complex fluid systems will have more meaning.

Petroleum fluids are complex mixtures of multiple component composition, among which are oil, water, salts, emulsions (oil in water and water in oil), waxes, organometallics, gases, and solids. The interactions of these multiple components are also very complex and can include a variety of physical changes. These changes include:

- Liquids that cool to solids
- Gases that condense to liquids

- Dispersions that aggregate to form solids
- Emulsions that resolve into separate phases and reform
- Acid and base equilibria
- Ionic salt equilibria
- Soluble and insoluble salts

Thus, the application of chemicals to affect this complex system is a very tricky matter. As we have seen, the chemistry of corrosion and scale inhibition is dictated by the ends that one is trying to achieve. This criterion, in itself, is difficult to accomplish without the added complications arising from phase behavior of the inhibitor within the complex petroleum fluid systems. Thus effective chemical placement is critical to the success or failure of a specific application. If corrosion inhibition requires adsorption of chemical at a metal's surface, and the chemical is soluble in the oil phase, how can one be sure that it will get to the metal's surface? Thus, the question of whether or not the "grease is getting to the squeak" is still a major concern.

Corrosion Inhibitor Solubilities

As we have seen, some of the corrosion inhibitor products consist of heteroatomic aromatic structures, which include pyridine, quinoline, isoquinoline, phenanthridine, acridine, and fatty imidazoline structures. These products represent only a few of the chemical types used for corrosion inhibition. In addition to these types, there are acetylinic alcohols (e.g., propargyl alcohol), quaternium salts (e.g., alkyl pyridinium chloride), and alkyl oxazolidines. Among this long list of corrosion inhibitors, the degree of water dispersibility ranges from very low in the case of the heteroatomic aromatics to highly dispersible for the quaternium salts.

Corrosion is a process that requires the metallic surface to be in intimate contact with water. Consequently, it follows that the chemical inhibitor should be able to traverse the phase boundary between the water and the metal surface. If the corrosion inhibitor is predominantly soluble in the oil phase of the petroleum fluid, then its ability to reach the metal interface is severely hampered. The quinoline and isoquinoline structures possess a slight polarity associated mainly with the unshared pair of electrons on the nitrogen. However, their dispersibility in the aqueous phase is quite small. Figure 14–1 illustrates some of the properties of these structures.

The water solubility or dispersibility of the quinoline structures is quite low. The ability of these chemicals to traverse the interface between the water and the metal interface is also expected to be quite low. The fatty imidazolines are bipolar, but the polarity of these structures is also small. They are expected to emulsify when added to systems containing water in oil, and field experience has shown this to be the case. It is often necessary to add de-emulsifier products to the fatty imidazolines to avoid the formation of difficult field emulsions.

The quaternium derivatives of the pyridines (quinolines and isoquinolines) add considerable water dispersibility to these chemicals. The increased dispersibility carries with it the drawback that increased water dispersibility can cause emulsions. Figure 14–2 shows some of these structures and some of their properties before and after quaternization.

In addition to enhancing the aqueous solubilities of these chemicals, another effect must be considered. Does quaternization change the corrosion control mechanism of the product with respect to surface passivation of the metal? The answer to this question comes from an examination of the difference in the complexing ability of the quaternary salt versus that of the free amine. Recall that the complex form of the metal and the aromatic amines involved a pi-bonding complex between the metal and the aromatic amine. When the unpaired electrons on the amine are reacted with a methyl group and the quaternary salt is formed, the participation of the amine nitrogen in the metal complex is forbidden. If the amine is unable to interact with the ligand groups of the metal, the mechanism of protection is altered.

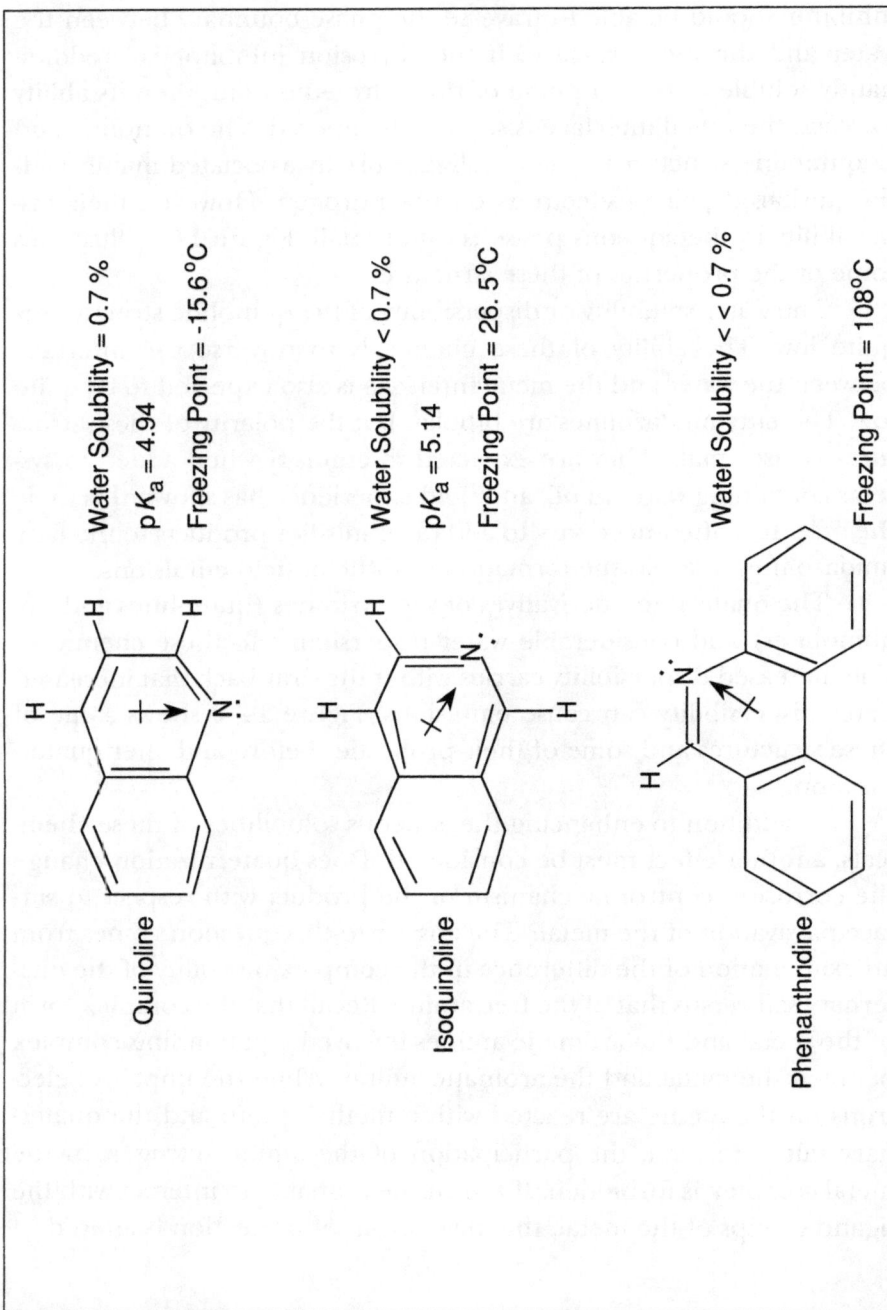

Fig. 14-1 Properties of the quinoline structures

Quinoline
Water Solubility = 0.7 %
$pK_a = 4.94$
Freezing Point = -15.6 °C

Isoquinoline
Water Solubility < 0.7 %
$pK_a = 5.14$
Freezing Point = 26.5 °C

Phenanthridine
Water Solubility << 0.7 %
Freezing Point = 108 °C

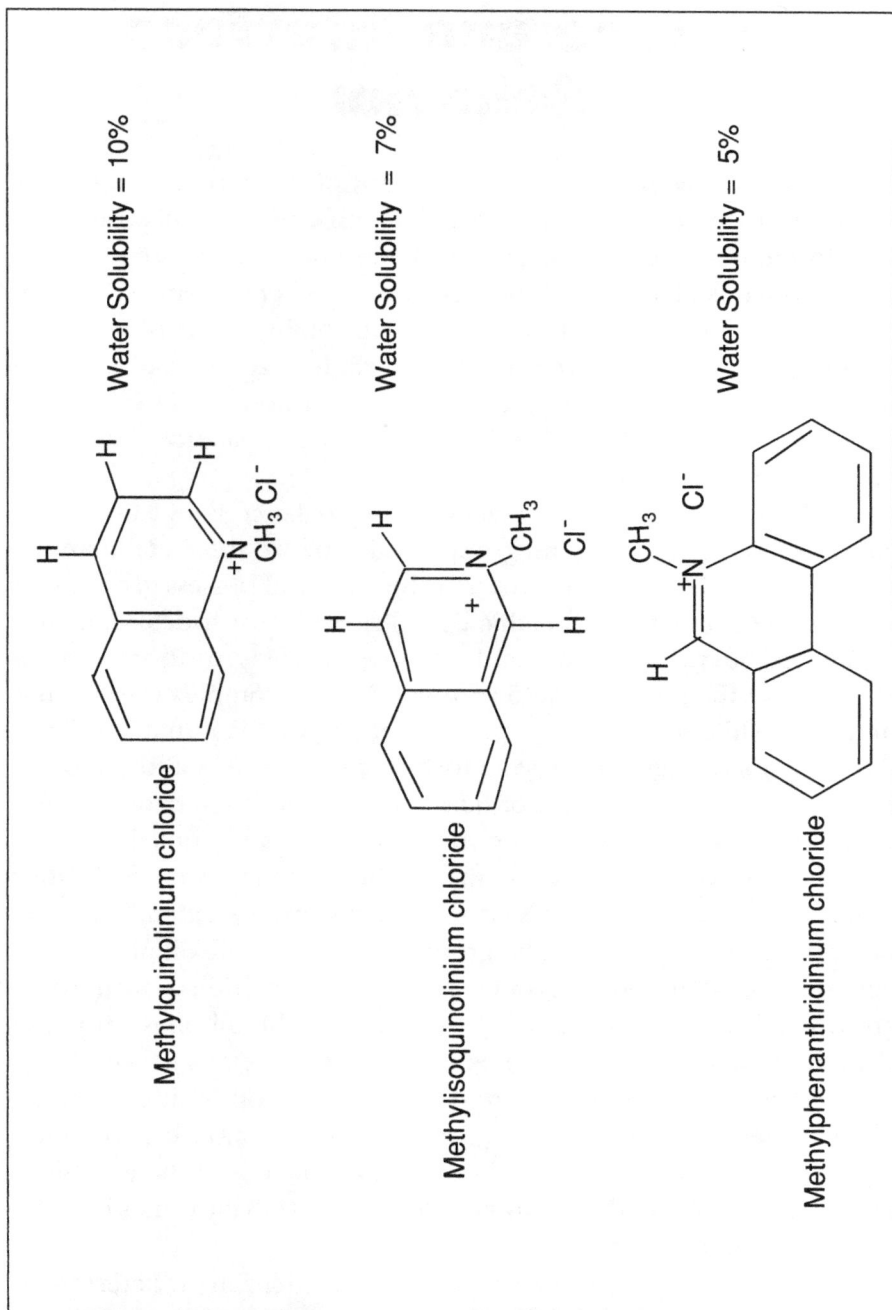

Fig. 14-2 Enhanced water solubility of the quaternary ammonium chloride salts of the quinoline type structures

Surface and Interface Behavior

In some ways it is fortunate that many of the corrosion inhibitor chemicals exhibit greater oil dispersibility than water. Since the complex formed between the metal and the inhibitor should resist attack by the aqueous phase, water solubility would be a problem. Thus, film persistency is often studied to determine the ability of the film formed by the inhibitor to resist removal by a turbulent aqueous phase. The need for film persistency and the need to traverse the interfacial water/metal surface boundary is then a tenuous balance that is frequently difficult to achieve.

The primary means of achieving this balance is by choosing a chemical with a higher ligand group activity toward the metal than the corroding species. This requires that the chemical possess groups that successfully compete for metallic ligand sites against hydroxyl, hydrosulfide, carboxyl, carbonate, and chloride anions. An obvious disadvantage that the inhibitors suffer from is that the complex they form is nonionic, while the attacking species is often anionic in nature. Thus, the complex strength is a driving force for successful ligand site occupation. This added strength of ligand group site interaction is often gained by pi complexes between the inhibitor and the metal.

Because of the bipolar nature of the corrosion chemicals, they orient their structures such that their polar groups extend into the water phase and their nonpolar groups extend into the oil phase. This tendency to align polar groups (the amine fractions) with polar groups and their nonpolar (alkyl) groups with the oil phase at phase boundaries allows polar group exposure to the metal surface. These alignments result from the need to minimize discontinuities between phase surfaces. In the emulsion form of the inhibitor, it is oriented with the polar fractions directed to the outer surface of the emulsion, making it available to the metal surface. Figure 14–3 illustrates how the corrosion inhibitor covers the metal surface.

Much of the activity of corrosion control chemistry is determined by its contact with the metal surface to be protected. If the product partitions into the oil phase of a mixed-phase fluid system, it may never

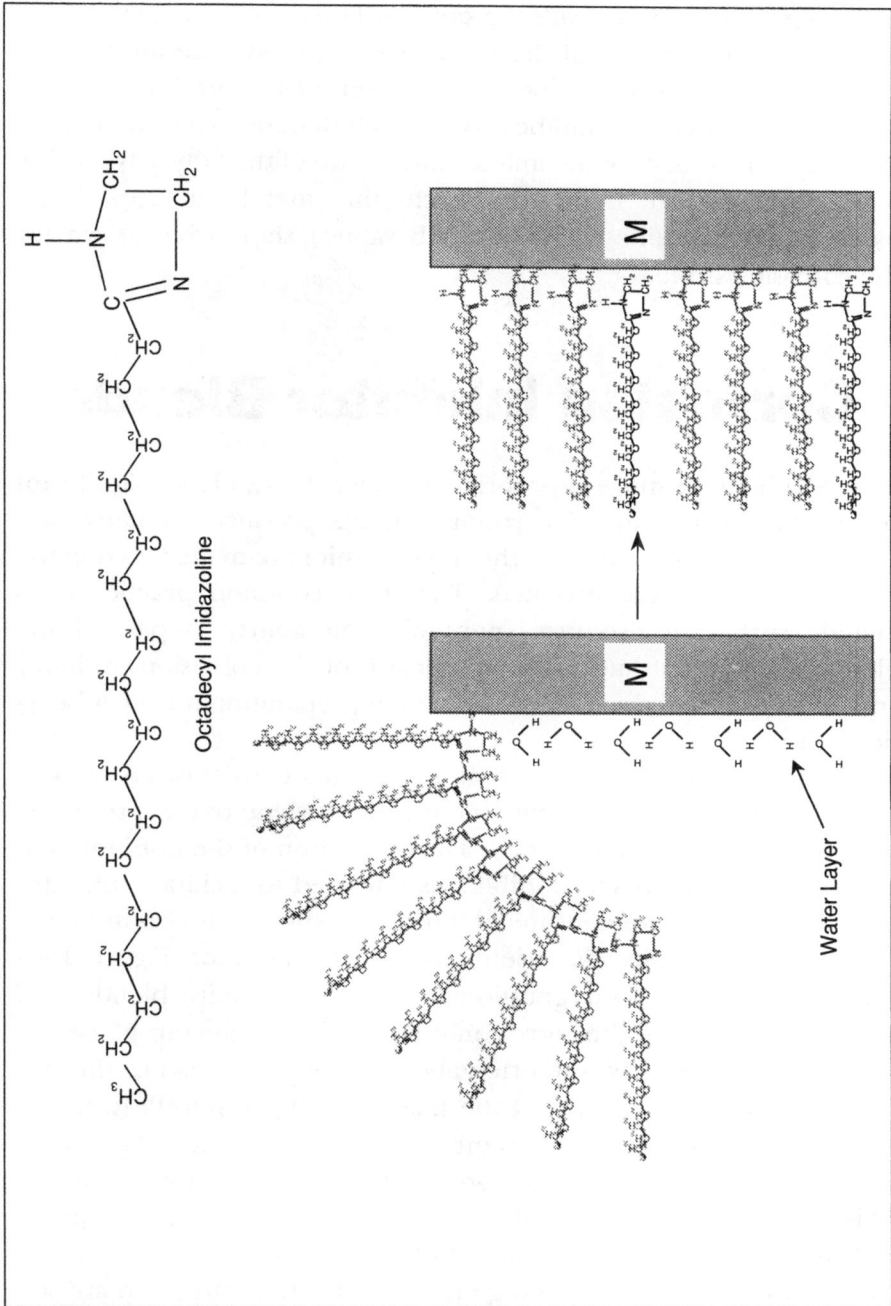

Fig. 14-3 Imidazoline corrosion inhibitor at metal surface, showing displacement of water

make significant contact with the surface. Thus, careful control of the solubility characteristics of the product is required. The imidazoline alkyl groups must provide for emulsification of the product, oil, and water. The emulsion should be formed such that the external phase of the emulsion consists of the imidazoline group of the fatty alkyl derivative; otherwise, it may not reach the metal surface. Thus, hydrophilic/lipophilic balances (HLB values) should be known for the corrosion inhibitor.

Corrosion Inhibitor Blends

Solubility, or more appropriately dispersibility, plays a significant role in the effectiveness of corrosion inhibitor products. In many cases cooperative blends (some call them synergistic) are made to aid in the dispersibility of these products. The most common practice is to include surfactant products (nonionic detergents) as part of the blend. These agents aid in the dispersion of the corrosion inhibitor and, in some cases, clean metal surfaces in preparation for the adsorption of the inhibitor.

The most common nonionic surfactants consist of an anionic nucleus molecule with varying amounts of ethylene oxide, propylene oxide, or mixture of both added. The association of the nonionic surfactant and the corrosion inhibitor is intended to facilitate inhibitor passage across the aqueous metal interface region. It is also intended to assist in the molecular orientation of the inhibitor. Figure 14–4 shows the expected configuration of these cooperative blends, and how they help orient the corrosion inhibitor in an aqueous phase.

Figure 14–4 shows the orientation of the polar head of the imidazoline toward the center of the macro-aggregate micelle with the fatty alkyl groups of the surfactant and inhibitor aligned. The water-dispersible ethylene oxide fraction is at the exterior of the aggregate. This arrangement is expected to be accommodated by an oil phase and disperse readily in an aqueous phase. In each case of surfactant blends discussed we have used generic chemistries, but there are several combinations of these methods that are employed by specialty chemical manufacturers.

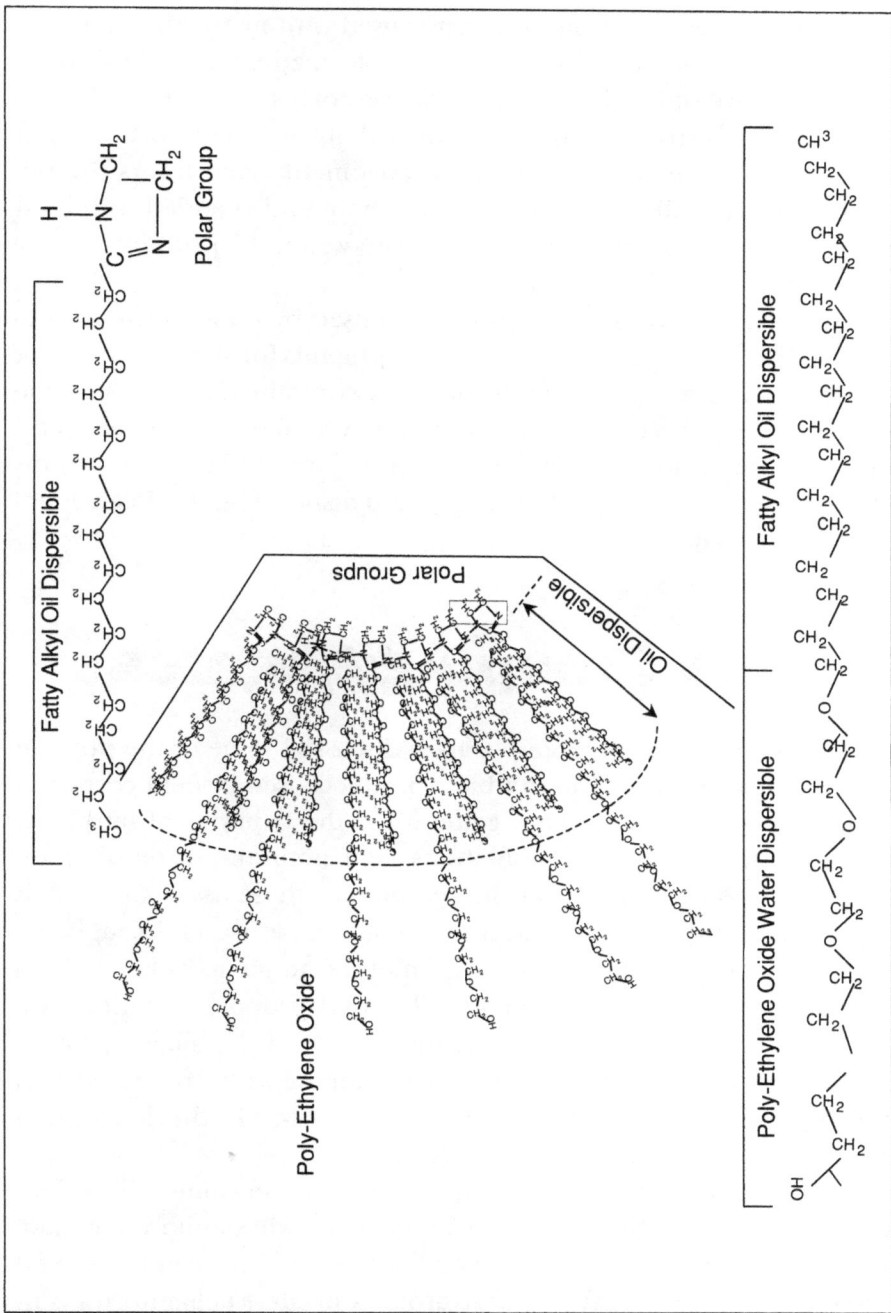

Fig. 14-4 Ethylene oxide nonionic surfactant blend with imidazoline corrosion inhibitor

Another type of surfactant that is used with nearly the same frequency as the non-ionic type involves the formation of alkyl-substituted aromatic sulfonic acid salts of the amine corrosion inhibitor. Figure 14–5 shows this structure and its suspected arrangement with the imidazoline structure. The molecular arrangement again favors the oil-soluble external alkyl projections when the blend is added to the oil phase, but when it comes into contact with water, the product is easily dispersed.

Fatty alkyl carboxylate salts are also used by some in the formulation of corrosion product blends. These blends form a weaker salt of the amine and have a lower dissociation constant than the sulfonic acid salts. They also undergo self-assembly (e.g., fatty alkyls of acid and amine interact through London inductive forces). The alkyl groups are projected to an external oil matrix and disperse readily in an aqueous environment.

Like Group Aggregates

Throughout this chapter we have discussed some of the physical properties of the corrosion inhibitor products used in field corrosion product formulation. We have generalized the behavior of fatty alkyl associations, amine salt combinations, and nonionic dispersibilities, but we have only touched on the reasons for these associations. Salt formation of the sulfonic acid amine combination acts through the normal mechanism of acid protonation of the amine to yield products with improved water dispersibility. The alkyl group associations are more complicated and involve the interaction of the alkane groups, which possess no dipole moment. When amine salts are added to a mixed solvent system, they migrate to regions within the system that best accommodate their structural groups.

In aqueous-containing petroleum systems, the amine will be best accommodated by the oil phase, while the salt will be more compatible with the aqueous phase. Thus addition of the amine salt to the oil phase produces an aggregate that projects its alkyl tails into the surrounding media. The polar head is directed toward the center of the

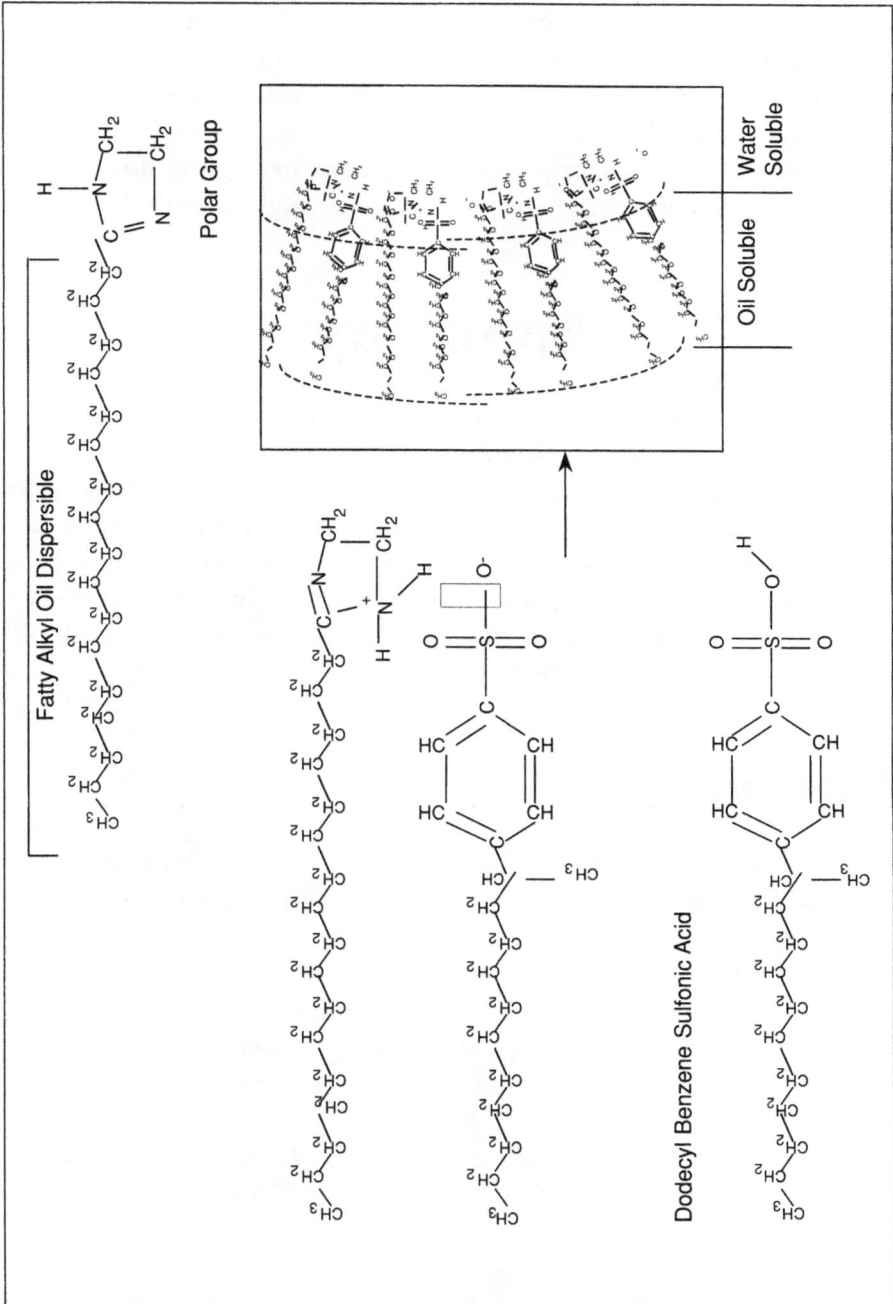

Fig. 14-5 Sulfonic acid salt of an alkyl imidazoline corrosion inhibitor showing its micelle arrangement

micelle. Conversely, if the amine salt is added to a predominately aqueous environment, the nonpolar alkyl groups will associate at the center of the micelle. The polar heads will project into the aqueous phase. This behavior is not surprising, since the compatibility of one phase with that of another is largely determined by like group association. Powerful driving forces are the result of attaining minimized entropy.

Summary

The preceding chapter discussed the complex nature of petroleum systems and their interaction with corrosion products. A general discussion of some of the physical properties associated with aromatic amines and their salts illustrated how these forms are employed to produce chemicals that can reach the metallic surface. In connection with facilitating inhibitor metallic contact, product blends were discussed and views of their orientation in complex multiple phase systems were developed.

References

Barrow, Gordon M. 1966. *Physical Chemistry.* 2d ed. New York, St. Louis, San Francisco, Toronto, London, and Sydney: McGraw Hill Book Company.

Dickerson, R. F., H. B. Gray, and G. P. Haight, Jr. 1970. *Chemical Principles.* 1st ed. New York: W. A. Benjamin, Inc.

Huheey, James E. 1978. *Inorganic Chemistry Principles of Structure and Reactivity.* 2d ed. New York, Hagerstown, San Francisco, and London: Harper & Row.

Jones, Loyd W. 1988. *Corrosion and Water Technology for Petroleum Producers.* 1st ed. Tulsa: OGCI Publications, Oil & Gas Consultants International, Inc.

Handbook of Chemistry and Physics. 1976. 56th ed. Cleveland: CRC Press.

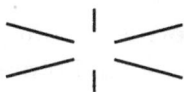

15
Phase Behavior of Scale Chemicals

Scale Inhibitor Solubilities

Scale inhibitors are generally intended to partition into the aqueous phase of petroleum fluid systems. Because of this, they are not formulated with surfactant packages as extensively as the corrosion inhibitors described earlier. Very often the commercial inorganic scale inhibitors are sold as aqueous dilutions of anionic salts. Thus the sodium and potassium salts are quite common in this product area. The polyamine type products are often neutralized with mineral acids (hydrochloric and sulfuric) to produce highly water-soluble products. Free acids are also produced (e.g., polyacrylic, phosphonic, sulfonic, and sulfonamides).

However, formulations with nonionic and bipolar organic acids are sometimes necessary, particularly when the system in which it is applied consists mainly of oil. Thus, some of the same rules of blending and types of products used in corrosion inhibitor formulations are employed in the formulation of scale blends.

Scale Inhibitor Blends

Formulation of scale compounds with surfactants proves to be considerably different than blending corrosion products. Choosing the polyamines as a starting point for blending procedures, a picture of the differences between corrosion inhibitor and scale inhibitor surfactant packages can be developed. Polyamines, used for scale inhibition, are frequently composed of repeating amine nitrogens separated by ethyl groups. These polyamines typically contain from two to five amine nitrogen atoms and exhibit various ranges of water solubility. When a system to which these amines are to be added is primarily oil, surfactant packages are added to enhance their oil dispersibility. Figure 15–1 shows a few of the polyamine surfactant combinations.

The macro-aggregate is organized such that the polar salts tend to concentrate in oil systems in a certain orientation. The surfactant alkyl group is extended into the oil phase with the polar heads in the center of the micelle. Thus, the molecular arrangements of the oil-dispersible scale blends are able to be suspended in oil as macro-aggregates and in water as freely dissociated salts.

Anionic Salts

Figure 2–6 illustrated some of the polar groups associated with scale inhibitor chemistry that are usually blended as sodium or potassium salts in an aqueous solvent.

The polar groups used for scale inhibitor chemicals are highly anionic in character, and as a consequence, they undergo considerable dissociation when placed in water. The macro-aggregate form taken by these products is a function of both the media into which it is placed and the nature of its nonpolar substituent. Thus, blending of these products can also be conducted based on the solubility traits of the nonpolar fraction of the anion. In some cases the acid form of these anionic products can be blended with the polyamines to produce salts with activities of both the polyamine and the anion.

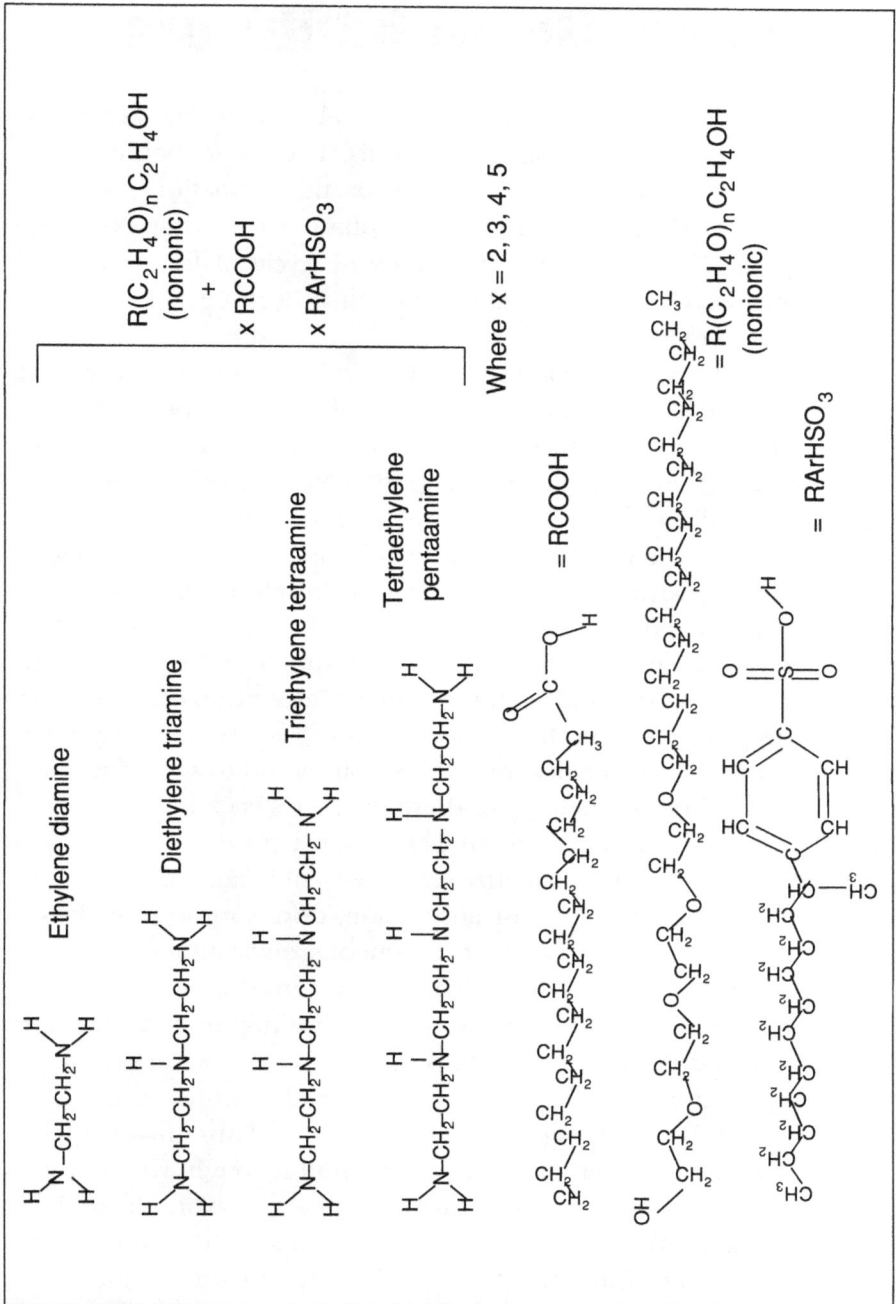

Fig. 15-1 Polyamines used for scale control with surfactants added to produce enhanced oil-phase dispersibility

Acidic Scale Inhibitors

Protonated forms of the anionic scale inhibitors are often employed as "squeeze treatment" chemicals. The theory behind these inhibitors is that they can be adsorbed by the formation rock and released slowly over time to the aqueous phase, where they can interfere with scale formation. A very interesting cycle of liquid to solid adsorption and reversion to solid to liquid de-adsorption is involved in the scale squeeze technique.

Usually the squeeze chemical is forced back into the formation of a well, where it deposits on the silicates of the sandstone. The well is brought back on production, and the chemical is monitored from time to time. The adsorbed chemical gradually de-adsorbs from the silicate rock and disperses in the water phase of the production fluids. As the temperature and pressure change within the wellbore, scale begins to form. As the scale forms, the free scale inhibitor adsorbs to the forming crystals and alters the crystal form. The altered crystal forms cause disruptions in the normal scale formation, and the scale is rendered incapable of depositing in the well tubing or transfer lines.

This squeeze technique can be considered phase behavior, since the adsorption/de-adsorption process is from liquid to solid and solid back to liquid. The reversibility of these processes is extremely important to the effectiveness of squeeze treatments. Figure 15–2 is a depiction of polyacrylic acids in the process of reversible adsorption on silicate rock. The driving force for adsorption/de-adsorption involves a diffusion gradient that arises as the aqueous environment changes with the reestablishment of fluid production from the formation.

The amount of scale inhibitor released is a function of the equilibrium concentration of the inhibitor in the connate water (capillary water-silicate coating). It is also a function of the diffusion gradient between it and the changing external "free water." Thus, the changing external environment acts as a dialysis medium through which differential concentrations of inhibitor diffuse, seeking a favorable equilibrium concentration. The squeeze process application is more commonly performed with phosphonates and phosphate esters, since they can be monitored quite readily by spectroscopic methods. The polyacrylic acid applications are more difficult to monitor.

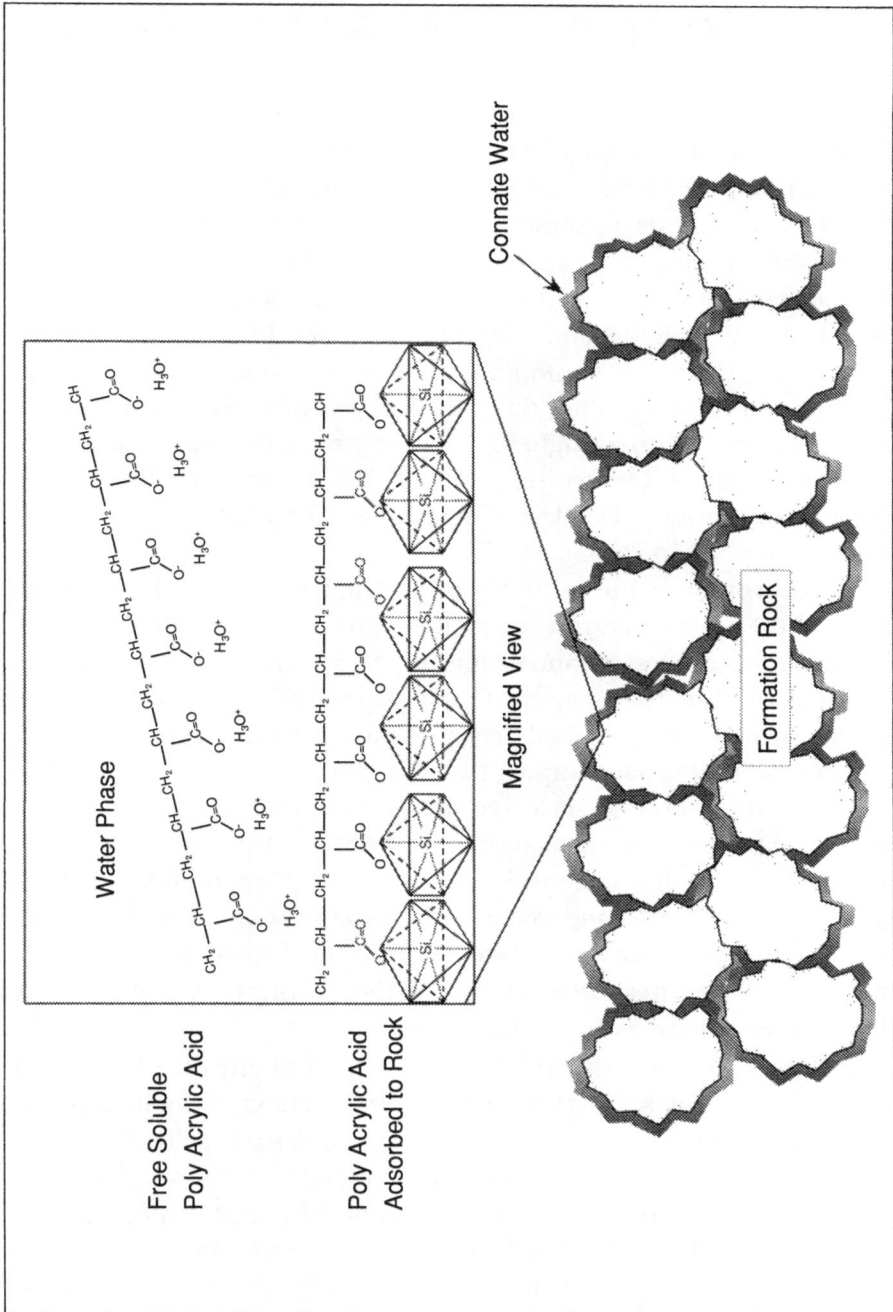

Fig. 15-2 The adsorption and de-adsorption of polyacrylic acid with silicate rock

Scale Inhibitor Adsorption

The initial driving force for scale and inhibitor interaction is determined by the strength of association of the ligand group (scale inhibitor) with the ligand site (potential scale complex). In the case of ionic scale inhibitors, the anionic forces between the ligand group and the ligand site must be competitive with the forces existing between the natural ligand groups. (These include carbonates, sulfates, and sulfides.) Alternatively, the concentration of the inhibitor must be greater than that of the natural anionic ligand groups present in solution. In the case of the polyamines, the statistically favorable combinations of the amine groups on a single ligand group must overwhelm the natural ligand groups. Otherwise, it will be less likely to exchange with other ligand groups. This last effect is called the chelate effect, which has been described earlier.

Adsorption of the inhibitor to a forming scale must take place prior to the scale's network formation. It follows that scale formation is a multistep process. If one examines the formation of scale, it soon becomes obvious that scale is a macro-aggregate form of individual crystal forms. The individual crystal forms, in turn, are macro-aggregates of the individual complex metallic salts.

The formation of scale requires a minimum of three steps of aggregate interaction, nucleation, and primary and secondary crystal interaction. The first stage is the formation of the complex metal salt (nucleation). The second stage is the organization of the complex metal salts into a crystal structure, and the third stage is the networking of the crystals to form scale. Figure 15–3 illustrates this three-stage development process of scale formation.

It is possible for the inhibitor to interact at any of the stages of aggregation of the scale. However, the effectiveness of the interaction and subsequent crystal deformation is greatest early in the stages of scale formation. One interesting consequence of crystal formation involves the increase in local concentration of scale-forming species, which is a result of the crystallization. A concentration gradient (diffusion force) arises as a consequence of crystallization that in turn acts to attract the scale inhibitor molecules as well as the natural ligand groups.

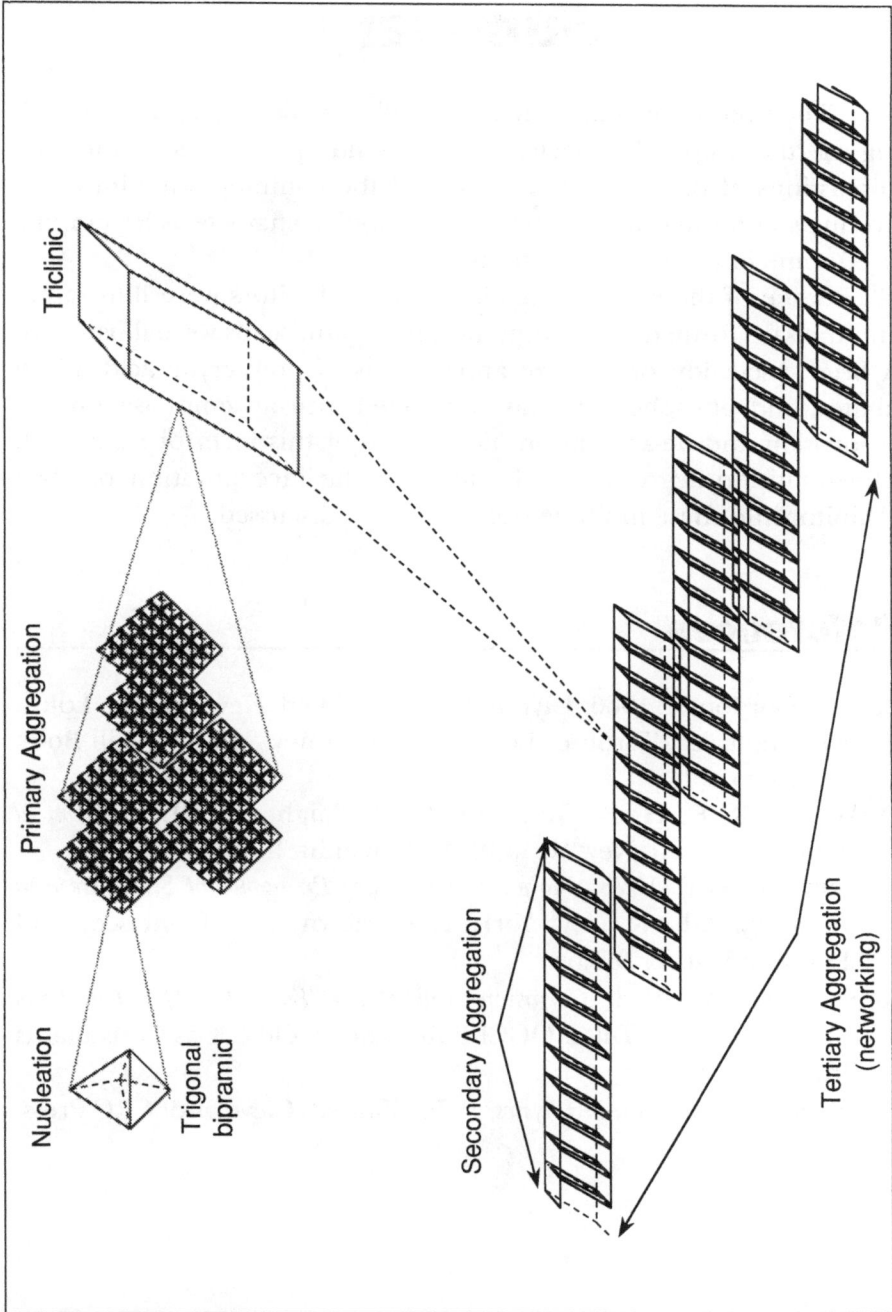

Fig. 15-3 Three-state development process for scale

Summary

The preceding chapter was directed at achieving an understanding of the physical aspects of the blending of scale inhibitors. Discussions of the general chemistry of the common scale inhibitor products and their active polar and nonpolar characteristics emphasized some of their solubility traits.

Some of the types of blends of scale inhibitors were illustrated, and the salts formed with inorganic and organic acids were illustrated. A brief discussion of squeeze applications of polyacrylic acid, phosphates, and phosphonates was conducted with an emphasis on the adsorption and de-adsorption mechanism of this form of treatment. The multistep aggregation of scale and the incorporation of scale inhibitor into these multiple steps were also discussed.

References

Barrow, Gordon M. 1966. *Physical Chemistry*. 2d ed. New York, St. Louis, San Francisco, Toronto, London, and Sydney: McGraw Hill Book Company.

Dickerson, R. F., H. B. Gray, and G. P. Haight, Jr. 1970. *Chemical Principles*. 1st ed. New York: W. A. Benjamin, Inc.

Huheey, James E. 1978. *Inorganic Chemistry Principles of Structure and Reactivity*. 2d ed. New York, Hagerstown, San Francisco, and London: Harper & Row.

Jones, Loyd W. 1988. *Corrosion and Water Technology for Petroleum Producers*. 1st ed. Tulsa: OGCI Publications, Oil & Gas Consultants International, Inc.

Handbook of Chemistry and Physics. 1976. 56th ed. Cleveland: CRC Press.

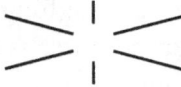

16
Petroleum Fluids Corrosion and Scale Testing

General Corrosion Testing

In the oilfield, the types of metal employed in equipment are numerous, but because of the cost versus performance properties of mild steel, it is the most commonly used. Thus, it is frequently employed as the measuring gauge of the effectiveness of corrosion protection. Weight loss is probably the most common form of measurement of corrosion control effectiveness.

Mild steel coupons are used in a large number of these test procedures and are placed in the production, transport, and storage areas of petroleum production facilities. They are measured periodically to determine weight loss. In addition to the placement in actual operating environments, they are also used for laboratory testing procedures. The degree of sophistication of these laboratory tests ranges widely. These vary from simple exposure to an aqueous mixture of mineral acid to electron microscopy of the metal surface after exposure to various corrosive environments. One such measurement involves precoating the coupon with an inhibitor, exposing the treated coupon to a corrosive mixture (e.g., aqueous mineral acids), and determining the weight loss.

Another method involves adding the corrosion inhibitor to the mineral acid solution and subjecting the pre-weighed coupon to this mix for a period of time, then re-weighing the coupon to determine weight loss. Frequently the temperature is elevated by the placement of the corrosive blend and coupon in an elevated temperature bath. Several variations of this test procedure are employed. Often it is necessary to determine the effects of agitation on the coupon, and this is accomplished by methods such as stirring, fluid circulation, and pumping.

Increased Severity Corrosion Testing

In addition to mineral acid exposure testing, other tests involving corrosive mixtures (e.g., hydrogen sulfide, carbonic acid, and hydrogen cyanide) are conducted in the laboratory. Many of these procedures are conducted in high-pressure, high-temperature autoclaves in an effort to reproduce some of the field conditions. In these tests, coupons with metal composition representative of the field equipment are placed in an aqueous environment in an autoclave. The autoclave is pressurized to thousands of pounds per square inch with the offending gas (e.g., hydrogen sulfide, carbon dioxide, or hydrogen cyanide). It then is heated to field temperatures and measured for weight loss at the conclusion of the test.

Electrical Methods

With the advent of the potentiostat, field methods and laboratory methods of corrosion analysis have been employed. In many tests, actual field fluids can be used. Additionally, approximations to the corrosive environment prevalent under field conditions are often made. The field unit is actually placed in the environment to be monitored, while in the laboratory a considerably more sophisticated set of instruments is employed. Field measurement requires that a unit be attached to the probe that may be remote from the central unit. Laboratory measurements are generally conducted with the test probe immersed in either the actual field fluid or a synthetic approximation. Figure 13–1 previously showed a schematic of the laboratory unit.

Field probe units are designed to withstand operating pressures and temperatures and are necessarily more compact than the laboratory units. Several specialty chemical companies produce their own versions of electrical corrosion meters, and the designs are quite varied. A quite general design is based on the concept of current imposition in opposition to the corrosion current. In order for this technique to be used, there is a requirement that the instrument be capable of balancing its input current to that of the corrosion current. Thus, most of these devices require a three-electrode arrangement in the probe. This three-electrode arrangement is illustrated in Figure 16–1.

The three-probe potentiostat device is mounted inside the tubing, flow lines, or storage vessels, while the instrument is situated in a remote location. The instrument readings can be recorded or stored digitally in a computer, to be examined at some point after the data have been acquired. Several advantages are obtained by the employment of these corrosion probes. The most obvious advantage is the ability to determine corrosion potentials existing at the time of measurement, and the ability to increase or change chemical inhibitor treatment in response.

Ultrasonic Thickness Testing

Recent developments in electronics have made it possible to use ultrasonic measurement techniques to examine corrosion damage to tubing, pipes, and storage vessels. This instrument uses ultrasonic signals to measure changes in pipe wall thickness. Figure 16–2 illustrates the principle of thickness measurement using ultrasonic testing.

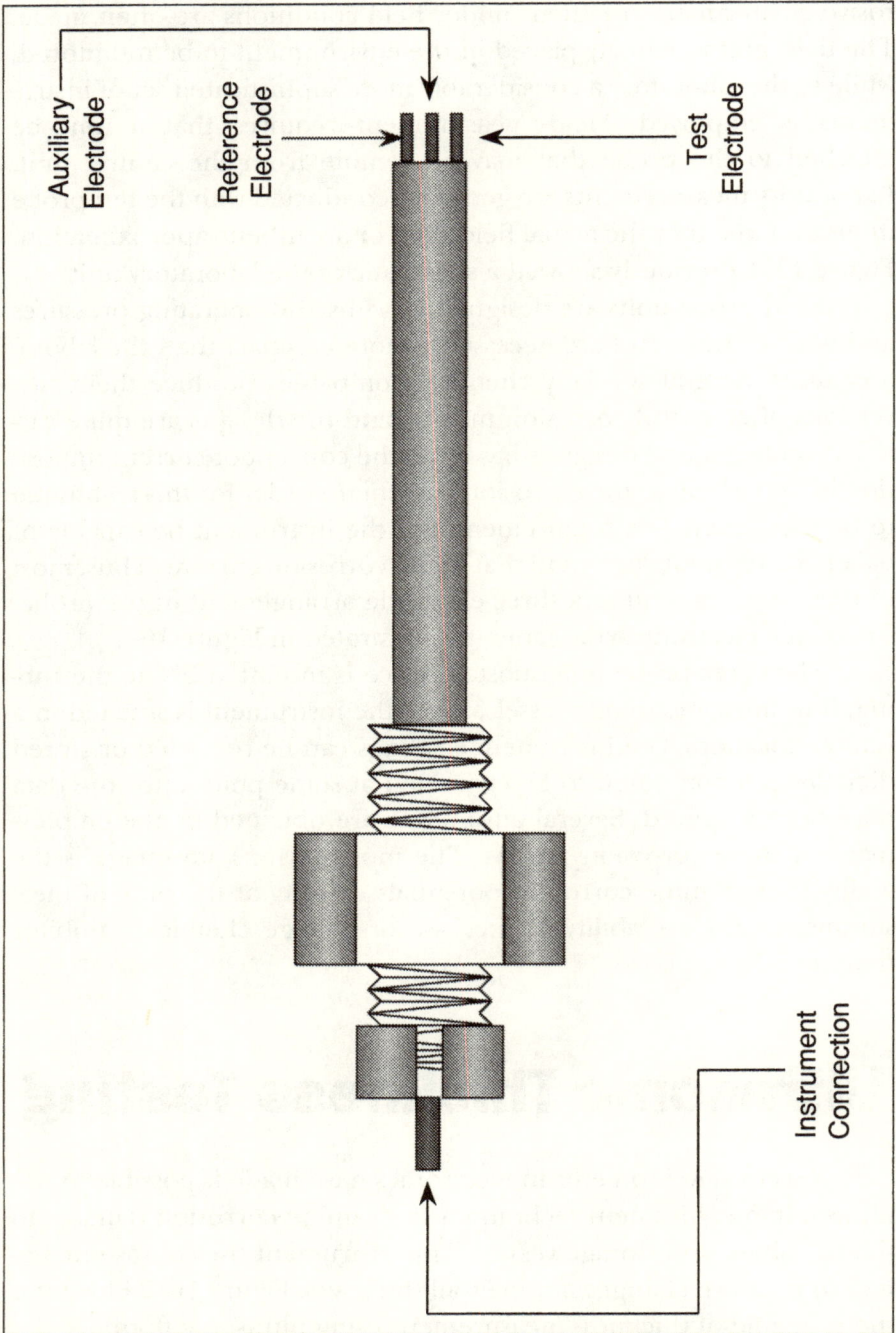

Fig. 16-1 The potentiostat test probe for field applications

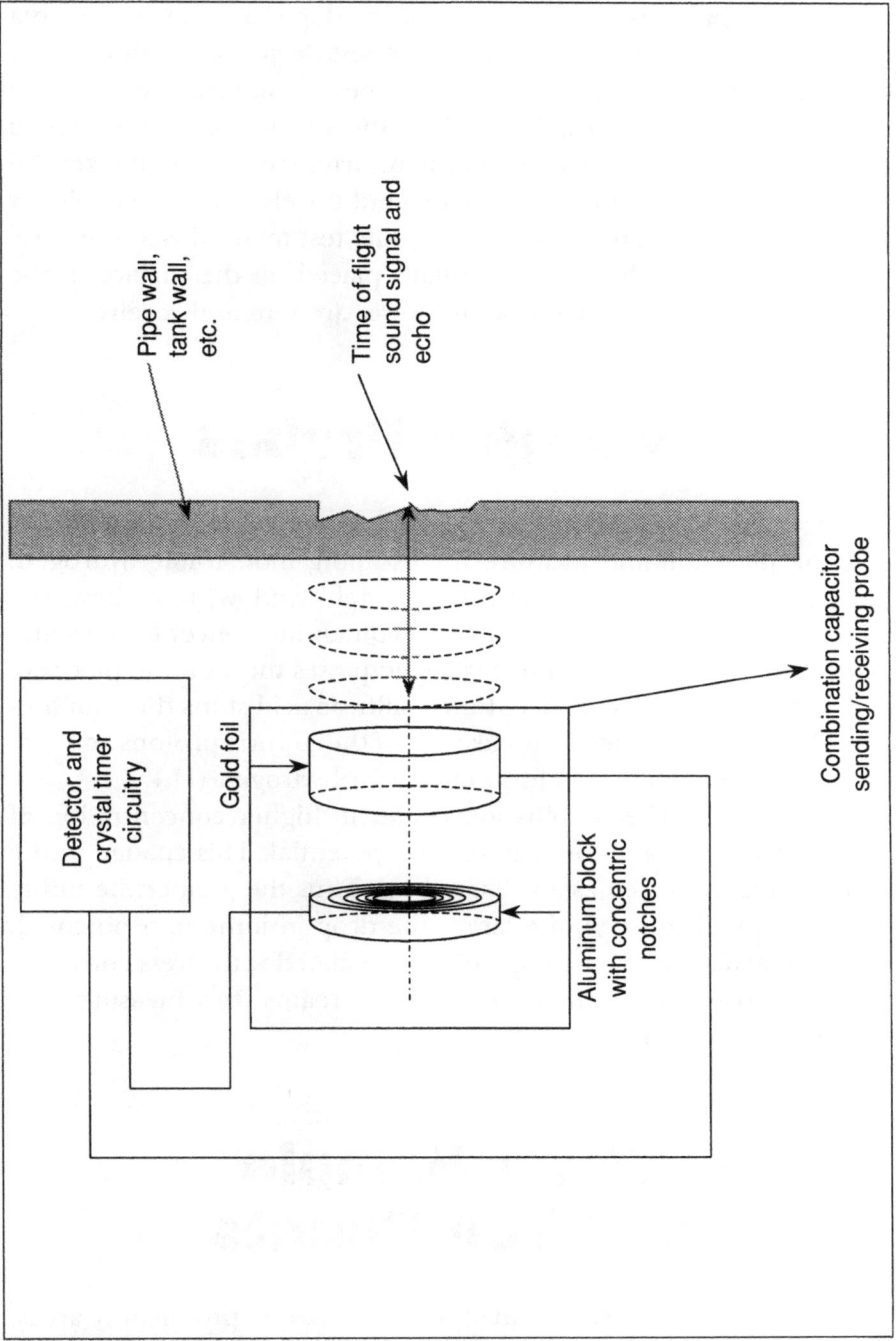

Fig. 16-2 Ultrasonic thickness testing device for testing wall thickness

Sound propagates at various speeds depending on the media through which it travels. The ultrasonic test device sends short ultrasonic sound wave bursts through a given media and measures the echo return time of the signal. In general, the shorter the time interval between the signal being sent and the returning echo, the thicker the material. The reason for this is that sound travels at a higher velocity in material with greater density. The sonic test method is a nonintrusive method, since the probe is usually placed on the surface of the pipe wall, and nondestructive sound waves are sent and received.

Hydrogen Probes

Hydrogen probes and patches are made that are mounted flush with the pipe wall and measure the escaping monatomic hydrogen. This technology uses a very thin film metal "window" that allows the monatomic hydrogen to enter a cavity containing concentrated sulfuric acid. When the monatomic hydrogen passes through the pipe wall and into the probe, the concentrated sulfuric acid shifts the equilibrium. This occurs by heterolytic cleavage of the sulfuric protons and radical-radical combination to form molecular hydrogen (H_2).

More radical recombination results in higher concentration of the sulfurous acid and lower ionization potential. This change in the ionization is measured as a voltage drop. Thus the greater the influx of monatomic hydrogen, the larger the drop in ionization potential. Since molecular hydrogen is generally associated with stress corrosion that takes place in highly corrosive gas streams, this measurement reflects the corrosion rate.

Water Sample and Iron Counts

Water samples are routinely collected from production areas, storage vessels, distillation units, heater treatment devices, transfer lines, and settling tanks. The samples are either tested in the field and/or sent to a laboratory for iron analysis. Field testing is usually done by forming colored complexes with the iron and other metallic species and measuring their absorbance of filtered light. Standard curves of filtered light absorbance versus iron, calcium, barium, and strontium content are then compared to those obtained from the field test sample.

Laboratory samples are usually stabilized by acidification with hydrochloric acid and sent to a central testing laboratory for later analysis. Laboratory analysis is usually performed using atomic absorption (AA analysis) or atomic emission (AE analysis) testing methods. Atomic absorption methods atomize the aqueous sample and measure the specific absorbance of a spectral frequency specific to metals in the sample. Atomic emission measures the electromagnetic emission spectrum specific to the various metals.

With computer interface capabilities and plasma temperature generators, AE has taken over as the preferred testing method. These devices are called inductively coupled plasma (ICP) units and measure specific emission spectra for several atoms simultaneously. The spectra obtained are then compared electronically with internally stored computer standards, and a measure of the concentration of a specific species is obtained. Thus, a complete array of metallic atom composition is obtained, and both corrosion and scale samples can be characterized using this instrument.

Anion Determination

ICP analysis is very accurate for the determination of the metallic species present in water samples. However, the anions associated with these metals are combinations of atoms (e.g., sulfate, carbonate, hydrosulfide, carboxylate, hydroxyl, bicarbonate, etc.) that are not easily detected by this method. Additionally, the combinations of these anions with metal complexes can be extremely confusing; for example, sulfate anions can be associated with barium, iron, calcium, and strontium, as can carbonate, carboxylate, or hydroxyl. Consequently, separate methods for the determination of anions must be employed.

During the last several years, methods employing specific ion electrodes have been developed that can distinguish between different anions in solution. These specific ion electrodes are also capable of measuring the concentrations of these anions. Thus, the specific ion electrode technology is generally used to determine the concentration of the anionic species present in either corrosion or scale samples.

X-ray Crystallography

Very often solid samples of scale are collected from petroleum fluid processing areas, and analysis is requested to determine the composition of the scale. Neither the ICP nor the combination of specific ion and ICP methods can reveal the composition of the scale. These methods can be used as a starting point, but full structural elucidation of the scale form must be acquired by other means. X-ray crystallography is the method of choice in these determinations.

First, a sample crystal of the scale is placed in a beam of X-rays and irradiated at a specific angle. The reflected X-ray pattern is collected as a set of spots on an exposed area of film, or a charge-coupled device (CCD). The x and y absorbance positions are digitized and saved in a computer matrix for further analysis. The image is transformed using internal algorithms involving Fourier analysis. This is based on the realization by Sir William Bragg that X-ray scattering patterns derived from crystals can be viewed as a three-dimensional distribution of varying electron densities. The transformed patterns are then a depiction of the actual scale's molecular arrangement.

Physical Methods of Scale Testing

Scale testing in the laboratory is often conducted by mixing brine waters from the field, heating them together for specific periods, filtering the precipitate, drying it, and weighing the amount produced. In this way chemicals can be screened to determine their effectiveness in scale prevention. Additionally, synthetic brines are produced by combining various compositions of metallic salts in aqueous blends containing a series of inhibitor chemicals. Measurements are used to determine the inhibitor's effectiveness.

These tests are commonly termed jar tests and represent quick and simple methodology. As field conditions become more severe, and scaling tendencies need more detailed analysis, increasingly sophisticated methods of analysis are required. Thus instruments that attempt to duplicate as many existing physical conditions of field operations as possible have been developed.

Scale Loop Testing Device

In an effort to duplicate as many existing physical field variables as possible, equipment and methods have been developed to test scale build-up under dynamic conditions of flow. One such device circulates the brine solution (field sample or synthetic brine) through a capillary tube and measures pressure as the scale deposits in the tubing. This unit can operate at pressures approximating those of the field by employing high-pressure positive displacement pumps such as HPLC pumps—high-pressure liquid chromatography pumps. Additionally, the appropriate temperature can be approximated by heating or cooling the capillary loop while circulating the scaling liquid brine. Figure 16–3 is a schematic of this device and shows how this instrument may be interfaced with a computer.

Core Testing Device

Frequently customers want to know how effective scale inhibitors are in squeeze applications. The only satisfactory way to show meaningful data to customers is through core testing. Core testing can be

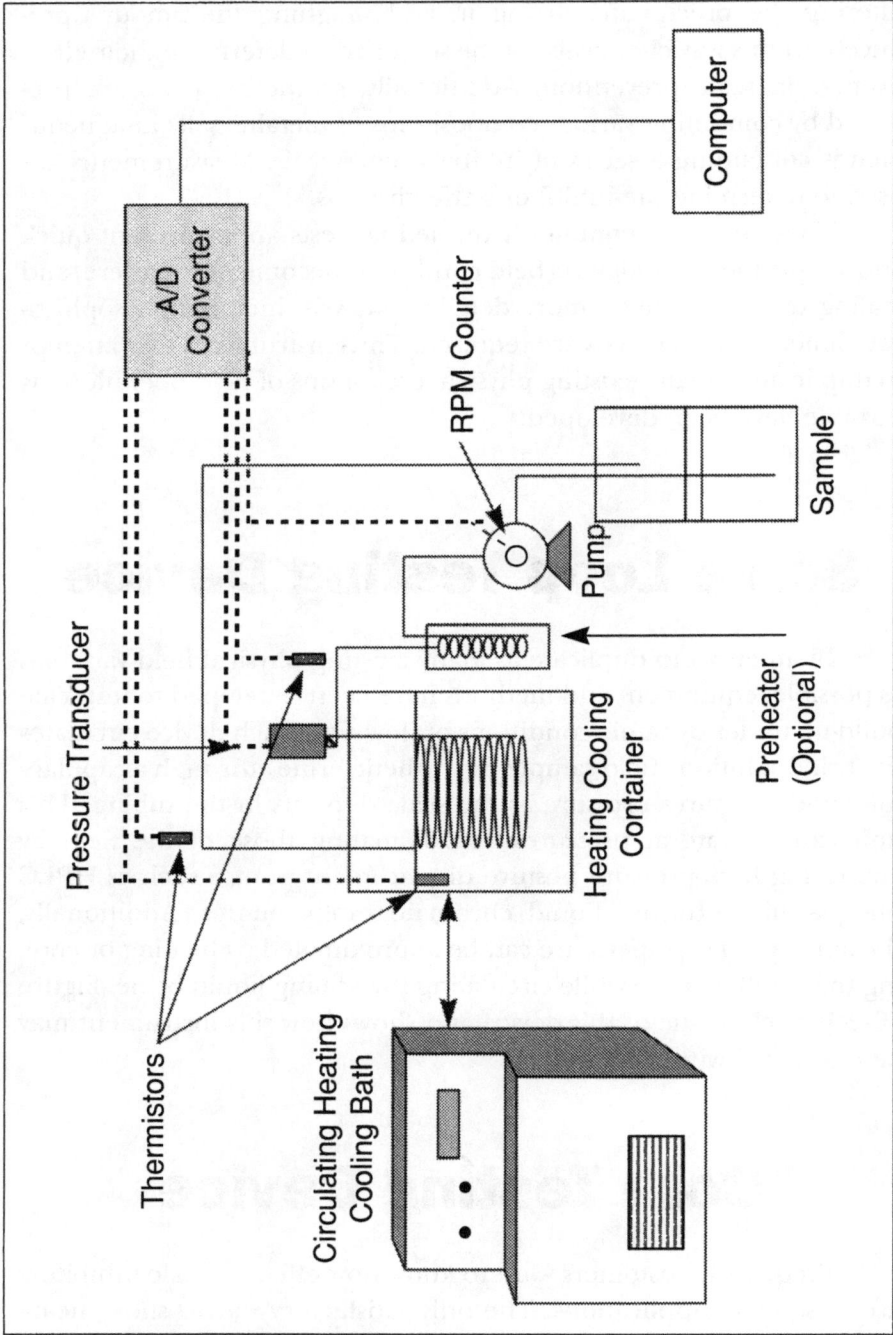

Fig. 16-3 A typical scale test loop

performed on synthetic Brea stone cores or on actual field core samples. Testing methodology varies, but the basic techniques are similar. The core sample is placed in a high pressure Hastler core holder. Next the inhibitor is placed in a reservoir and pumped on the core, just as it would be used in the field. Scaling brine is added to the reservoir and circulated through the core; the core can be heated to simulate down-hole conditions. Pressure is monitored, and the test continues until pressure profiles of inhibited and uninhibited samples have been collected. Figure 16–4 is a schematic diagram of the basic instrumental design.

Sonic Testing Device

The Sonic Portable Laboratory is an instrument that consists of a collection of electromechanical sending and receiving devices that are interfaced to a computer through an analog to digital converter (A/D converter). The concepts of operations performed by this unit are based on many of those employed by the devices mentioned above, with two notable exceptions. First, sound is used as a means of imparting shear forces to the sample, and a kinetic model is used in conjunction with temperature measurements to generate physical profiles of the system. Second, the device is designed to be portable and operate in a wide range of remote environments.

Because the unit employs probes for measurement, several measurements can be conducted remotely, under prevailing conditions of temperature and pressure. The unit is also modular and can be used as part of some complementary package including such other devices as the scale test loop described above. Figure 16–5 shows a schematic of this device.

Summary

The preceding chapter presented a very brief overview of the methods of testing corrosion and scale problems encountered in

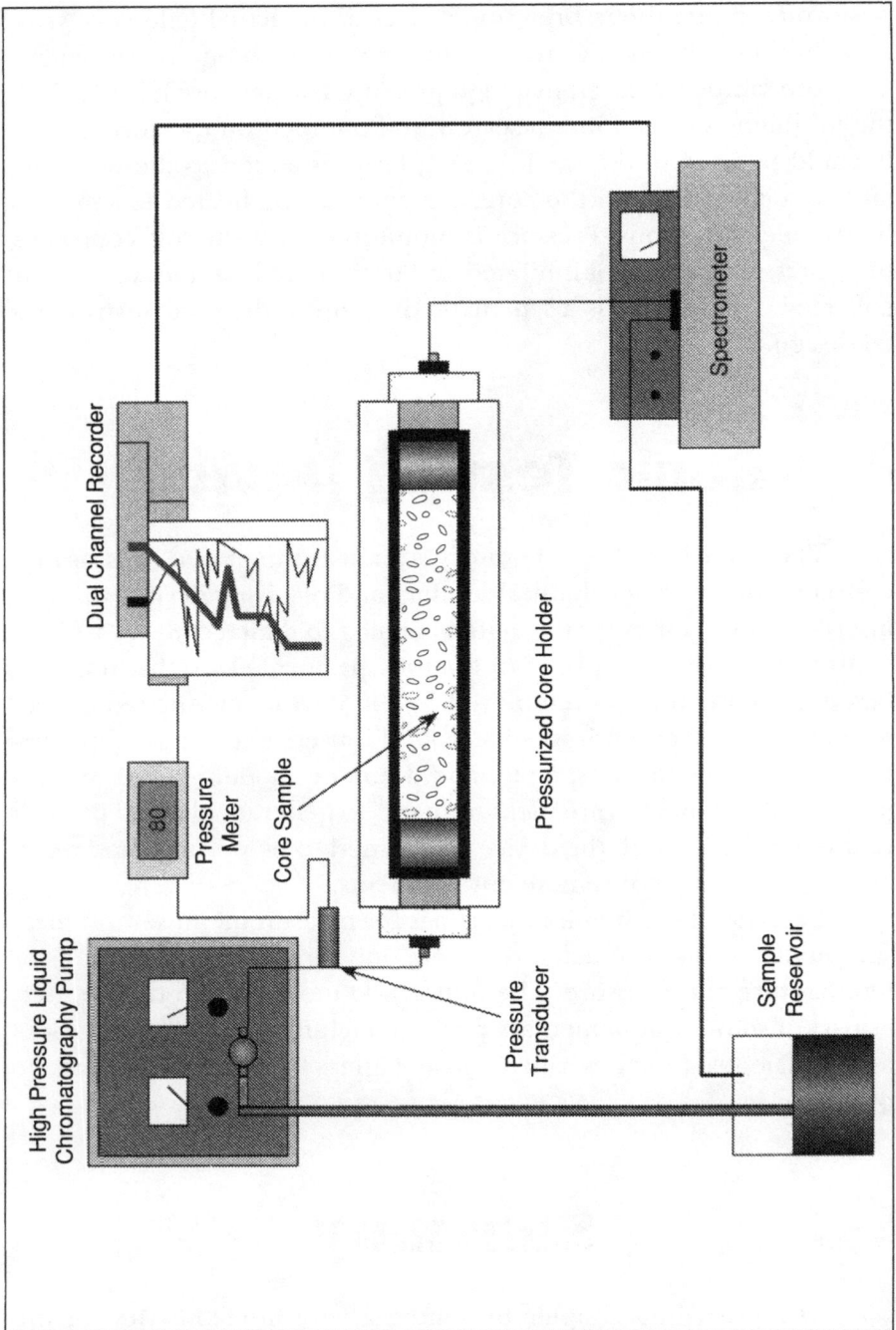

Fig. 16-4 The basic configuration of a core test apparatus

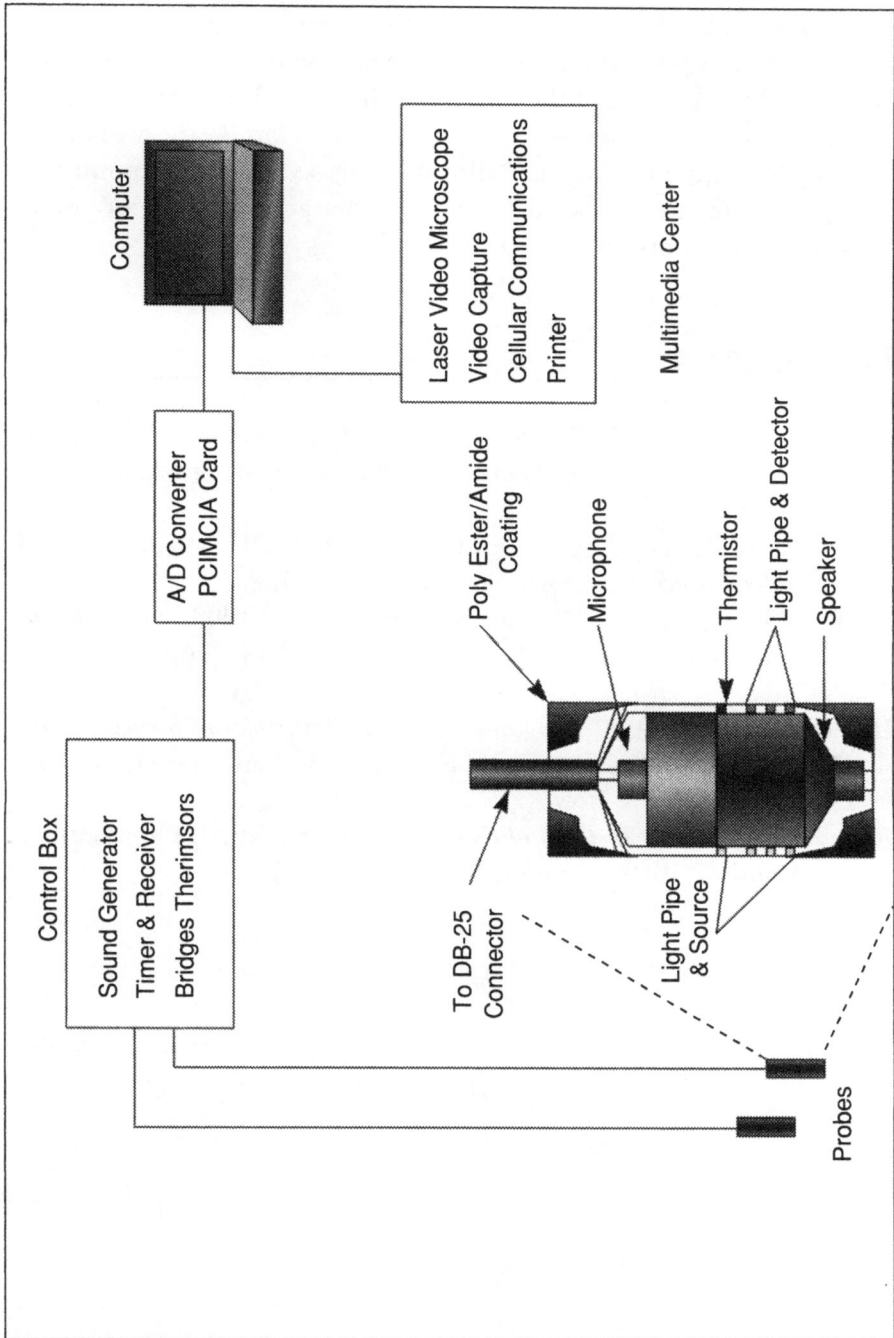

Fig. 16-5 The sonic portable testing device

petroleum fluids. Most of these test methods are available in standard format as ASTM procedures, and can be examined in more detail in this series. The specifics of the analysis techniques discussed are left as an exercise for the reader, since they appear in a large variety of other sources. Consequently, the main intent of this section was to point out the variety of techniques available for the characteristics of corrosion and scale as they appear in the field.

References

Barrow, Gordon M. 1966. *Physical Chemistry*. 2d ed. New York, St. Louis, San Francisco, Toronto, London, and Sydney: McGraw Hill Book Company.

Dickerson, R. F., H. B. Gray, and G. P. Haight, Jr. 1970. *Chemical Principles*. 1st ed. New York: W. A. Benjamin, Inc.

Hamill, William H. and Russell R. Williams, Jr. 1966. *Principles of Physical Chemistry*. 2d ed. Englewood Cliffs, New Jersey: Prentice Hall.

Huheey, James E. 1978. *Inorganic Chemistry Principles of Structure and Reactivity*. 2d ed. New York, Hagerstown, San Francisco, and London: Harper & Row.

Noller, Carl R. 1966. *Textbook of Organic Chemistry*. 3d ed. Philadelphia and London: W. B. Saunders Company.

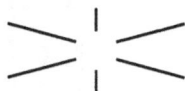

17
Synthetic Routes to Some Scale and Corrosion Chemicals

Alkyl Phosphonates

The alkyl phosphonates have been made for many years and have been found to be quite effective in the inhibition of scale formation in carbonate scale systems. Figure 17–1 shows how these products interact with calcium carbonate scale.

The phosphonates have typically been made in a reaction in which the primary alcohols are reacted with phosphorus trichloride. However, the phosphorus trichloride can be generated in situ by the reaction of phosphorous acid and hydrochloric acid. Thus, the overall reaction is as follows:

$$2\ ROH + H_3PO_3 + HCl \longrightarrow (RO)_2POH + 2\ H_3O+ + Cl^-$$

There are some important considerations in this reaction that involve the purity of the phosphorous acid. The phosphorous acid

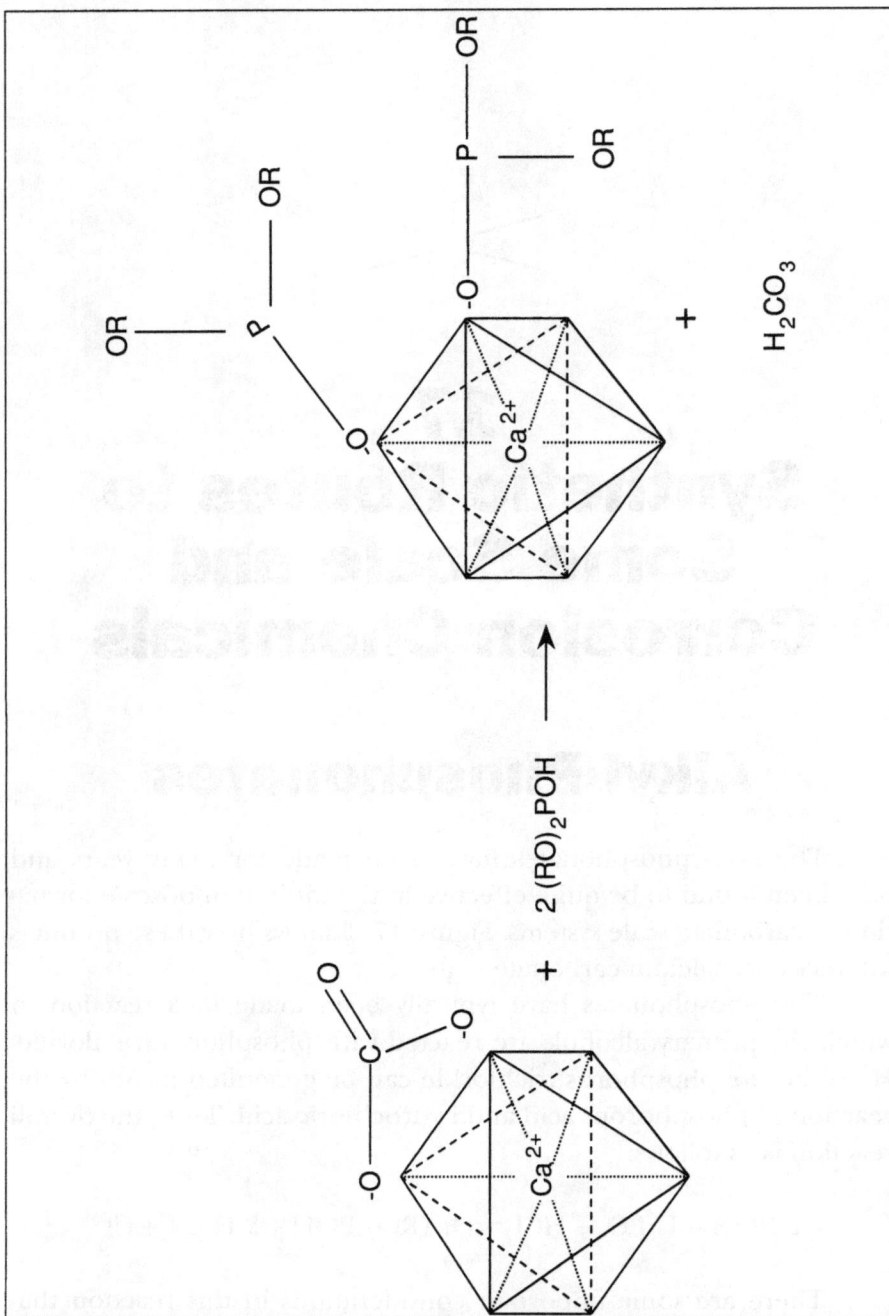

Fig. 17-1 Calcium carbonate undergoing phosphonate anionic ligand exchange

(H_3PO_3) must be as free of phosphoric acid as possible. It is generally analyzed and certified by the supplier prior to use in this reaction. Further, the alcohol (ROH) used in the synthesis must be a primary alcohol $(C_nH_{2n+1}OH)$. The primary alcohol requirement does not preclude the use of ethoxylated or propoxylated products such as $RO(C_2H_4O)_{n2}C_2H_4OH$ or $RO(C_3H_7O)_{n2}C_3HOH$. However, the hydrochloric acid can cause cleavage of the ether linkages if the reaction conditions are sufficiently severe.

The formation of phosphonate ester requires protonation of the primary alcohol, attack by the phosphorous anion, elimination of water, and protonation of the exiting water by the excess phosphorous acid.

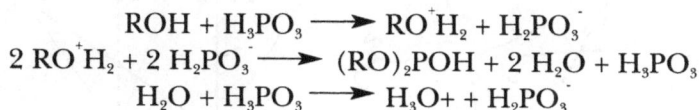

$$ROH + H_3PO_3 \longrightarrow RO^+H_2 + H_2PO_3^-$$
$$2\ RO^+H_2 + 2\ H_2PO_3^- \longrightarrow (RO)_2POH + 2\ H_2O + H_3PO_3$$
$$H_2O + H_3PO_3 \longrightarrow H_3O+ + H_2PO_3^-$$

The ethoxylated and propoxylated phosphonate esters are particularly active as chelating groups, since they possess unshared electrons on the ether oxygen. Thus, if the primary alcohol is the terminal alcohol of a polyether, the complexing abilities of the phosphonate are considerably increased. Figure 17–2 illustrates the complexing abilities of a polyether phosphonate.

Polyamine Derivatives

The polyamines are commercially available from the ethylene diamine through pentaethylene tetraamine. They are produced by the reaction of aziridine (the cyclic three-membered amine analog of ethylene oxide) and fractionally distilled to provide the products. These polyamines are superior complexing agents (chelating products) by themselves. However, they can be given anionic character by attaching anionic functional groups to the terminal primary amines. Figure 17–3 shows an example of the attachment of acetyl groups through the reaction of the polyamine with chloroacetic acid ester.

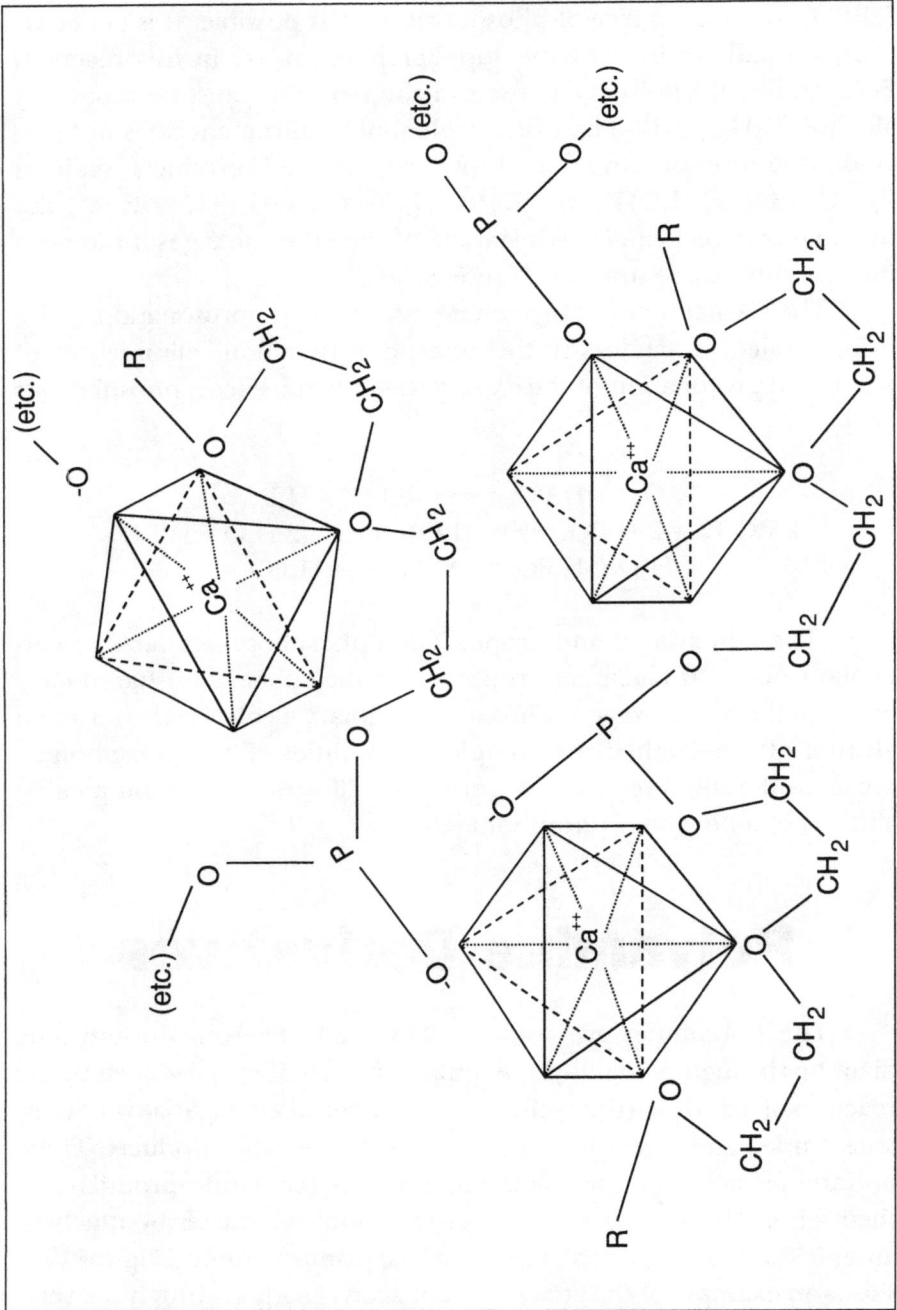

Fig. 17-2 The polydentate nature of the polyether phosphonate complex

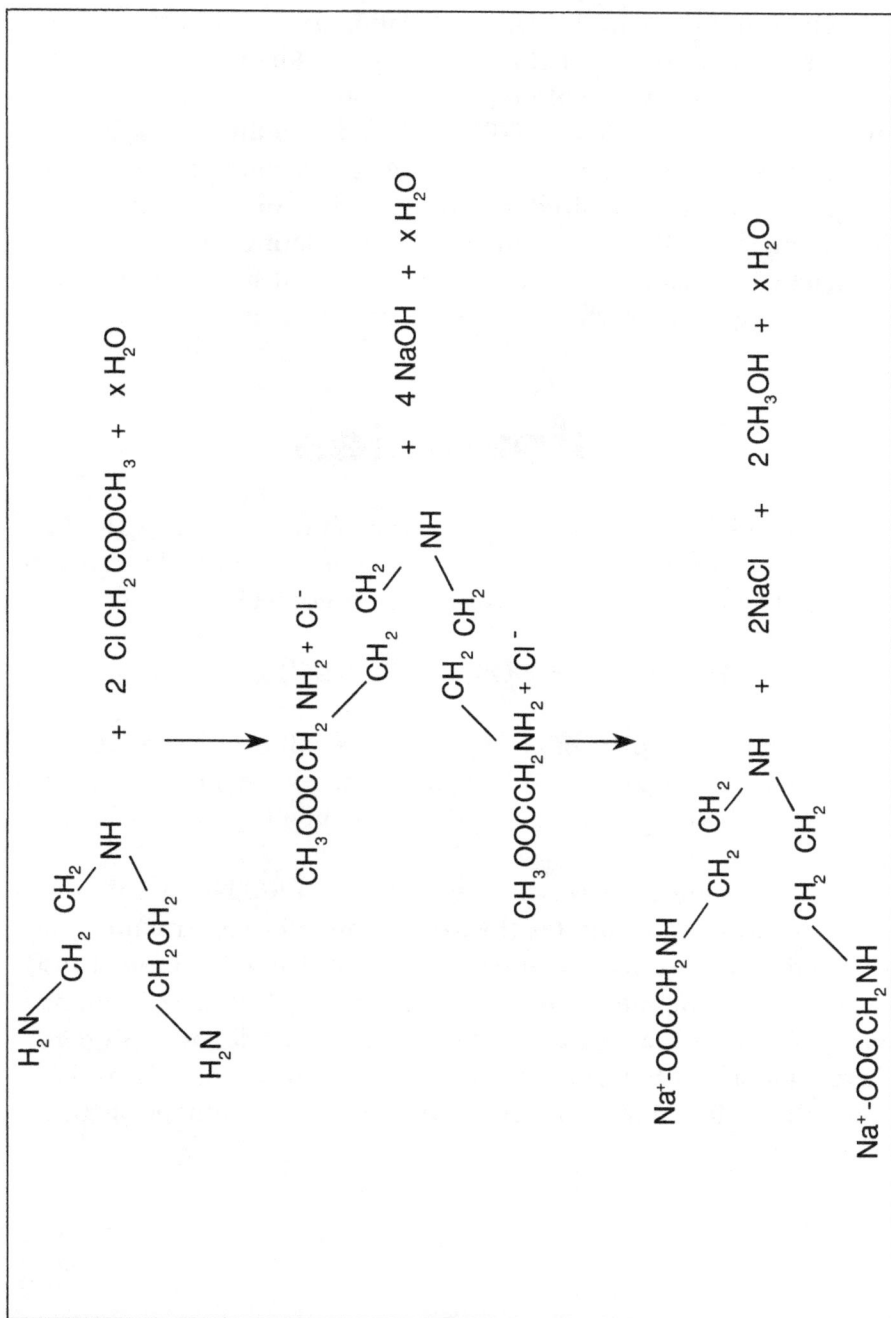

Fig. 17-3 Reaction of diethylene triamine with chloroacetometholate to form ethylenediaminedisodiumacetate

These products are known as chelating agents because they are able to interact with several of the ligand sites of metals. Further reaction of the product with chloroacetometholate can produce the very powerful chelating agent EDTA (ethylenediaminetetrasodiumacetate). These salts are particularly strong complexing agents because they possess several unpaired electrons capable of occupying ligand sites on the metal. Further, they exhibit the chelating effect, since they are attached as a single molecule to several ligand sites on the central metal ion. Figure 17–4 shows one such arrangement.

Thioamides

Thioamides are prepared by the reaction of an amide with phosphorous pentasulfide, and they are tautomeric (the amide nitrogen exchanges a proton with the sulfur in one tautomer).

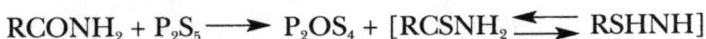

$$RCONH_2 + P_2S_5 \longrightarrow P_2OS_4 + [RCSNH_2 \rightleftharpoons RSHNH]$$

The unshared pairs of electrons on the sulfur and the nitrogen are capable of complexing with metallic cations. Figure 17–5 shows two tautomeric complex forms possible with an alkyl thioamide structure.

The thioamide are very interesting molecules, because they are capable of forming clathrates (hydrogen-bonded cage structures, see Fig. 17–6), as are thiourea, urea, and water. These clathrate (host) structures are complex geometric aggregates of hydrogen-bonded repeating units that are capable of including molecules or ions (guest) of a specific size within a hole their aggregation produces. Figure 17–6 shows the thioamide clathrate form and the methane hydrate (clathrate) structures.

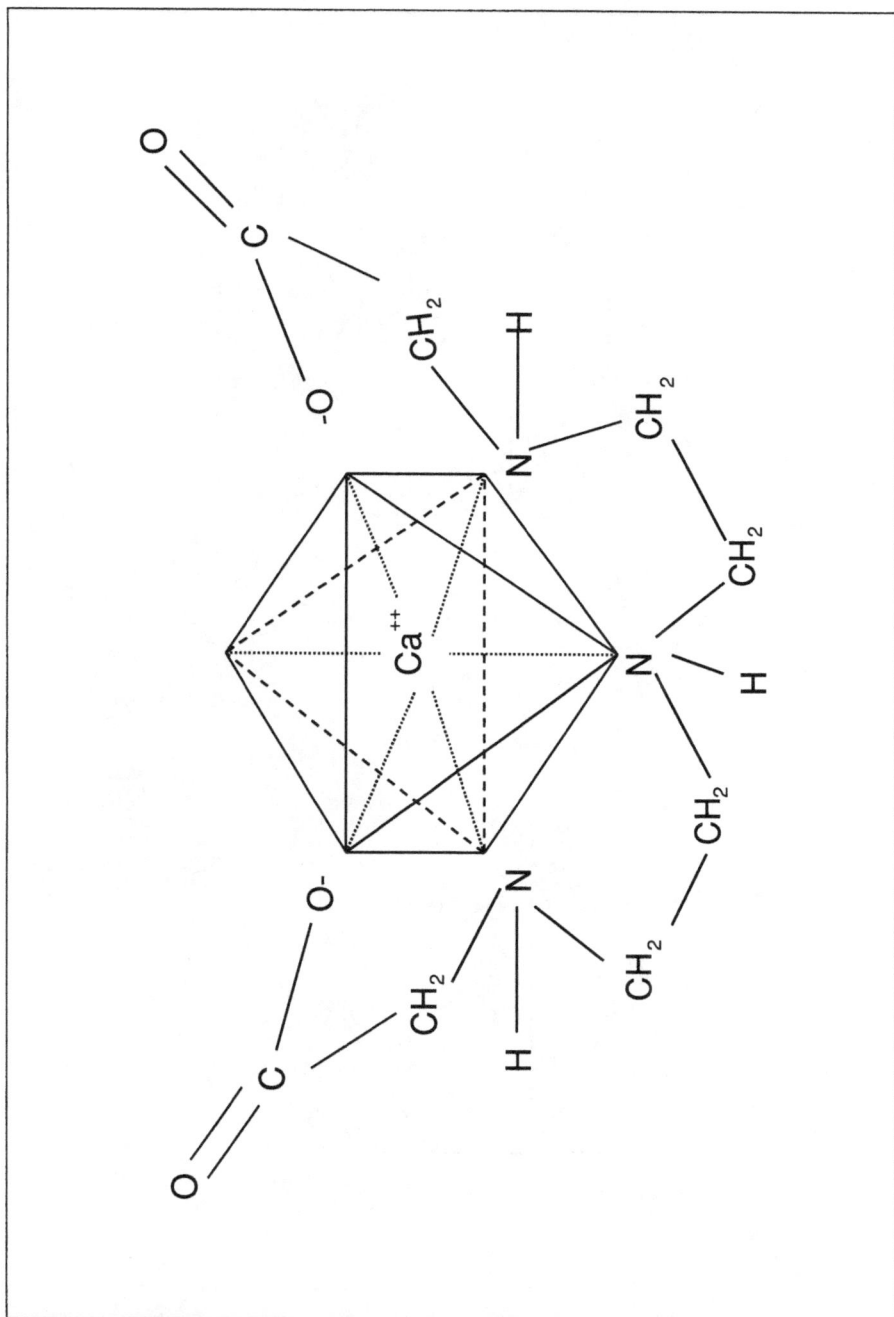

Fig. 17-4 Diethylenetriaminodiacetate complex of calcium

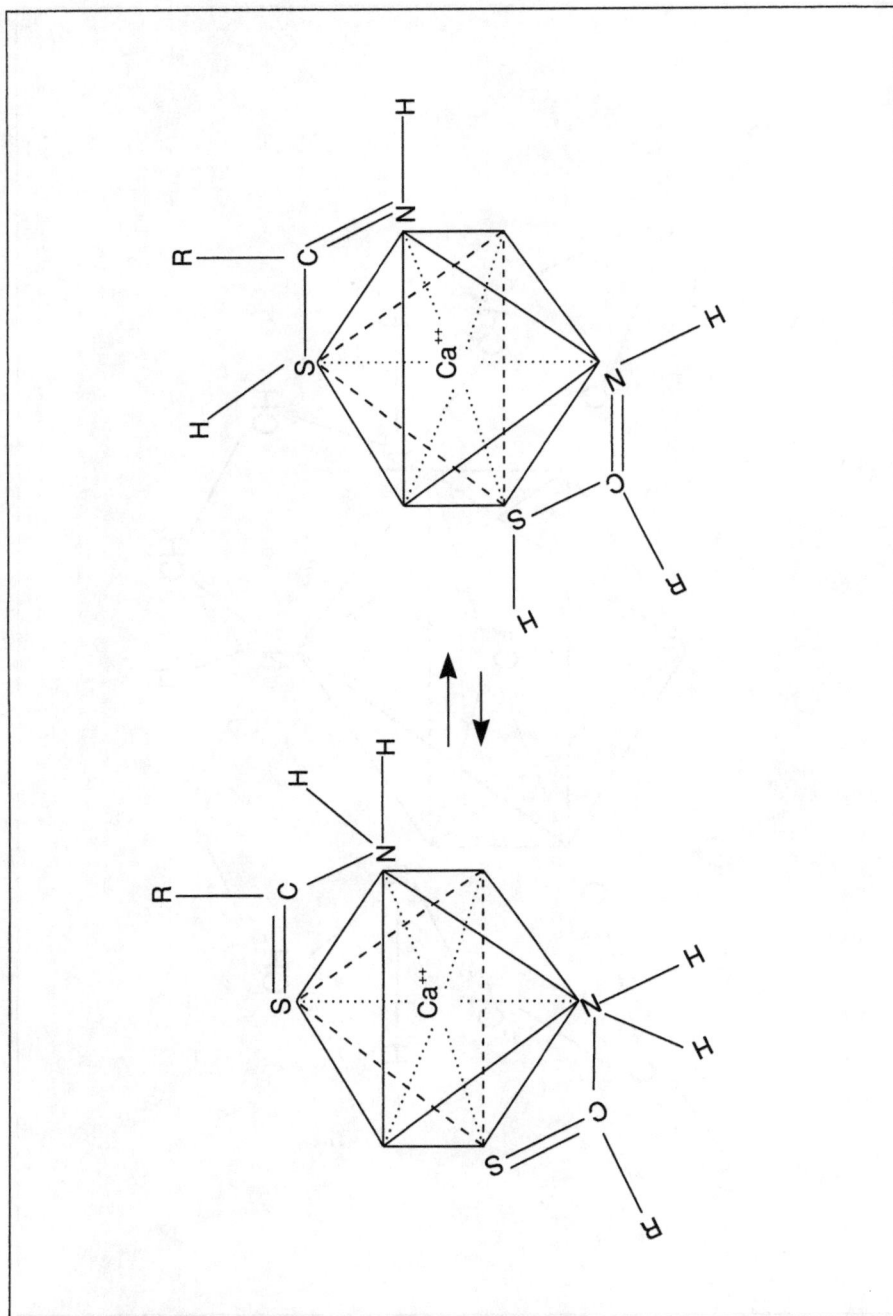

Fig. 17-5 Tautomeric alkyl thioamide molecules forming a cationic complex with a calcium ion

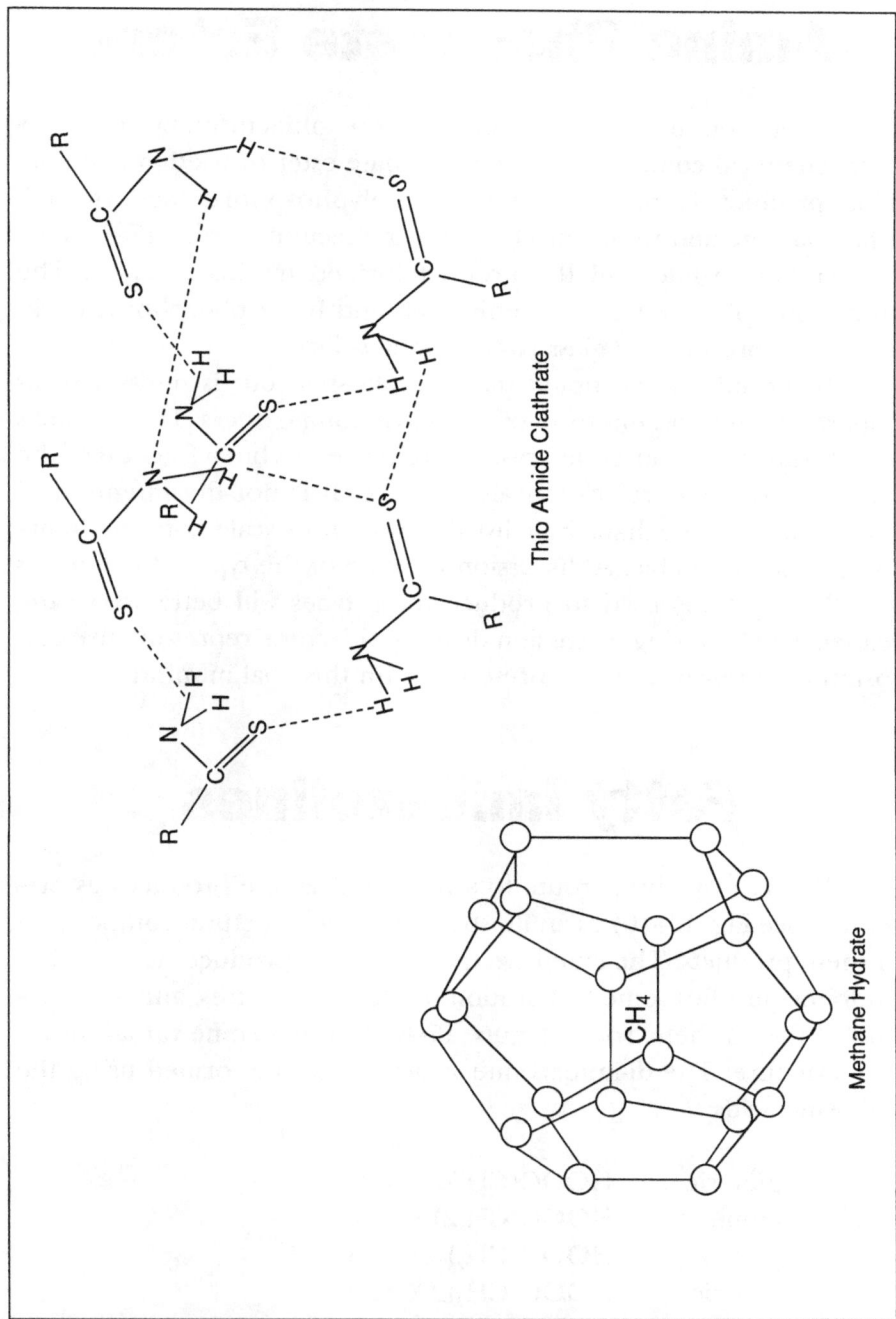

Fig. 17-6 Two clathrate structures, in which holes accommodate guest molecules (e.g., methane in the methane hydrate)

Amino Phosphate Esters

A very common scale product that is manufactured by many specialty chemical companies is the phosphate ester of triethanolamine. This product is made by adding polyphosphoric acid to triethanolamine and removing the water of reaction. Figure 17–7 shows the synthetic route and the product formed in this reaction. The amine phosphate ester is an amine salt and has a phosphorous acid anion; the product is a glass (free-flowing solid).

It should also be noted that the phosphorous is hexavalent as opposed to the trivalent form of the phosphonate esters. The products and the synthetic routes described here represent but a fraction of the different scale control chemicals available. It is not the intention of this discussion to exhaustively list the numerous scale control chemicals that are available. A discussion of some of the types of chemicals and the reactions used to produce these types will better serve the reader. The following discussion dealing with some representative corrosion inhibitor products is presented with this goal in mind.

Fatty Imidazolines

Earlier a synthetic route to a fatty imidazoline product was presented (see Fig. 13–10) as an illustration of the structural composition of these products. The extensive usage of these products as corrosion inhibitors justifies some elaboration on these structures, and a reiteration of the synthetic route. Figure 17–8 illustrates some variations on this structure. The di-imidazoline structure can be formed using the following acids:

- glutaric $HOOC(CH_2)_3COOH$
- adipic $HOOC(CH_22)_4COOH$
- pimelic $HOOC(CH_2)_5COOH$
- suberic $HOOC(CH_2)_6COOH$
- azelaic $HOOC(CH_2)_7COOH$
- sebacic $HOOC(CH_2)_8COOH$

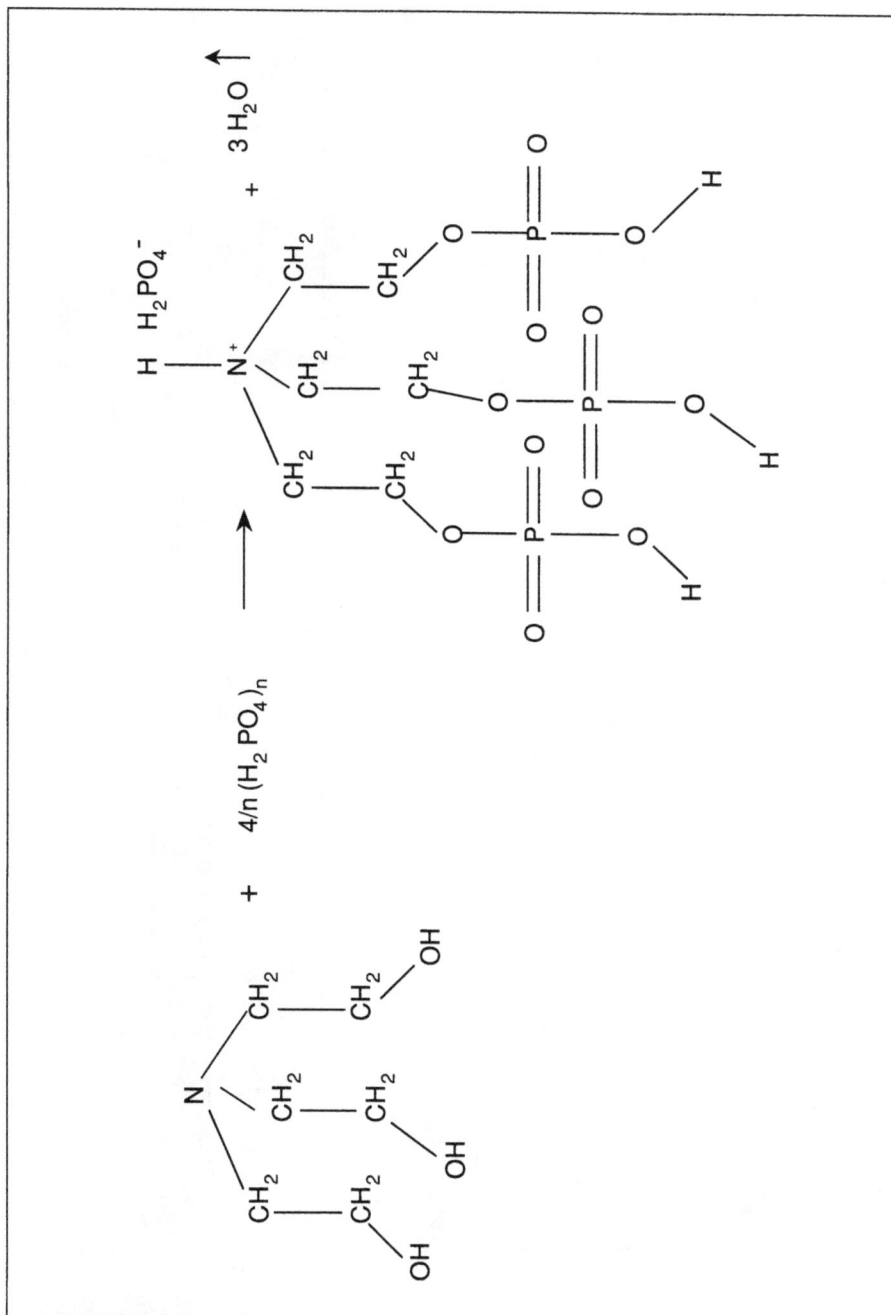

Fig. 17-7 The reaction product of triethanolamine and polyphosphoric acid

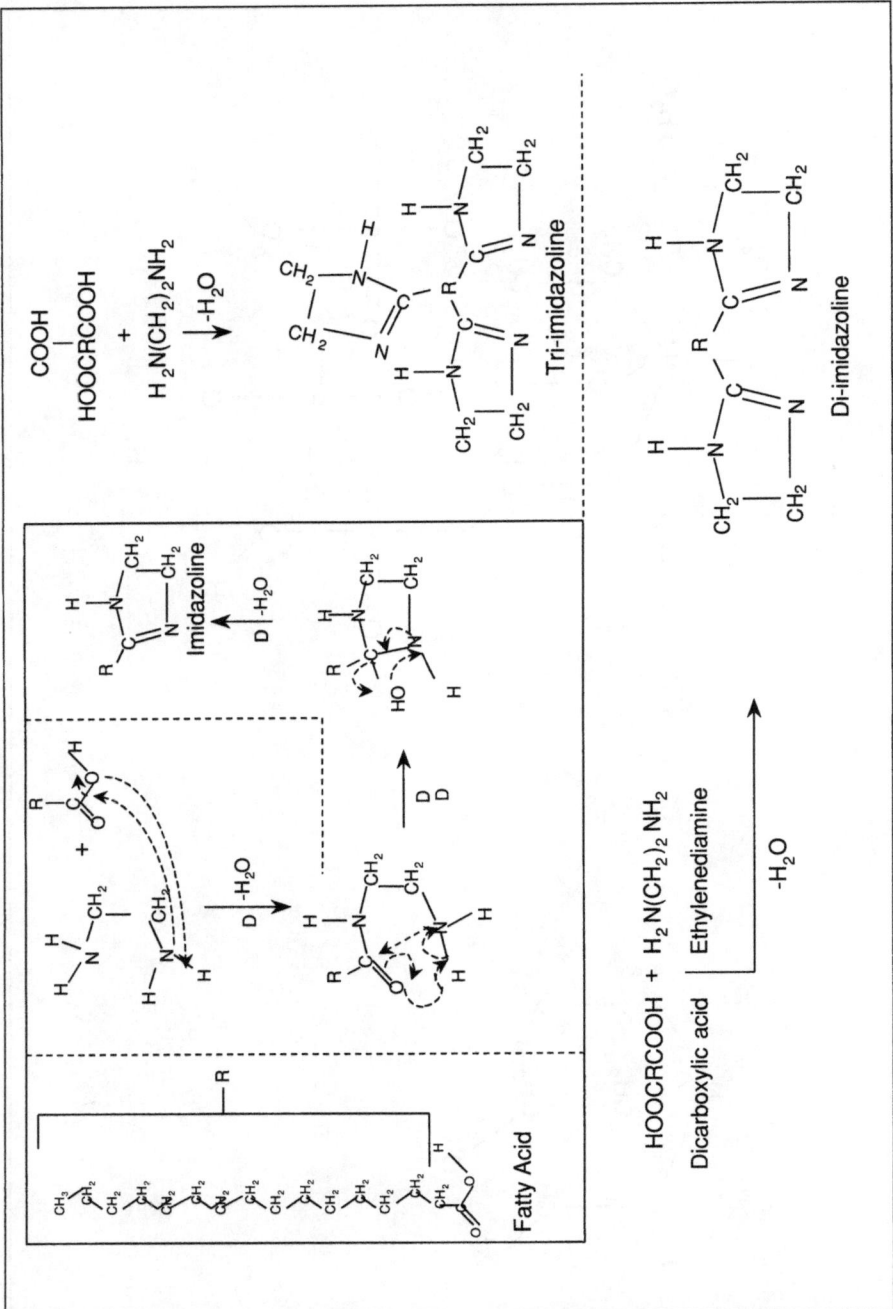

Fig. 17-8 Mono-, di-, and tri-imidazoline from carboxylic acids

The tri-imidazoline products can be manufactured from tricarboxylic acids. One tricarboxylic acid of particular interest is citric acid: $HOOCCH_2C[(COOH)(OH)]CH_2COOH$. This product can be acquired readily from citrus fruit. Citric acid is a hydroxyl tricarboxylic acid but is suspected to remain stable at the temperatures required for the cycylization of the imidazoline product.

Quinolines

The quinolines are best prepared in the Skraup synthesis. A mixture of glycerol and aniline is heated with concentrated sulfuric acid in the presence of a mild oxidizing agent, such as nitrobenzene. The quinoline produced in the Skraup synthesis can then be reacted with methyl chloride or dimethyl sulfate to form the quaternary derivative. The quaternary derivative is then susceptible to nucleophilic attack at position two.

The eight-hydroxy quinoline is a valuable reagent in inorganic chemistry because it forms strong complexes with metallic cations. It can be made using a modification of the Skraup synthesis, where the aniline is replaced by orthoamino phenol. The space relationship of the hydroxyl group and the unshared pair of electrons on the nitrogen atom is such that insoluble chelate coordination complexes are formed. It is also possible to react the hydroxyl group of the hydroxy quinoline with ethylene oxide to further functionalize the product, giving it interesting scale-inhibiting qualities. Figure 17–9 shows the Skraup synthesis and some modifications of the quinoline product.

Other derivations are possible both to the quaternary product and the hydroxy quinoline. Nucleophilic reaction as the methyl quinolinium chloride salt (in polar, aprotic solvent) leads to the two and four substitution products.

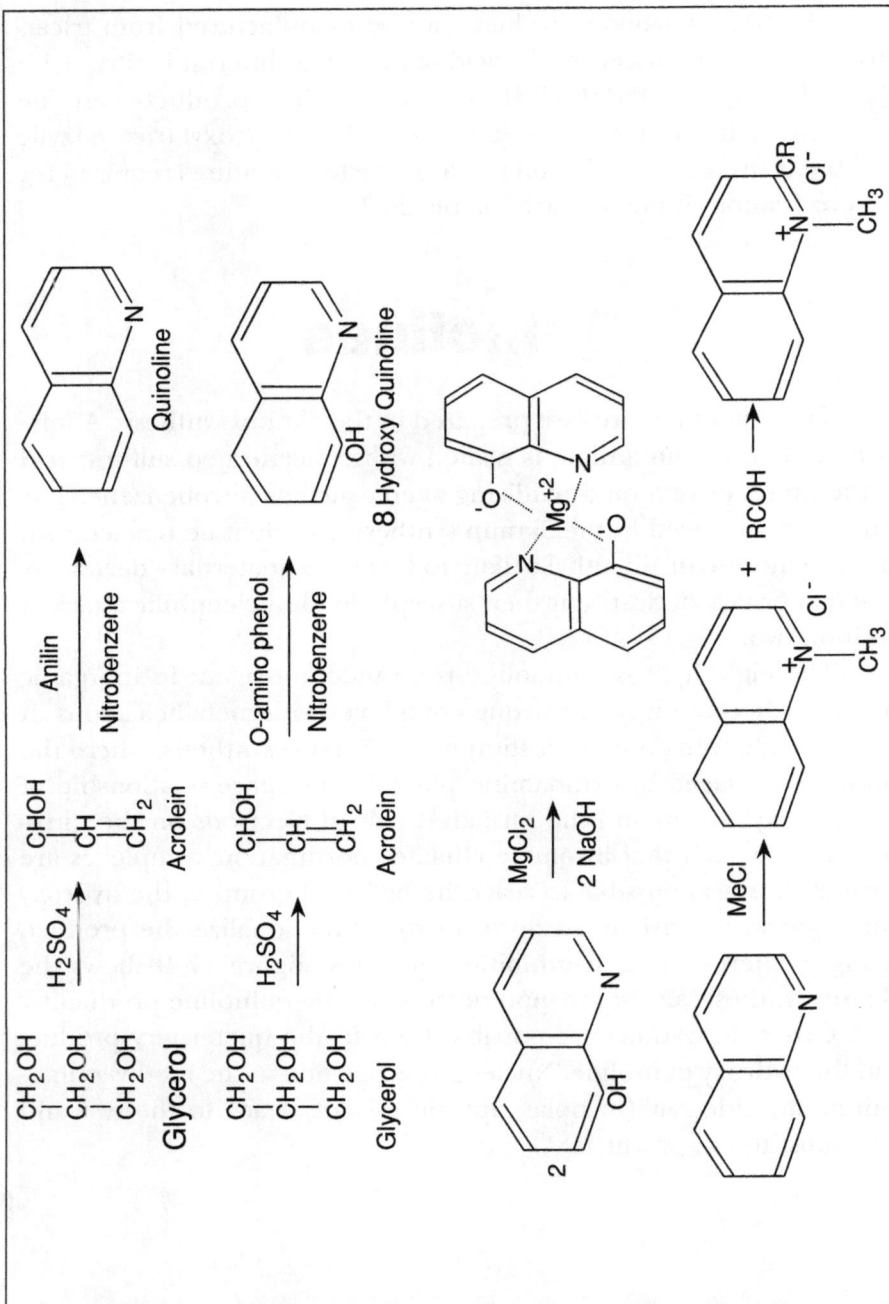

Fig. 17-9 The Skraup synthesis and some modifications of the quinoline product

Alkyl Pyridines

Alkyl pyridines are usually purchased by specialty chemical companies and quaternized to form the methyl chloride salts, which are freely soluble in water. Additionally 2-vinylpyridine is made from the reaction of 2-methyl pyridine with formaldehyde to form the 2-hydroxyethyl pyridine, which dehydrates to the vinyl pyridine. Figure 17–10 shows some of the derivatives of pyridine used as corrosion inhibitors.

The alkyl pyridines possess an unshared pair of electrons on the nitrogen, giving it properties of a tertiary amine. Pyridine is also aromatic, and the combination of the unshared pair of electrons on the nitrogen and its aromatic qualities make it a good pi-complexing species. The pi-complexing capabilities give the alkyl pyridines superior metals adhesion properties.

Summary

The preceding chapter gave a few selected synthetic routes to representative scale and corrosion inhibitors. Some complexed metal structures were included to show the interaction of the complexing agent with the metallic cations. This listing is not intended to be exhaustive, and consequently provides only an overview of a few classes of compounds.

References

Barrow, Gordon M. 1966. *Physical Chemistry.* 2d ed. New York, St. Louis, San Francisco, Toronto, London, and Sydney: McGraw Hill Book Company.

Bockris, J. O'M. and A. K. N. Reddy. 1973. *Modern Electrochemistry.* New York: Plenum Publishing Corporation.

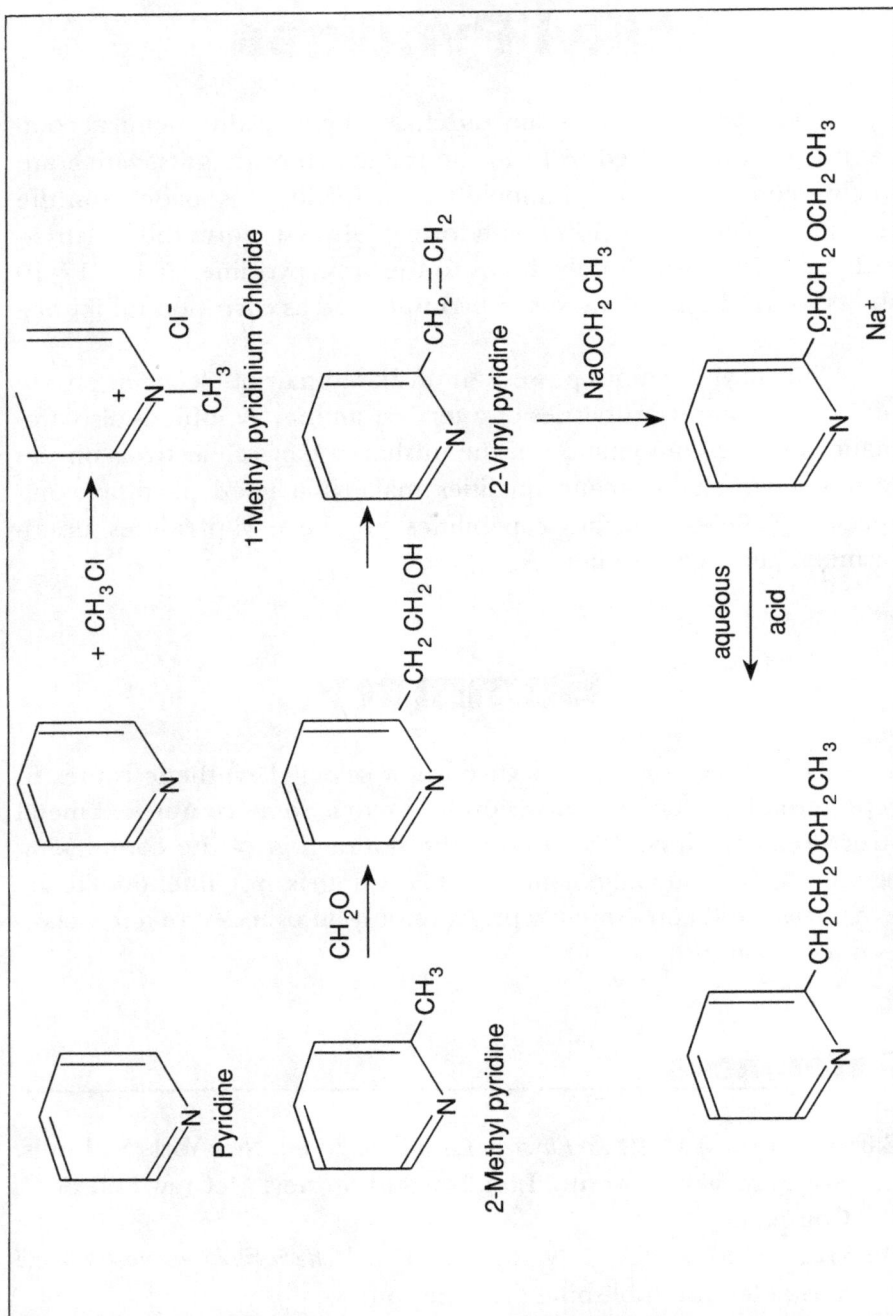

Fig. 17-10 Reactions of pyridine and alkyl pyridine with vinyl pyridine as a starting material for further substituted products

Dickerson, R. F., H. B. Gray, and G. P. Haight, Jr. 1970. *Chemical Principles.* 1st ed. New York: W. A. Benjamin, Inc.

Hamill, William H. and Russell R. Williams, Jr. 1966. *Principles of Physical Chemistry.* 2d ed. Englewood Cliffs, New Jersey: Prentice-Hall.

Huheey, James E. 1978. *Inorganic Chemistry Principles of Structure and Reactivity.* 2d ed. New York, Hagerstown, San Francisco, and London: Harper & Row.

Noller, Carl R. 1966. *Textbook of Organic Chemistry.* 3d ed. Philadelphia and London: W. B. Saunders Company.

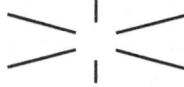

18
Corrosion and Scale Inhibitors and Native Petroleum Surfactants

Native Surfactants in Petroleum Fluids

Petroleum fluids (unrefined production fluids) can contain a considerable number of surface-active chemicals. These surface-active chemicals combine with dissimilar phases present in the petroleum fluids to produce emulsions, suspensions, aerosols, micelles, and colloids. The activity of these surfactants is determined by their interaction with chemistries and phases contained in the fluid system, as well as the strength of their polar and nonpolar interactions. Further, the extent of their interaction is determined by their concentration and the ratio of dissimilar phases present in the system.

Most crude oils contain significant amounts of water (0.5% to 50% by volume) and solids (fine dispersions of sand, asphaltene, paraffin, scale, and corrosion products). They also contain significant amounts of gas (lighter hydrocarbon fractions of methane, ethane, propane, hydrogen sulfide, and carbon dioxide). The physical condition of the phases is determined by the temperature, pressure, agitation, and more generally, the dynamics of the system.

Native surfactants play a key role in the physical state manifested by the petroleum system, because they provide the stabilizing influences and provide pathways for the interchange of components. Some of the types of surfactants present in crude oil systems were listed in Figure 7–2, and it should be noted that several possibilities are represented in this listing. Each chemical listed possesses the capability of containing several other groups. These groups are indicated by the numerous arrows in the illustration.

The enormous number of possible combinations of the components listed in Figure 7–2 indicates the possibility of interactions with heterogeneous phases present in crude oil systems. These heteroatomic species exhibit varying degrees of potential surface activity, since they possess properties such as dipole moments, hydrogen bonding capability, acid and base character, and London interaction capabilities. These chemistries are all possible native components of crude oil systems. When they combine with dissimilar phases, they can produce emulsions, suspensions, aerosols, micelles, and colloids. Although the percentages of these components are generally low, the powerful effects they impart to crude oil systems can be very significant.

Native Surfactants and Scale Inhibitors

As we have seen, the formulation of scale inhibitor chemicals is generally intended to facilitate their water solubility. Thus, these chemicals are usually prepared as aqueous blends of their salt forms, and are expected to partition mainly into the aqueous regions of petroleum fluids. However, considering the previous discussion of the surface-active chemicals (surfactants) present as natural components in crude oils, the possibility of their inclusion in emulsions is high.

In fact, scale inhibitors containing alkyl groups in excess of two carbons can interact through London forces of dispersion with a number of the species listed in Figure 7–2. Additionally, the possibilities for hydrogen bonding and ionic interactions with fatty acids can present a significant impediment to the distribution of the scale chemical in the free aqueous phase of the system. Thus, a considerable percentage of the scale-inhibitor chemical can be incorporated into emulsions within the petroleum fluid system.

Once incorporated as an emulsion, or occluded by crystallizing paraffin or destabilized asphaltenes, the scale product is often unavailable to forming scales in the bulk water phase. There is, however, one saving grace to this situation; scales do form in emulsions, so the partitioning of the scale chemical into the emulsion phase is not a total loss. The occlusion of scale chemical by crystallizing paraffin or destabilized asphaltene does remove the scale chemical from the system and results in loss of activity. The general composition of scale inhibitor products is quite fortunate, since their activity probably suffers the least as a function of their removal from actively scaling sites. Such is not the case for corrosion chemicals, as we shall see in the next section.

Native Surfactants and Corrosion Chemicals

Corrosion inhibitor chemistry is dramatically affected by the presence of native surfactants in mixed phased (water and oil) systems. Figure 18–1 shows how a fatty imidazoline corrosion inhibitor can be removed from its primary function by phase partitioning.

Figure 18–1 also shows how the fatty imidazoline corrosion inhibitor can form an emulsion in the presence of water in oil. The native surfactants also interact with these bipolar molecules as can be seen from Figure 18–2.

Of course imidazoline corrosion inhibitors are perfectly set up to form emulsion, since they possess both a nonpolar and polar function

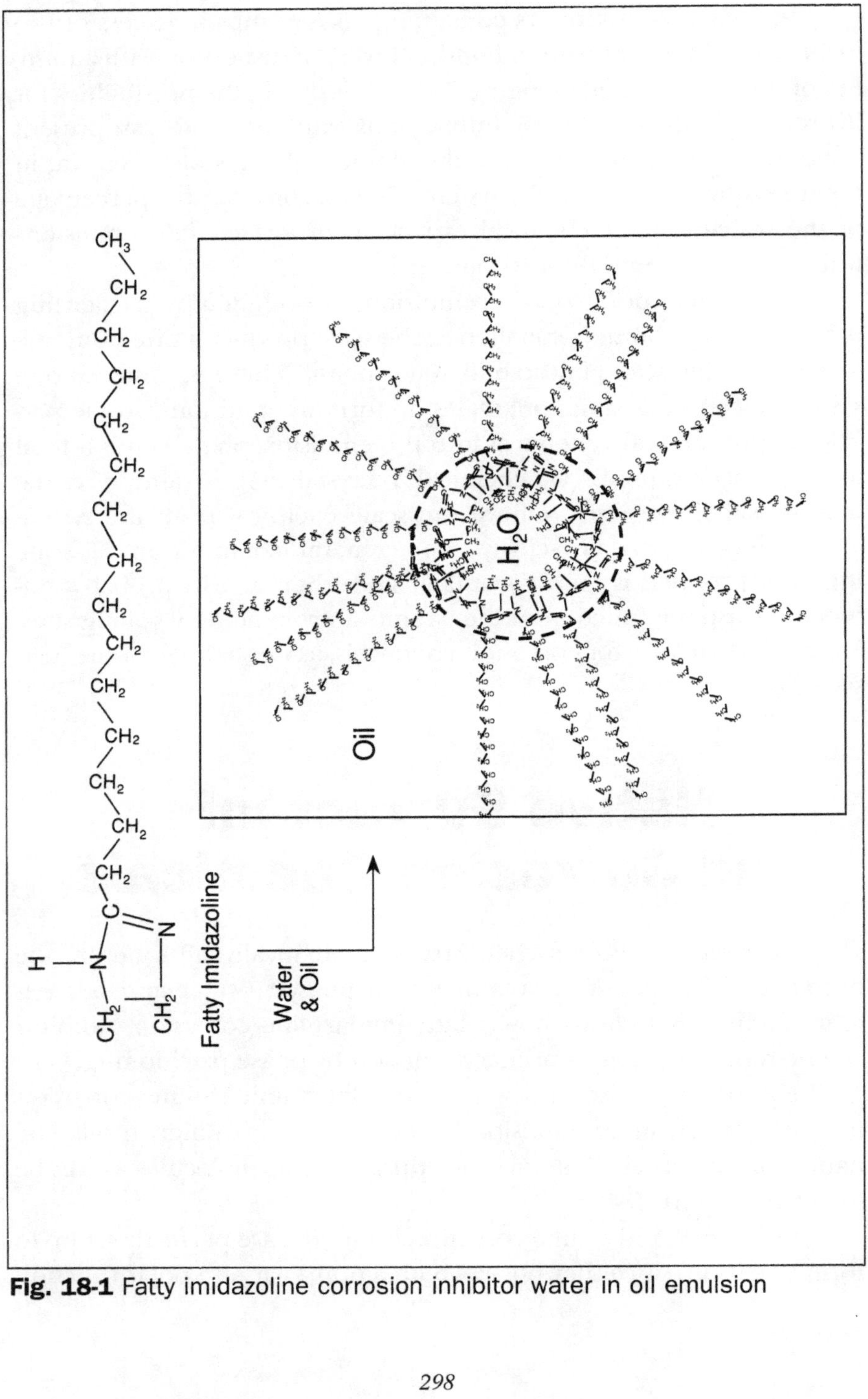

Fig. 18-1 Fatty imidazoline corrosion inhibitor water in oil emulsion

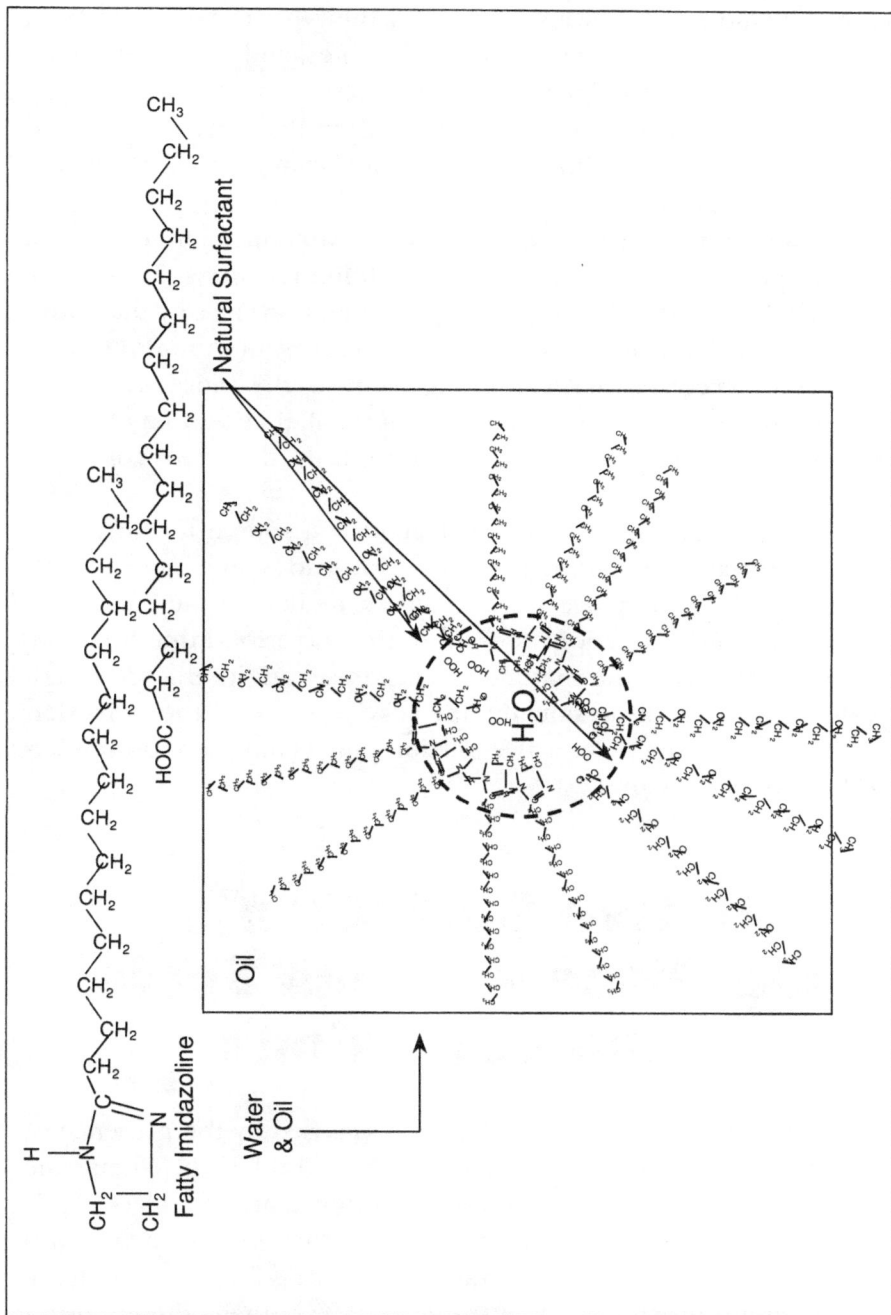

Fig. 18-2 The natural surfactant's interaction with fatty imidazoline corrosion inhibitors to assist in forming emulsions

in the same molecule. The alkyl pyridines, their quaternary derivatives, and the quinolines, alkyl-substituted quinolines, and their quaternary derivatives all possess polar and nonpolar attributes. These attributes combine to produce chemicals that emulsify dissimilar phases and facilitate interactions with the native surfactants present in the petroleum fluid systems.

In addition to the problems associated with emulsion formation, there are problems associated with competition reactions. These competition reactions involve the coverage of metal surfaces by the native surfactants, thus making it difficult for the corrosion inhibitor chemical to get to the corrosion site. Although some of the native surfactants present can have corrosion protection potential, they seldom offer the same degree of protection as the specifically designed corrosion inhibitor product.

In some cases the natural surfactant is a source for corrosion (e.g., some naturally occurring fatty acids and some mercaptans). Further, some of the native surfactants can neutralize the base properties of the amine functions present in the corrosion inhibitor product. In some cases the natural surfactants can interact through London forces of dispersion on the alkyl groups of the corrosion inhibitor, forcing the polar active portion into a configuration rendering the product inactive.

Native Surfactants and Discontinuous Scale and Corrosion

Among the many detrimental effects caused by the presence of native surfactants, discontinuous surface effects rank high. When scale forms in an emulsion, the crystal forms produced are often less highly organized than those formed in the free water phase (e.g., aggregates of fewer steps in crystal associations). As the scale forms in the emulsion, the emulsion changes to a suspension. Eventually the emulsion breaks or the suspension becomes unstable, depositing formed crystals to the system. The crystals thus formed combine with those formed in the free water phase, and the combined scale forms on surfaces and in constricted areas in the system.

In addition to the formation of (and rupture by) scale crystals in emulsions, the crystals released become oil coated as they pass from the water phase. This oil coating also adds extraneous materials to the scale forming in the free water. The extraneous materials produce scales that have different adhesion properties than the bulk scale forms and tend to cause spotty scale accumulations.

The situation with corrosion inhibitors is even more noticeable, since corrosion inhibitors tend to partition as emulsions if appropriate additions of surfactants are not added to prevent them. If the corrosion inhibitor partitions into emulsion forms, either by the action of their own surfactant properties or in combination with native surfactants present in the oil, then metal surfaces will exhibit spotty coverage. This occurs because the emulsions containing the corrosion inhibitor may or may not rupture at the metal/water interface to form a uniform coverage of the surface. This spotty coverage is even more damaging than no corrosion control, since the metal forms additional electron sinks and sources between areas where the chemical has successfully contacted the metal.

Phase Transfer by Native Surfactants

Organic chemists have known for a long time that sometimes functional salts can be made to go into the oil phase. This can occur if the water-soluble anion or cation of the salt can be interchanged with an organically soluble anion or cation. A good example would be the exchange of the acetate anion in tetramethyl ammonium acetate by sodium benzene sulfonate. Figure 18–3 shows this salt interchange and the resultant oil-soluble tetramethyl ammonium benzene sulfonate.

Similar ionic exchanges can occur with salts of the aromatic amines, sulfhydrides, and carboxylates present in the organic and

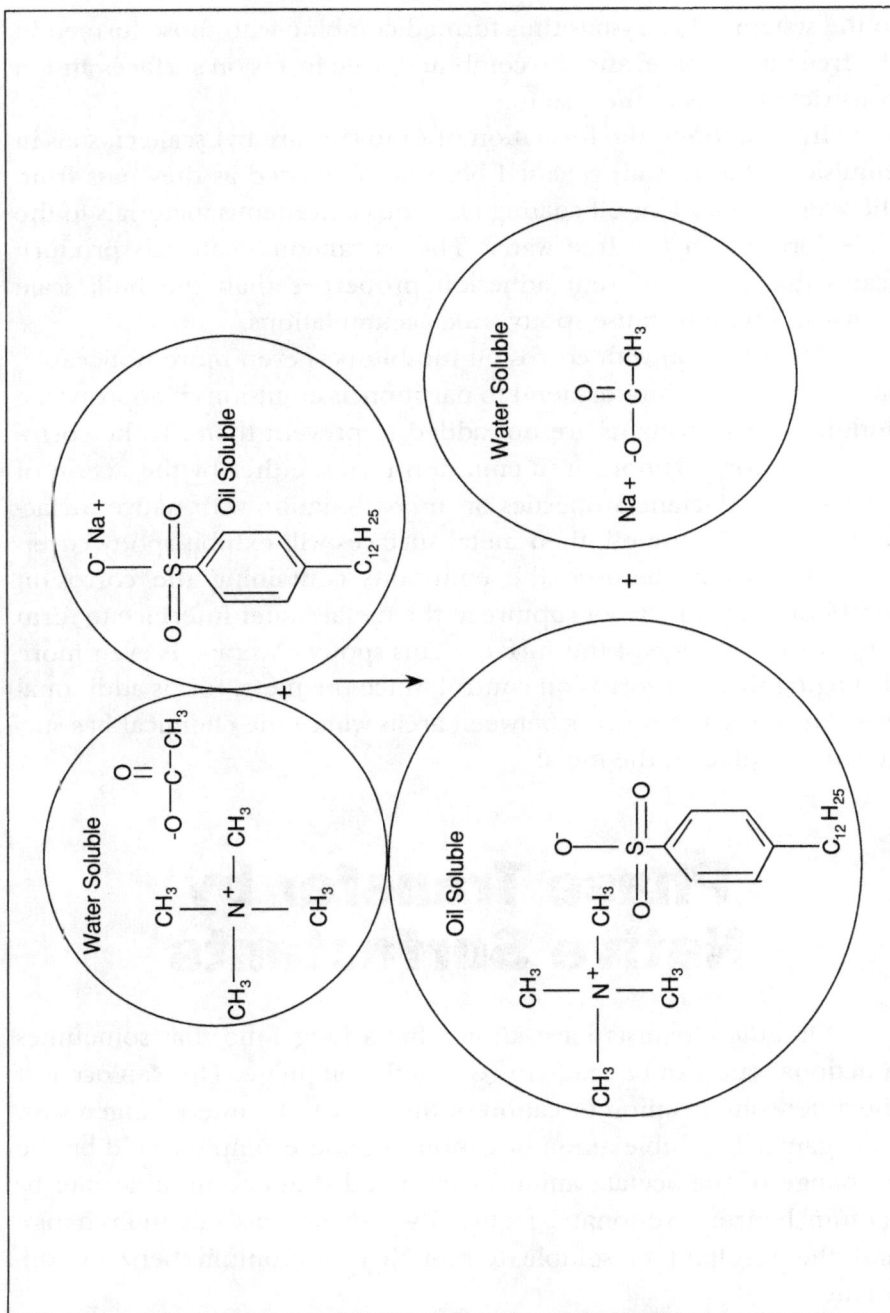

Fig. 18-3 Transfer of phase behavior by the interchange of anions and cations

aqueous phases of the petroleum fluids. Likewise, corrosion inhibitor molecules and scale inhibitor molecules can form phase transfer products. The phase transfer of the active products (corrosion and scale inhibitors) presents a real problem in certain crude oil systems, since the products never get to the site they were intended to reach. Phase transfer reactions also affect the scale products formed in the system. Sometimes aromatic amines, fatty acids, and thio compounds interact with ligand sites of the metallic salt complexes, and transfer them from the aqueous phase to the oil phase. Figure 18–4 shows how this phase transfer can destabilize some scale inhibitor products.

The influences of native surfactants can be overcome by judicious choices of added surfactant packages. The forces directing the aggregation of these species are a function of the concentration of the native surfactants and their reactivity with the added product or the metal complex.

Summary

Native surfactant species present in petroleum fluid systems produce complications that change the effectiveness of products primarily intended to perform in the aqueous phase. The concentration and chemical reactivity of these native surfactants can, and often do, compete with the chemicals added to perform corrosion and scale alterations to the system. The formation of emulsions, phase transfer salts, and macro-aggregate species in the presence of these native surfactants can be a significant factor in the loss of chemical effectiveness.

Addition of counteractive surfactant products can overcome some of these effects, but the formulation of such effective counteractive measures is highly dependent on the characteristics of the particular system. In addition to the liquid phase behavior of these native surfactants, liquid/solid effects are also affected by these agents. Competition for active sites on the metal surface or ligand groups in metallic salt complexes also results in the loss of activity by the corrosion or scale product. In general this discussion is intended as a listing of some of the pitfalls encountered in the treatment of problems in complex heterophased petroleum fluid systems.

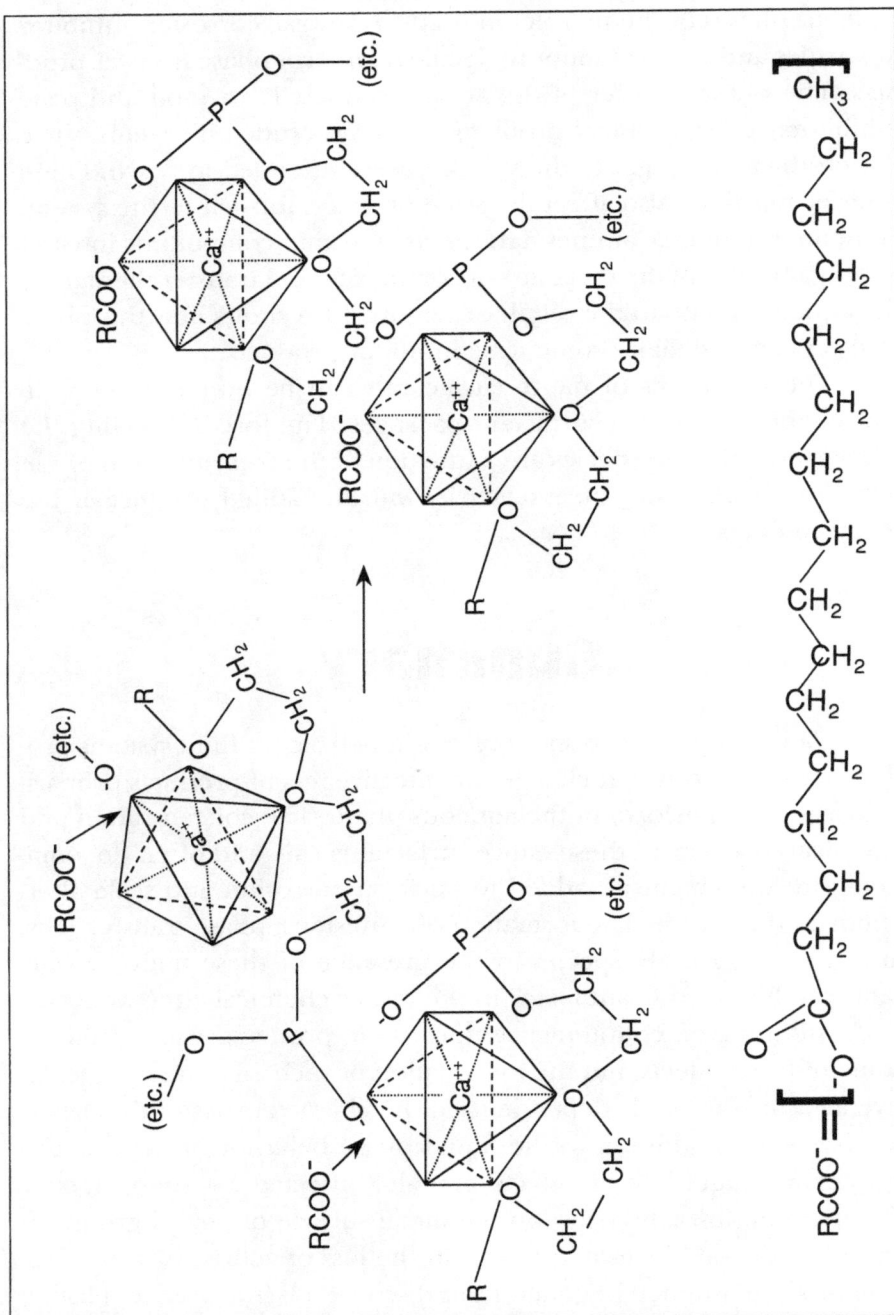

Fig. 18-4 The competition for ligand sites by native fatty acid surfactants, and the possibility for phase transfer by the fatty tail of the acid ligand

References

Barrow, Gordon M. 1966. *Physical Chemistry*. 2d ed. New York, St. Louis, San Francisco, Toronto, London, and Sydney: McGraw Hill Book Company.

Dickerson, R. F., H. B. Gray, and G. P. Haight, Jr. 1970. *Chemical Principles*. 1st ed. New York: W. A. Benjamin, Inc.

Huheey, James E. 1978. *Inorganic Chemistry Principles of Structure and Reactivity*. 2d ed. New York, Hagerstown, San Francisco, and London: Harper & Row.

Jones, Loyd W. 1988. *Corrosion and Water Technology for Petroleum Producers*. 1st ed. Tulsa: OGCI Publications, Oil & Gas Consultants International, Inc.

Handbook of Chemistry and Physics. 1976. 56th ed. Cleveland: CRC Press.

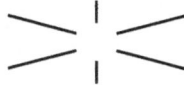

19
Crude Oil Production/
Refining Corrosion and
Scale Problems

Throughout the previous discussions, our aim was to develop an understanding of the mechanisms leading to corrosion and scale problems, and the design of chemical treatments to address these problems. We have examined some of the complex phase behavior of scale and corrosion phenomena along with their chemical treatments, but we have not detailed their occurrence in crude oil production streams. This section is intended to address some of these issues and describe some of the field conditions leading to scale and corrosion problems. Figure 19–1 shows some of the facilities in place in a modern production area.

Reservoir Conditions

Crude oil production is not just a simple matter of drilling a hole in the earth's crust, tapping a pool of oil and gas, and allowing the gas

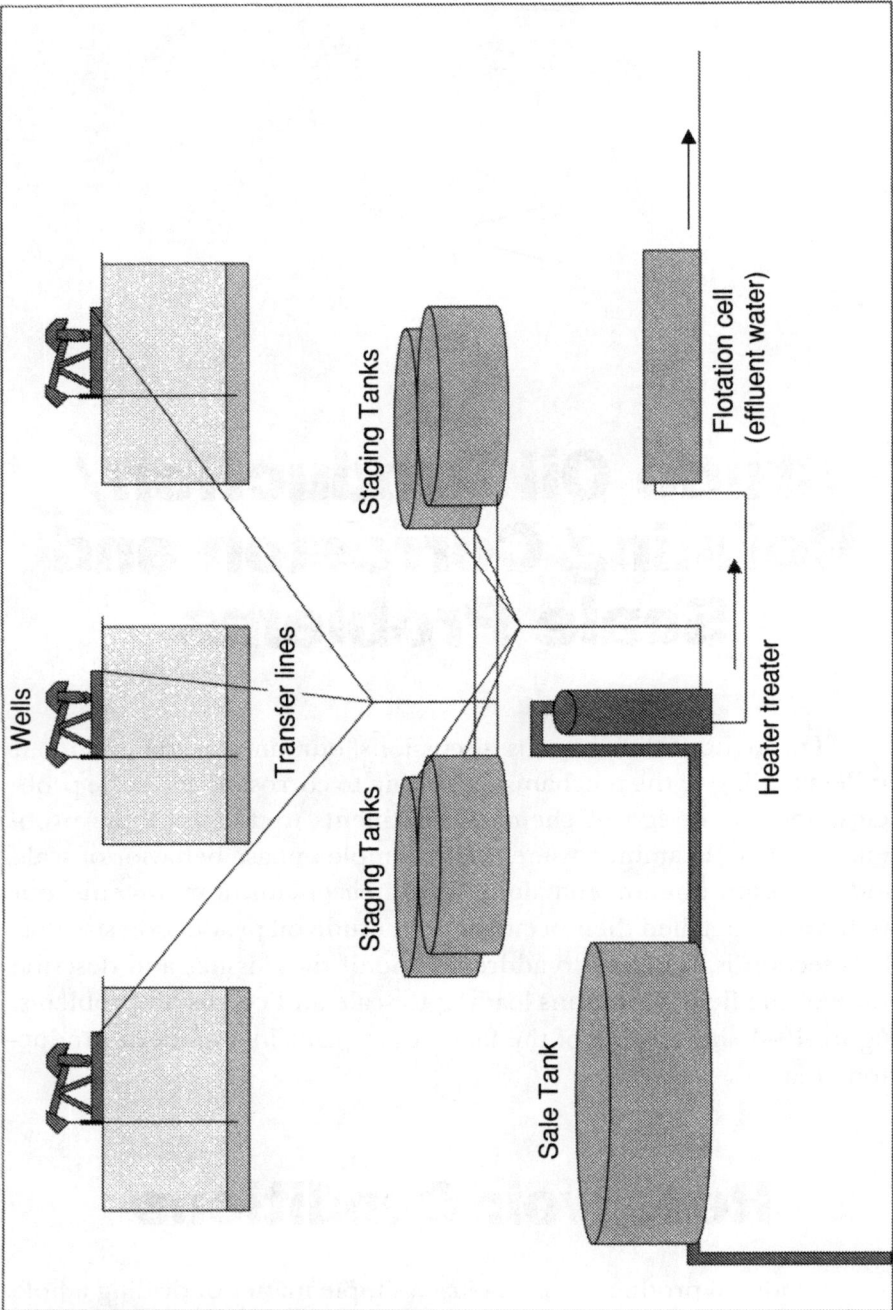

Fig. 19-1 Oil production facilities

pressure to push up the oil. More often than not, the producing interval consists of highly compressed and dense reservoir rock sandwiched between rock of even greater density.

During vast periods of geologic time, the percolation of surface water and products of organic decay are entrained by the dynamic shifts of the earth's crust. Slow but inexorable forces thrust large segments of the crust over and under each other, trapping and compressing the sands and silts that contain the products of organic decay. The tremendous forces of compression compacted the organic-containing sands and silts to produce dense rock from which oil and gas are extracted.

Generally the water that has carried the products of organic decay to the sand and silt is preferentially adsorbed to their polar surfaces. This adsorbed water is called connate water, and its polar nature acts as a barrier to the nonpolar oil phases. Thus, the compressed sands and silts act in concert with their water coatings to provide storage vesicles for the nonpolar oil phase. With this description, it can be seen that the processes required for oil production are vastly more complex than the simple act of drilling a well.

Well Completion

Now that we understand that the matrix containing crude oil is very often highly compressed rock with varying densities, it can be seen that drilling the hole is only the beginning. Once it has been determined that a zone of oil-containing rock has been reached in the drilling process, a means of increasing the available surface area for oil exit must be provided. During the drilling process, well tubing must be lowered into the hole to prevent collapse of the upper well regions. Additionally, cement is pumped between the hole and the tubing to further strengthen the well walls.

Once the tubing and casing are placed, the well must be opened to the producing interval. This is accomplished by placing a device in the prospective producing interval of the well. This shotgun device contains shaped explosives that propel projectiles that penetrate the tubing and casing and lodge deep within the producing interval. The holes thus produced provide paths for the subsequent pumping of weighted synthetic muds that provide the hydraulic pressures required to further fracture the producing rock interval. The muds are complex mixtures of water, swollen clays, and additives such as hydroxyethyl cellulose and other synthetic derivatives. Quite often they contain porous additives known as proppants, which are left behind to prevent the fractures from closing.

Once the fracturing process is completed, the mud is pumped from the well, allowing the pressure of the reservoir to produce oil. It is often necessary to perform an initial acid job to provide a clean up after the drilling and fracturing processes. Acid jobs consist of following the mud removal with weighted acid to counterbalance reservoir pressure. The acid used is most often a mixture of hydrochloric acid containing metallic salts for additional weight. The acid helps to flush the residual mud and attacks scales such as calcium carbonate that might act to impede flow.

Production

Once the well has been completed, the light hydrocarbon species begin to expand in response to the reduced confining pressure provided by the well opening. These light hydrocarbons lose solubility in the oil phase as they reach pressures below a critical value (where they are compressed to the liquid state). As these liquids convert to gases, the reservoir pressure increases, and liquid crude oil fractions are driven to the perforations and up the well tubing. When this process acts to push oil from a reservoir, it is called a natural gas drive. Each oil pocket contained in the spaces between the water-coated rock face is driven from its position by the increasing gas pressure.

In this process, turbulent forces act to mix some of the polar components of the crude oil with their companion water and dissolved salts. Because the perfect gas law ($PV = nRT$) demands that the temperature increase as pressure increases, the temperature of the reservoir must increase. The increased temperature of the reservoir assists the processes of salt hydration, and concentrated water and salt solutions are formed. Additionally, the combined turbulence and temperatures produce emulsions from the bipolar crude oil and water.

Production Decline

After a period of time, the natural gas drives become depleted, and artificial methods must be employed to lift the petroleum to the surface. At this stage of a well's life, it may be practical to use pumps to decrease the well-tubing pressure and lift the oil from the reservoir. The practicality of such methods is determined by the nature of the crude oil remaining in the formation.

If the viscosity of the oil is low, surface pumps are acceptable; if it is high, then electronic submersible pumps might be required. These methods add to the turbulence experienced by the fluids as they move up the well, thereby exacerbating problems of emulsion formation and tubing corrosion and scale. Once these methods decline to a point where lifting costs exceed profitability, new methods must be employed to produce the oil.

Secondary Recovery Methods

Often secondary recovery methods are employed to produce the considerable amounts of crude oil left behind by depleted natural gas drives and artificial lift methods. These methods include water flood, steam flood, and nitrogen and carbon dioxide floods. These methods involve the injection of water, steam, nitrogen, and carbon dioxide into wells situated near the producing wells.

The injected water or gases act as sweeping fronts that push the oil from the formation and up the producing well. Water floods displace fluids from the producing zone and very often increase corrosion and scale problems by leaching soluble minerals from the surrounding rock. Steam flooding, with its elevated temperature, is even more effective in carrying scaling minerals and corrosive fluids to the producing areas.

The use of nitrogen and carbon dioxide has increased in importance from 1985 to the present. Although carbon dioxide floods have proven more effective than nitrogen floods, nitrogen is still used in some systems. Carbon dioxide floods are performed by injecting super critical carbon dioxide into an injection well and sweeping fluids toward the producing well. (The pressure and temperature of the carbon dioxide are maintained so that it is in the liquid state.)

The liquid carbon dioxide mixes with the hydrocarbon fractions in the oil and reestablishes formation pressures that are capable of lifting the crude oil to the surface. Thus, the carbon dioxide acts to replenish the natural gas drive and assists in the lifting of the crude. As the carbon dioxide changes from liquid to gas, and exits the well, Joule Thompson cooling occurs. This cooling effect changes the ability of the fluids being produced to solubilize salts. The decreased temperature thus enhances the process of scale formation within the reservoir near the wellbore.

Tertiary Recovery

Tertiary recovery methods involve polymer and surfactant floods. When the price of crude oil justified the additional cost, the production of heavier crude oil was practical. Since 1981, these methods generally have proven too costly. As a consequence, much of the recoverable reserves remain in these reservoirs. Therefore, discussion of these methods will remain as a topic for future discussion.

The Wellbore and Beyond

The problems of scale and corrosion attending the lifting of crude from the reservoir are just the beginning of the story. As these fluids travel up the tubing, through transfer lines, and into treating and storage facilities, they are continuously subjected to pressure and temperature changes. These changes, in turn, change the physical and chemical properties of these fluids. Shearing forces and surface interactions occur within tubing and transfer lines, producing conditions favorable to the processes of corrosion and scale.

The pressure drops and the attendant temperature declines act to create conditions that alter the type of corrosion and scale problem experienced. The placement of corrosion and scale probes in subsurface areas is not very feasible. Therefore, one must be content to place these probes after the wellhead. These methods give only relative information about the nature and extent of problems occurring downhole.

Transfer Lines

As fluids are delivered from the wellhead to the transfer lines, the insulation properties change, and in some cases, these transfer lines are exposed to very cold temperatures. The decreased temperature reduces the solubility of most scaling salts and aids in their deposit. Thus, placement of scale coupons in the transfer lines will help to evaluate the effectiveness of scale treatment programs. The lower temperatures prevalent in transfer lines generally result in lower corrosion rates than those that occur in well tubing and pumping equipment.

Thus it can usually be assumed that transfer line corrosion takes place at a slower rate than tubing corrosion or corrosion taking place under conditions of elevated temperatures. In many cases the crude oil is sent to a heated treatment vessel, where water is removed and discharged to additional treatment facilities. There the water is further cleaned prior to its release to the environment. These transfer lines are subjected to higher concentrations of soluble salts because of the added heat. This then is a good area for the monitoring of scale and corrosion.

Settling Tanks and Storage Vessels

After the crude has been heat treated, and water has been removed, it is transferred to settling tanks where continued water separation occurs. These vessels provide an environment of relatively low turbulence, where solids, scales, and conditions of high ionic strength exist. This set of physical conditions favors corrosion and scale processes, and provides a good spot to place monitoring equipment. Significant corrosion areas include the overhead regions of the storage vessels, where condensation occurs, and settling tanks; these areas should be monitored.

Valves and Pumps

The stability of scale-producing fluids is highly dependent upon the physical conditions of the system through which these fluids pass. Pressure drops occurring after valves and pumps can be significant, and the disturbances they cause in the equilibrium of salt solvation can lead to conditions favoring enhanced corrosion and scale rates. Increased turbulence resulting from pressure drops across pumps and valves can also result in mechanical disruption of emulsions.

This emulsion disruption often produces discontinuities in the nature of fluid wetting properties. Metal and brine water contact is increased, which promotes both scaling and corrosion effects; monitoring of each effect should be conducted. However, the practicality of measurements must also be considered. The ease of access to pumps and valves often dictates the placement of monitoring equipment.

Pipelines

The transportation of crude oil through pipelines to refineries is an important part of an integrated oil company's operation. The inflow of crude oil to pipelines is highly regulated, and stringent restrictions are placed on the water content of the crude oil being shipped through these lines. The size of these transport lines must be capable of accommodating the production of several producing areas, and consequently the widely variable physical characteristics of several crude oil types. Although the water content is tightly monitored in the pipelines, the volumes it must accommodate can lead to significant problems of corrosion and scale. Thus, monitoring for corrosion and scale in this conduit is critical for the prevention of failure and the disastrous consequences that could occur.

Refinery Storage

Refineries constitute a plexus for the mingling of several sources of crude oil production. Large storage tanks receive the effluent of several pipelines and provide staging areas for the crude oils that will later be refined. Just as the scale of fluids transported through pipelines magnifies the problems of scale and corrosion, the admixture of many pipelines acts to intensify these effects in storage areas of a refinery. Thus, this area is of considerable importance and should be monitored for corrosion and scale. Figure 19–2 shows many of the areas of a modern refinery.

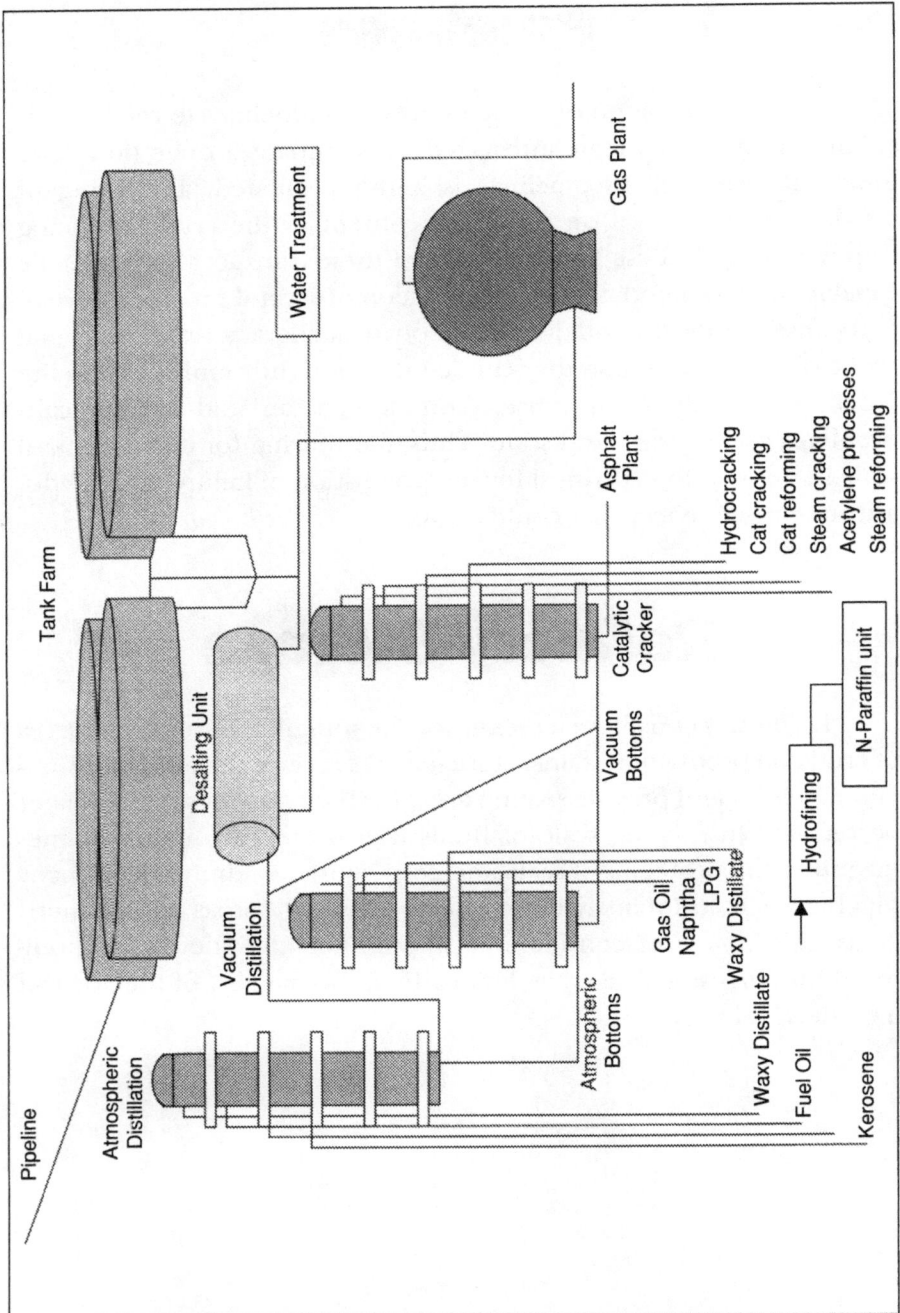

Fig. 19-2 Some refinery processes

Desalting Unit

As the crude oil is introduced into the refinery, it is subjected to a process in which 5% water is added. This mix of water and crude oil is exposed to a combination of heat and high-voltage alternating current to remove salts. The exiting wash water is then treated to remove oil and salts. The dehydrated, low-salt crude is then fed to an atmospheric distillation unit. The water fraction is high in dissolved salts and presents an exaggerated potential for scale formation. It should be monitored at this point. Likewise the overhead condensation water presents an increased potential for corrosion and should also be monitored.

Atmospheric Distillation Unit

The dehydrated, desalted crude oil leaves the desalting unit and enters the atmospheric distillation unit, where heating separates the different boiling point fractions of the raw crude oil. Several light-end fractions are recovered at this point, as well as some corrosive gases. Additionally, some corrosive gases are produced in the process of heating. Corrosion coupons are often placed in the distillation towers and in their overheads to monitor the effects of these corrosive gases. Although the dehydration process performed by desalting units removes much of the water, some water invariably remains. The water that remains is flashed from the crude as the distillation process proceeds.

Summary

The preceding chapter gave a view of some of the operations and physical structure of production areas and refinery systems. The purpose of this chapter was to familiarize the reader with the fundamental construction of production and refinery systems. It is intended to provide a picture of the areas in which corrosion and scale processes are important.

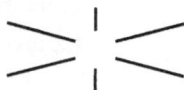

Index

A

Acid corrosion, 129-130
Acidic conditions, 129-130
Acidic scale inhibitors, 258-259
Adsorption, 5, 50, 260-261
Aggregate structures, *xix*, 3-4, 8-9, 252, 254
    architecture, 8-9
    like-group, 252, 254
Aliphatic-substituted aromatics, 9
Alkaline earth metal salt, 25, 28
Alkyl phosphonates, 277-280
Alkyl pyridines, 291-292
Amino acids, 8
Amino phosphate esters, 286-287
Amorphous solid, 114
Anion determination, 269-270
Anion/cation interactions, 3-4, 7-8, 10, 20
Anionic ligand site occupation, 42, 45
Anionic salts, 256
Anisotropic behavior, 114

Applications elsewhere, 116, 118
Aromatic amines, 235-239
    quinoline structures, 235-239
Asphaltenes, *xix*, 7, 14, 17, 50-53, 90, 94-99, 101
Atmospheric distillation unit, 317
Atmospheric factors, 46-47, 50
Atomic absorption analysis, 269
Atomic emission analysis, 269
Autoclaves, 264

B

Band theory, 238, 241
Barium carbonate, 20, 29
Barium sulfate, 20, 29, 33, 76
Basic conditions, 130-132
Biodegradation, 3, 19, 31, 309
Biological system, 7-9, 17
Bipolar molecules/species, 9-10, 17, 28, 50-51, 53, 99-100, 248
    surfactant, 99-100

Embrittlement, 4-5

Emulsions, *xix*, 5, 7-13, 15-17, 28, 51, 53-54, 298-300
  combined forms, 15-17
  oil in water, 12-13
  water in oil, 9-12, 298-300

Enforced passivation, 224

Ethylenediaminetetrasodiumacetate. *SEE EDTA.*

Evans diagram, 206-216, 220, 223, 241

External coatings, *xx-xxi*

External current, 191-192

Extrusion, 62

F

Fat/glyceride, 8

Fatty acids, 9, 12, 51

Fatty imidazolines, 238, 240, 286, 288-289

Ferric ion, 87-88, 122-123

Fluid phases (petroleum), 243-244

Forms (corrosion/scale), 4-5, 19-20, 169-183
  corrosion, 4-5, 169-183
  filling nonionic ligand sites, 179, 181-182
  mixed metallic salts, 179-180
  oxidation films and corrosion, 174-175
  radical-induced complex geometry, 175-178
  radicals, 173-183
  scale, 4-5, 19-20, 169-183

sequence of formation, 170-173

Free radicals. *SEE Radicals.*

G

Geometrical distortion, 25, 27

Glycerides, 9

Grain boundaries, 59-61, 63, 65, 69

Group 2A elements, 23, 31, 34, 37, 68-70, 103, 142, 147, 161-166, 169-170
  sulfur, 68-70

Group concentration, 45

H

Hard/soft acids and bases, 143-145

Hemin complex, 97

Hexagonal crystal, 74-75

High-pressure liquid chromatography, 271

Hydrated complexes, 12, 14, 23, 68-69

Hydride mechanism, 155, 160

Hydrogen and corrosion, 149-153

Hydrogen and scale, 161, 163

Hydrogen case, 149-167
  corrosion, 149-153
  group 2A elements, 161, 163-166
  metallic de-electronation, 152, 154

V

Valves/pumps, 314
van der Waals radius, 90, 92-93,
    171, 189

W

Water in oil emulsions, 9-12,
    298-300
Water sample, 268-269
Weight loss, 263-264
Welded pipe, 62
Well completion, 309-310
Wellbore and beyond, 312-313

X

X-ray crystallography, 270